HANDBOOK OF RESEARCH ON FOOD PROCESSING AND PRESERVATION TECHNOLOGIES

Volume 2

Nonthermal Food Preservation and Novel Processing Strategies

Handbook of Research on Food Processing and Preservation Technologies, 5 volume set:

Volume 1: Nonthermal and Innovative Food Processing Methods

Volume 2: Nonthermal Food Preservation and Novel Processing Strategies

Volume 3: Computer-Aided Food Processing and Quality Evaluation Techniques

Volume 4: Design and Development of Specific Foods, Packaging Systems, and Food Safety

Volume 5: Emerging Techniques for Food Processing, Quality, and Safety Assurance

Innovations in Agricultural and Biological Engineering

HANDBOOK OF RESEARCH ON FOOD PROCESSING AND PRESERVATION TECHNOLOGIES

Volume 2

Nonthermal Food Preservation and Novel Processing Strategies

Edited by

Preeti Birwal, PhD

Megh R. Goyal, PhD, PE

Monika Sharma, PhD

AAP | APPLE
ACADEMIC
PRESS

First edition published 2022

Apple Academic Press Inc.
1265 Goldenrod Circle, NE,
Palm Bay, FL 32905 USA

4164 Lakeshore Road, Burlington,
ON, L7L 1A4 Canada

CRC Press
6000 Broken Sound Parkway NW,
Suite 300, Boca Raton, FL 33487-2742 USA

4 Park Square, Milton Park,
Abingdon, Oxon, OX14 4RN UK

© 2022 Apple Academic Press, Inc.

Apple Academic Press exclusively co-publishes with CRC Press, an imprint of Taylor & Francis Group, LLC

Library and Archives Canada Cataloguing in Publication

Title: Handbook of research on food processing and preservation technologies / edited by Preeti Birwal, PhD, Megh R. Goyal, PhD, PE, Monika Sharma, PhD.
Names: Birwal, Preeti, editor. | Goyal, Megh R., editor. | Sharma, Monika (Food scientist), editor.
Series: Innovations in agricultural and biological engineering.
Description: First edition. | Series statement: Innovations in agricultural and biological engineering | Includes bibliographical references and indexes. | Incomplete contents: Volume 1. Nonthermal and innovative food processing methods-- Volume 2. Nonthermal food preservation and novel processing strategies
Identifiers: Canadiana (print) 20210143002 | Canadiana (ebook) 2021014307X | ISBN 9781771889827 (v. 1; hardcover) | ISBN 9781774638514 (v. 1 ; softcover) | ISBN 9781774630037 (v. 2 ; hardcover) | ISBN 9781774638521 (v. 2 ; softcover) | ISBN 9781774630334 (v. 3 ; hardcover) | ISBN 9781774638538 (v. 3 ; softcover) | ISBN 9781774630341 (v. 4 ; hardcover) | ISBN 9781774638545 (v. 4 ; softcover) | ISBN 9781774630358 (v. 5 ; hardcover) | ISBN 9781774638552 (v. 5 ; softcover) | ISBN 9781003153221 (v. 1 ; ebook) | ISBN 9781003161295 (v. 2 ; ebook) | ISBN 9781003184591 (v. 3 ; ebook) | ISBN 9781003184645 (v. 4 ; ebook) | ISBN 9781003184720 (v. 5 ; ebook)
Subjects: LCSH: Food industry and trade. | LCSH: Food—Preservation.
Classification: LCC TP371.2 .H36 2021 | DDC 664/.028—dc23

Library of Congress Cataloging-in-Publication Data

....

CIP data on file with US Library of Congress

....

ISBN: 978-1-77463-036-5 (5-volume set)
ISBN: 978-1-77463-003-7 (hbk)
ISBN: 978-1-77463-852-1 (pbk)
ISBN: 978-1-00316-129-5 (ebk)

ABOUT THE BOOK SERIES: INNOVATIONS IN AGRICULTURAL AND BIOLOGICAL ENGINEERING

Under this book series, Apple Academic Press Inc. is publishing book volumes over a span of 8–10 years in the specialty areas defined by the American Society of Agricultural and Biological Engineers (www.asabe. org). Apple Academic Press Inc. aims to be a principal source of books in agricultural and biological engineering. We welcome book proposals from readers in areas of their expertise.

The mission of this series is to provide knowledge and techniques for agricultural and biological engineers (ABEs). The book series offers high-quality reference and academic content on agricultural and biological engineering (ABE) that is accessible to academicians, researchers, scientists, university faculty and university-level students, and professionals around the world.

Agricultural and biological engineers ensure that the world has the necessities of life, including safe and plentiful food, clean air and water, renewable fuel and energy, safe working conditions, and a healthy environment by employing knowledge and expertise of the sciences, both pure and applied, and engineering principles. Biological engineering applies engineering practices to problems and opportunities presented by living things and the natural environment in agriculture.

ABE embraces a variety of the following specialty areas (www.asabe.org): aquaculture engineering, biological engineering, energy, farm machinery and power engineering, food, and process engineering, forest engineering, information, and electrical technologies, soil, and water conservation engineering, natural resources engineering, nursery, and greenhouse engineering, safety, and health, and structures and environment.

For this book series, we welcome chapters on the following specialty areas (but not limited to):

1. Academia to industry to end-user loop in agricultural engineering.
2. Agricultural mechanization.
3. Aquaculture engineering.
4. Biological engineering in agriculture.
5. Biotechnology applications in agricultural engineering.
6. Energy source engineering.

7. Farm to fork technologies in agriculture.
8. Food and bioprocess engineering.
9. Forest engineering.
10. GPS and remote sensing potential in agricultural engineering.
11. Hill land agriculture.
12. Human factors in engineering.
13. Impact of global warming and climatic change on agriculture economy.
14. Information and electrical technologies.
15. Irrigation and drainage engineering.
16. Micro-irrigation engineering.
17. Milk Engineering.
18. Nanotechnology applications in agricultural engineering.
19. Natural resources engineering.
20. Nursery and greenhouse engineering.
21. Potential of phytochemicals from agricultural and wild plants for human health.
22. Power systems and machinery design.
23. Robot engineering and drones in agriculture.
24. Rural electrification.
25. Sanitary engineering.
26. Simulation and computer modeling.
27. Smart engineering applications in agriculture.
28. Soil and water engineering.
29. Structures and environment engineering.
30. Waste management and recycling.
31. Any other focus areas.

Books published in the Innovations in Agricultural & Biological Engineering Series

- Biological and Chemical Hazards in Food and Food Products: Prevention, Practices, and Management
- Bioremediation and Phytoremediation Technologies in Sustainable Soil Management
    o Volume 1: Fundamental Aspects and Contaminated Sites
    o Volume 2: Microbial Approaches and Recent Trends
    o Volume 3: Inventive Techniques, Research Methods and Case Studies
    o Volume 4: Degradation of Pesticides and Polychlorinated Biphenyls

- Dairy Engineering: Advanced Technologies and Their Applications
- Developing Technologies in Food Science: Status, Applications, and Challenges
- Emerging Technologies in Agricultural Engineering
- Engineering Interventions in Agricultural Processing
- Engineering Interventions in Foods and Plants
- Engineering Practices for Agricultural Production and Water Conservation: An Interdisciplinary Approach
- Engineering Practices for Management of Soil Salinity: Agricultural, Physiological, and Adaptive Approaches
- Engineering Practices for Milk Products: Dairyceuticals, Novel Technologies, and Quality
- Field Practices for Wastewater Use in Agriculture: Future Trends and Use of Biological Systems
- Flood Assessment: Modeling and Parameterization
- Food Engineering: Emerging Issues, Modeling, and Applications
- Food Process Engineering: Emerging Trends in Research and Their Applications
- Food Processing and Preservation Technology: Advances, Methods, and Applications
- Food Technology: Applied Research and Production Techniques
- Handbook of Research on Food Processing and Preservation Technologies:
  - Volume 1: Nonthermal and Innovative Food Processing Methods
  - Volume 2: Nonthermal Food Preservation and Novel Processing Strategies
  - Volume 3: Computer-Aided Food Processing and Quality Evaluation Techniques
  - Volume 4: Design and Development of Specific Foods, Packaging Systems, and Food Safety
  - Volume 5: Emerging Techniques for Food Processing, Quality, and Safety Assurance
- Modeling Methods and Practices in Soil and Water Engineering
- Nanotechnology and Nanomaterial Applications in Food, Health, and Biomedical Sciences
- Nanotechnology Applications in Agricultural and Bioprocess Engineering: Farm to Table

- Nanotechnology Applications in Dairy Science: Packaging, Processing, and Preservation
- Novel Dairy Processing Technologies: Techniques, Management, and Energy Conservation
- Novel Strategies to Improve Shelf-Life and Quality of Foods: Quality, Safety, and Health Aspects
- Processing of Fruits and Vegetables: From Farm to Fork
- Processing Technologies for Milk and Milk Products: Methods, Applications, and Energy Usage
- Scientific and Technical Terms in Bioengineering and Biological Engineering
- Soil and Water Engineering: Principles and Applications of Modeling
- Soil Salinity Management in Agriculture: Technological Advances and Applications
- State-of-the-Art Technologies in Food Science: Human Health, Emerging Issues and Specialty Topics
- Sustainable Biological Systems for Agriculture: Emerging Issues in Nanotechnology, Biofertilizers, Wastewater, and Farm Machines
- Technological Interventions in Dairy Science: Innovative Approaches in Processing, Preservation, and Analysis of Milk Products
- Technological Interventions in Management of Irrigated Agriculture
- Technological Interventions in the Processing of Fruits and Vegetables
- Technological Processes for Marine Foods, from Water to Fork: Bioactive Compounds, Industrial Applications, and Genomics

OTHER BOOKS ON AGRICULTURAL AND BIOLOGICAL ENGINEERING FROM APPLE ACADEMIC PRESS, INC.

Management of Drip/Trickle or Micro Irrigation
Megh R. Goyal, PhD, PE, Senior Editor-in-Chief

Evapotranspiration: Principles and Applications for Water Management
Megh R. Goyal, PhD, PE, and Eric W. Harmsen, Editors

Book Series: Research Advances in Sustainable Micro Irrigation
Senior Editor-in-Chief: Megh R. Goyal, PhD, PE

Volume 1: Sustainable Micro Irrigation: Principles and Practices
Volume 2: Sustainable Practices in Surface and Subsurface Micro Irrigation
Volume 3: Sustainable Micro Irrigation Management for Trees and Vines
Volume 4: Management, Performance, and Applications of Micro Irrigation Systems
Volume 5: Applications of Furrow and Micro Irrigation in Arid and Semi-Arid Regions
Volume 6: Best Management Practices for Drip Irrigated Crops
Volume 7: Closed Circuit Micro Irrigation Design: Theory and Applications
Volume 8: Wastewater Management for Irrigation: Principles and Practices
Volume 9: Water and Fertigation Management in Micro Irrigation
Volume 10: Innovation in Micro Irrigation Technology

Book Series: Innovations and Challenges in Micro Irrigation
Senior Editor-in-Chief: Megh R. Goyal, PhD, PE

Volume 1: Management of Drip/Trickle or Micro Irrigation
Volume 2: Sustainable Micro Irrigation Design Systems for Agricultural Crops
Volume 3: Principles and Management of Clogging in Micro Irrigation
Volume 4: Performance Evaluation of Micro Irrigation Management

ABOUT THE EDITORS

Preeti Birwal, PhD
*Scientist (Processing and Food Engineering),
Department of Processing and Food Engineering,
College of Agricultural Engineering and
Technology, Punjab Agricultural University,
Ludhiana, Punjab, India*

Preeti Birwal, PhD, is working as a Scientist (Processing and Food Engineering) in the Department of Processing and Food Engineering, College of Agricultural Engineering and Technology, Punjab Agricultural University, Ludhiana, Punjab, India.

She holds a BSc (2012) in Dairy Technology from ICAR-National Dairy Research Institute (NDRI), Karnal; MSc (2014) in Food Process Engineering and Management from NIFTEM, Haryana; and PhD (Dairy Engineering) from ICAR-NDRI, Bangalore. She is a recipient of MHRD (2008), Nestle India (2009), GATE (2012–2014), and UGC-RGN fellowships (2014–2018).

She is currently working in the area of nonthermal food preservation, fermented beverages, food packaging, and technology of millet-based beer. She has served at Jain Deemed to be University, Bangalore as a member of the board of examiners and placements. She is advising several MTech scholars in food technology. She has participated in several national and international conferences and seminars. She has delivered lectures as a resource person on doubling farmers' income through dairy technology in training sponsored by the Directorate of Extension, Ministry of Agriculture and Farmers Welfare, Government of India. She has named outstanding reviewer of the month by the *Current Research in Nutrition and Food Science Journal*. She has successfully completed AUTOCAD 2D and 3D certification.

She has 18 research papers, one edited book, five book chapters, about 28 popular articles, five conference papers, 56 abstracts, and two editorial opinions to her credit. She has successfully guided five postgraduate students for their dissertation work. She is serving as the external examiner for various Indian state agricultural universities. She is also serving as editor and reviewer of several journals. She is a life member of IDEA. Readers may contact her at: preetibirwal@gmail.com.

Megh R. Goyal, PhD, PE
*Retired Professor in Agricultural and
Biomedical Engineering, University of Puerto
Rico, Mayaguez Campus; Senior Technical
Editor-in-Chief, Biomedical Engineering and
Agricultural Science, Apple Academic Press, Inc.*

Megh R. Goyal, PhD, PE, is a Retired Professor
in Agricultural and Biomedical Engineering
from the General Engineering Department in
the College of Engineering at the University of
Puerto Rico–Mayaguez Campus; and Senior Acquisitions Editor and Senior
Technical Editor-in-Chief in Agriculture and Biomedical Engineering for
Apple Academic Press, Inc. He has worked as a Soil Conservation Inspector
and as a Research Assistant at Haryana Agricultural University and Ohio
State University.

During his professional career of 52 years, Dr. Goyal has received many
prestigious awards and honors. He was the first agricultural engineer to
receive the professional license in Agricultural Engineering in 1986 from
the College of Engineers and Surveyors of Puerto Rico. In 2005, he was
proclaimed as "Father of Irrigation Engineering in Puerto Rico for the
Twentieth Century" by the American Society of Agricultural and Biological
Engineers (ASABE), Puerto Rico Section, for his pioneering work on
micro irrigation, evapotranspiration, agroclimatology, and soil and water
engineering. The Water Technology Centre of Tamil Nadu Agricultural
University in Coimbatore, India, recognized Dr. Goyal as one of the experts
"who rendered meritorious service for the development of micro irrigation
sector in India" by bestowing the Award of Outstanding Contribution in
Micro Irrigation. This award was presented to Dr. Goyal during the inaugural
session of the National Congress on "New Challenges and Advances in
Sustainable Micro Irrigation" held at Tamil Nadu Agricultural University.

Dr. Goyal received the Netafim Award for Advancements in Microir-
rigation: 2018 from the American Society of Agricultural Engineers at the
ASABE International Meeting in August 2018. VDGOOD Professional
Association of India awarded Lifetime Achievement Award at 12th Annual
Meeting on Engineering, Science and Medicine that was held on 20-21 of
November of 2020 in Visakhapatnam, India. A prolific author and editor, he
has written more than 200 journal articles and textbooks and has edited over
85 books. He is the editor of three book series published by Apple Academic

Press: Innovations in Agricultural & Biological Engineering, Innovations and Challenges in Micro Irrigation, and Research Advances in Sustainable Micro Irrigation. He is also instrumental in the development of the new book series Innovations in Plant Science for Better Health: From Soil to Fork.

Dr. Goyal received his BSc degree in engineering from Punjab Agricultural University, Ludhiana, India; his MSc and PhD degrees from Ohio State University, Columbus; and his Master of Divinity degree from Puerto Rico Evangelical Seminary, Hato Rey, Puerto Rico, USA.

Monika Sharma, PhD
Scientist, Dairy Technology Division, Southern Regional Station, ICAR-National Dairy Research Institute, Bengaluru, India

Monika Sharma, PhD, is working as a Scientist, Dairy Technology Division, Southern Regional Station, ICAR-National Dairy Research Institute, Bangalore, India.

She received a BSc (2006) in Food Science and Technology from Delhi University, New Delhi; MSc (2008) in Food Technology from Govind Ballabh Pant University of Agriculture and Technology, Pantnagar; and PhD (2015) in Dairy Technology from ICAR-National Dairy Research Institute (NDRI), Karnal, Haryana, India. She started her career as a Scientist in the Indian Council of Agricultural Research in 2010; as a Scientist in ICAR-Central Institute of Postharvest Engineering and Technology, Ludhiana, Punjab, for more than five years. She is now working as a Scientist at Southern Regional Station, ICAR-National Dairy Research Institute, Bangalore, and is actively involved in teaching and research activities. She has more than ten years of research experience. She has worked in the area of convenience and ready-to-eat foods, functional foods, quality evaluation, composite dairy foods, starch modification and its application in dairy food products, etc. Presently, she is working in the area of functional and indigenous dairy foods.

She has published 25 research papers in peer-reviewed journals, three edited books, four technical bulletins, two technology inventory books, six book chapters, more than 25 popular articles, and more than 20 conference papers. She has successfully guided 6 postgraduate students for their dissertation work. She has worked as Principal Investigator in several research projects and has developed various technologies. She has also conducted entrepreneurship development programs for some of the developed technologies. She has earned several awards, such as an ICAR-JRF award and fellowship (2006–2008), first rank in all India level Agricultural Research Services (2008) examination in the discipline of Food Science and Technology, ICAR-NET (2009), conference awards, institute awards, etc. She is a life member of the Indian Science Congress and the Association of Food Scientists and Technologists (India). Readers may contact her at: sharma.monikaft@gmail.com

CONTENTS

CONTRIBUTORS

Mohammed Shafiq Alam
Senior Research Engineer, Department of Processing and Food Engineering, Punjab Agricultural University, Ludhiana–141004, Punjab, India, Mobile: +91-9417188501, E-mail: ms_alam@pau.edu

Preeti Birwal
Scientist, Department of Processing and Food Engineering, Punjab Agricultural University, Ferozepur Road, Ludhiana–141004, Punjab, India, Mobile: +91 989669633, E-mail: preetibirwal@gmail.com

Prasad Chavan
PhD Research Fellow, Department of Processing and Food Engineering, Punjab Agricultural University, Ludhiana–141004, Punjab, India, Mobile: +91-8569030671, E-mail: erprasad.chavan@gmail.com

Swastika Das
Guest Faculty, Department of Agricultural Engineering, Assam University, Silchar, Assam–788011, India, Mobile: +91-9735836444, E-mail: swastika5396@gmail.com

Brindha Deivendran
MTech Research Scholar, Department of Food Technology, FET, Jain University (Deemed-to-be-University), Jain Global Campus, Jakkasandra Post, Kanakapura Taluk, Ramanagara District, Karnataka–562112, India, Mobile: +91-8883583777, E-mail: brindhadeivendran7@gmail.com

R. Devaraju
PhD Research Scholar in Dairy Engineering, ICAR-National Dairy Research Institute, Southern Regional Station, Adugodi, Bangalore–560030, Karnataka, India, Mobile: +91 8088168268, E-mail: draj.raju20@gmail.com

P. Devikrishna
MTech Research Scholar Dairy Engineering, ICAR-National Dairy Research Institute, Southern Regional Station, Adugodi, Bangalore – 560030, Karnataka, India, Mobile: +91 9496271085, E-mail: devikrishnap95@gmail.com

Neela Emanuel
Associate Professor, Department of Basic and Applied Science, National Institute of Food Technology Entrepreneurship and Management (NIFTEM), An Autonomous Institution under Ministry of Food Processing Industries (MoFPI), Government of India, Plot No. 97, Sector 56, HSIIDC Industrial Estate, Kundli–131028, Sonepat, Haryana, India, Mobile: +91-8222000399, E-mail: nemanuel99@hotmail.com

F. Magdaline Eljeeva Emerald
Senior Scientist, Dairy Engineering Section, ICAR-National Dairy Research Institute, Southern Regional Station, Adugodi, Bangalore – 560030, Karnataka, India, Mobile: +91 9481886670, E-mail: eljeeva1@gmail.com

R. S. Gaudham
Assistant Professor, Department of Food Technology, Sri Shakthi Institute of Engineering and Technology, Coimbatore – 641062, Tamil Nadu, India, Mobile: +91-9953580606, E-mail: gaudham.fpe@gmail.com

Megh R. Goyal
Senior Editor-in-Chief (Agriculture and Biomedical Engineering) for AAP; Retired Professor in
Agricultural and Biomedical Engineering, University of Puerto Rico-Mayaguez Jardines de Rincon,
E25 Calle 1, Rincon-PR–00677, USA, Mobile: 001-787-536-0039, E-mail: goyalmegh@gmail.com

Prerna Gupta
Assistant Professor, Department of Food Technology and Nutrition, Lovely Professional University
(LPU), Jalandhar-Delhi G.T. Road (NH1), Phagwara–144411, Punjab, India, Mobile: 91-8169252003,
E-mail: prerna.gskuastj@gmail.com

Gurjeet Kaur
PhD Research Scholar, Department of Processing and Food Engineering, Punjab Agricultural
University, Ferozepur Road, Ludhiana–141004, Punjab, India, Mobile: +91-6280295606,
E-mail: Gurjeet.jatana9@gmail.com

Arvind Kumar
Assistant Professor, Department of Dairy Science and Food Technology, Institute of Agricultural
Sciences, Banaras Hindu University, Varanasi–221005, Uttar Pradesh, India, Mobile: +919793583702,
E-mail: arvind00000@gmail.com

Rohit Kumar
PhD Research Scholar, Department of Food Engineering, National Institute of Food Technology
Entrepreneurship and Management (NIFTEM), An Autonomous Institution under Ministry of Food
Processing Industries (MoFPI), Government of India, Plot No. 97, Sector 56, HSIIDC Industrial Estate,
Kundli–131028, Sonepat, Haryana, India, Mobile: +91-9709250177, E-mail: rohit1995shahi@gmail.com

Priyanka Kundu
PhD Research Fellow, Department of Food Technology and Nutrition, Lovely Professional University
(LPU), Jalandhar-Delhi G.T. Road (NH1), Phagwara–144411, Punjab, India, Mobile: 8168634233,
E-mail: kundup36@gmail.com

Ronit Mandal
PhD Research Fellow, Food, Nutrition, and Health, University of British Columbia, 2205, East Mall,
Vancouver, BC, V6T 1W4, Canada, Mobile: +1-2368639270, E-mail: ronit.mandal@ubc.ca

Madhumitha Maran
MTech Research Scholar, Department of Food Technology, FET, Jain University (Deemed-to-
be-University), Jain Global Campus, Jakkasandra Post, Kanakapura Taluk, Ramanagara District,
Karnataka–562112, India, Mobile: +91-8056374858, E-mail: madhumithaaquarius@gmail.com

Rekha Ravindra Menon
Principal Scientist, ICAR-National Dairy Research Institute (NDRI), Southern Research Station (SRS),
Adugodi, Bangalore–560030, Karnataka, India, Mobile: +91-9916703069, E-mail: rekhamn@gmail.com

Sadhna Mishra
PhD Research Scholar, Department of Dairy Science and Food Technology, Institute of Agricultural
Sciences, Banaras Hindu University, Varanasi–221005, Uttar Pradesh, India, Mobile: +918953236018,
E-mail: sadhnamishra2649@gmail.com

Laxmana N. Naik
Scientist, Dairy Chemistry Section, ICAR-National Dairy Research Institute, Southern Regional Station
(SRS), Adugodi, Bangalore–560030, Karnataka, India, Mobile: +91 9996123653,
E-mail: laxmandcnaik@gmail.com

B. Surendra Nath
Principal Scientist, Dairy Chemistry Section, ICAR-National Dairy Research Institute, Southern
Regional Station, Adugodi, Bangalore–560030, Karnataka, India, Mobile: +91 9449028628,
E-mail: bsn_ndri@gmail.com

Shikha Pandhi
PhD Research Scholar, Department of Dairy Science and Food Technology, Institute of Agricultural Sciences, Banaras Hindu University, Varanasi–221005, Uttar Pradesh, India, Mobile: +91-9958382808, E-mail: shikhapandhi94@gmail.com

Pramod K. Prabhakar
Assistant Professor, Department of Food Science and Technology, National Institute of Food Technology Entrepreneurship and Management (NIFTEM), An Autonomous Institution under Ministry of Food Processing Industries (MoFPI), Government of India, Plot No. 97, Sector 56, HSIIDC Industrial Estate, Kundli–131028, Sonepat, Haryana, India, Mobile: +91-9800105812, E-mail: pkprabhakariitkgp@gmail.com

Heartwin A. Pushpadass
Principal Scientist, Dairy Engineering Section, ICAR-National Dairy Research Institute, Southern Regional Station, Adugodi, Bangalore–560030, Karnataka, India, Mobile: +91 9448927370, E-mail: heartwin1@gmail.com

Dinesh Chandra Rai
Professor, Department of Dairy Science and Food Technology, Institute of Agricultural Sciences, Banaras Hindu University, Varanasi–221005, Uttar Pradesh, India, Mobile: +919415225645, E-mail: dcrai@bhu.ac.in

Robina Rai
Guest Faculty, Department of Agricultural Engineering, Assam University, Silchar, Assam–788011, India, Mobile: +91-7086125023, E-mail: robinarai95@gmail.com

B. Rajunaik
PhD Research Scholar, Dairy Engineering Section, ICAR-National Dairy Research Institute, Southern Regional Station (SRS), Adugodi, Bangalore – 560030, Karnataka, India, Mobile: +91-9844779444, E-mail: raju.ndri@gmail.com

Rajasree Ranjit
PhD Research Scholar, Department of Food Engineering, National Institute of Food Technology Entrepreneurship and Management (NIFTEM), An Autonomous Institution under Ministry of Food Processing Industries (MoFPI), Government of India, Plot No. 97, Sector 56, HSIIDC Industrial Estate, Kundli–131028, Sonepat, Haryana, India, Mobile: +91-9496326711, E-mail: rajasreeranjit@gmail.com

B. G. Seethu
PhD Research Scholar, Dairy Engineering Section, ICAR-National Dairy Research Institute, Southern Regional Station (SRS), Adugodi, Bangalore–560030, Karnataka, India, Mobile: +91-9400877694, E-mail: seethubg2011@gmail.com

Arun Sharma
Assistant Professor, Department of Food Engineering, National Institute of Food Technology Entrepreneurship and Management (NIFTEM), An Autonomous Institution under Ministry of Food Processing Industries (MoFPI), Government of India, Plot No. 97, Sector 56, HSIIDC Industrial Estate, Kundli–131028, Sonepat, Haryana, India, Mobile: +91-9416656492, E-mail: arunsharma1712@gmail.com

Monika Sharma
Scientist (Food Science and Technology), Department of Dairy Technology, Southern Regional Station (SRS), ICAR-National Dairy Research Institute, Bangalore–560030, Karnataka, India, Mobile: +91-9915018948, E-mail: sharma.monikaft@gmail.com

Gagandeep Kaur Sidhu
Senior Research Engineer, Department of Processing and Food Engineering, Agricultural University, Ferozepur Road, Ludhiana–141004, Punjab, India, Mobile: +91 9815854300, E-mail: gagandeep@pau.edu

Anubhav Pratap Singh
Assistant Professor, Food, Nutrition, and Health, University of British Columbia, 2205, East Mall, Vancouver, BC, V6T 1W4, Canada, Mobile: +1-6048225944, E-mail: anubhav.singh@ubc.ca

Jashandeep Singh
PhD Research Fellow, Department of Processing and Food Engineering, Punjab Agricultural University, Ferozepur Road, Ludhiana–141004, Punjab, India, Mobile: +91 7973624393,
E-mail: sidhujash@gmail.com

Shyam Kumar Singh
PhD Research Scholar, Department of Food Science and Technology, The Ohio State University, Columbus–43210, Ohio, USA, E-mail: shyamsingh.iitkgp@gmail.com

S. Subhashini
Assistant Professor, Department of Food Technology, FET, Jain University (Deemed-to-be-University), Jain Global Campus, Jakkasandra Post, Kanakapura Taluk, Ramanagara District, Karnataka–562112, India, Mobile: +91-9962610287, E-mail: s.subhashini10@gmail.com

Aditya P. Sukumar
MTech Dairy Engineering Scholar, ICAR-National Dairy Research Institute, Southern Regional Station, Adugodi, Bengaluru–560030, Karnataka, India, Mobile: +91 8885345172,
E-mail: adityasukumar888@gmail.com

Rajasekhar Tellabati
Assistant Professor, Department of Dairy Engineering, College of Dairy Technology,
Sri Venkateswara Veterinary University (SVVU), Tirupati–517502, Andhra Pradesh, India,
Mobile: +91-9701048600, E-mail: rajasekhar.svvu@gmail.com

Aswin S. Warrier
Assistant Professor (Dairy Engineering), KVASU Dairy Plant, Kerala Veterinary and Animal Sciences University, Mannuthy Campus, Thrissur, Kerala–680651, India, Mobile: +91-7559067959,
E-mail: aswinswarrier@kvasu.ac.in

ABBREVIATIONS

1-MCP	1-methylcyclopropene
3D	three-dimensional
4D	four-dimensional
A	area m^2
a_w	water activity
Be	berry number
BHA	butylated hydroxyanisole
BHT	butylated hydroxytoluene
BSH	bile salt hydrolases
C_2H_4	ethylene
CA	controlled atmosphere
CAS	controlled atmospheric storage
CBDR	common but differentiated responsibilities
CCD	central composite design
CFCs	chlorofluorocarbons
CFD	computational fluid dynamics
CFU	colony forming units
CIP	cleaning in place
CO_2	carbon dioxide
COP	Coefficient of performance
De	Deborah number
DMSO	dimethyl sulfoxide
DNA	deoxyribonucleic acid
DON	deoxynivalenol
ECG	echocardiogram
etc.	Et cetera
EVOH	ethylene-vinyl alcohol
FAO	Food and Agriculture Organization
FCV	feline calicivirus
FDA	Food and Drug Administration
FEA	finite element analysis
FFA	free fatty acids
FOS	fructo-oligosaccharides
FSC	food supply chain

FSIS	Food Safety and Inspection Service
FTIR	Fourier transform infrared spectroscopy
g	grams
GHG	greenhouse gas
GI	gastrointestinal
GIT	gastrointestinal tract
GNP	gross national product
GRAS	generally recognized as safe
GWP	global warming potential
h	heat transfer coefficient w-m^{-2} k^{-1}
H	hydrogen
H_2O_2	hydrogen peroxide
HAV	hepatitis A virus
HCFC	hydrochlorofluorocarbons
HDPE	high-density polyethylene
HDPE	high-density polymers
HFCs	hydrofluorocarbons
HIFU	high intensity focused ultrasound
HP	high pressure
HPL	hydroperoxide lyase
HPP	high-pressure processing
Hz	hertz
I	current A
I_2	iodine
IARI	Indian Agricultural Research Institute
IDF	International Dairy Federation
IJAM	International Journal of Ayurvedic Medicine
IPP	inspection program personnel
K	Kelvin
k	thermal conductivity, W-m^{-1} K^{-1}
kHz	kilohertz
L	liter
LAB	lactic acid bacteria
LDPE	low-density polyethylene
LOX	lipoxygenases
MAP	modified atmosphere packaging
MB	methyl bromide
mg	milligrams
MHz	megahertz
ml	milliliter

MNV-1	murine norovirus 1
MPa	mega Pascal
MR	moisture ratio
MS	manosonication
MTS	manothermosonication
MVD	microwave vacuum dehydration
N_2	nitrogen
nm	nanometer
O/W	oil in water
O_2	oxygen
O_3	ozone
ODP	ozone depletion potential
OH	hydroxide
Oh	Ohnesorge number
O-W-O	oil-water-oil
Pa	Pascal
PCL	poly-e-caprolactone
PE	polyethylene
PEF	pulsed electric field
PEO	polyethylene oxide
PET	polyethylene terephthalate
PG	polygalacturonase
pH	potential of hydrogen
PLA	polylactic-acid
PLGA	polylactide-co-glycolide
PME	pectin methylesterase
POD	peroxidase
PP	polypropylene
ppm	parts per million
PPO	polyphenol oxidase
psi	pound per square inch
PTFE	polytetrafluoroethylene
PVA	polyvinyl alcohol
PVDC	polyvinylidene chloride
PVP	polyvinylpyrrolidone
Q_c	cooling capacity
qi	cooling rate
R&D	research and development
Re	electrical resistance of thermoelectric module Ω
RH	relative humidity

RNA	ribonucleic acid
RSM	response surface methodology
RTD	residence time distribution
RTE	ready-to-eat
SEM	scanning electron microscopy
SFE	supercritical fluid extraction
SGF	simulated gastric fluid
SIF	simulated intestinal fluid
SLNs	solid lipid nanoparticles
SPI	soy protein isolate
SV	sous vide
T_c	cold side temperature
TCD	tip to collector distance
TEMP	temperature
T_h	hot side temperature
TMV	tobacco mosaic virus
TPC	total phenol content
Tr	Trouton ratio
TS	thermo sonication
UN	United Nations
UNFCCC	United Nation Framework Convention on Climate Change
US	ultrasonication
USFDA	United States Food and Drug Administration
UV	ultraviolet
V	voltage
VCRS	vapor compression refrigeration system
VSV	vesicular stomatitis virus
W	watt
W/cm^2	watt per square centimeter
W/O	water in oil
We	Weber number
WHO	World Health Organization
W-O-W	water-oil-water
WPC	whey protein concentrate
WPI	whey protein isolate
ZEA	zearalenone
ZT	thermoelectric figure of merit
Mm	micrometer

SYMBOLS

%	percentage
ΔT	temperature difference between hot and cold side of thermoelectric module
°C	degree Celsius
η	efficiency %
α	Seebeck coefficient V-K^{-1}
θ	$T_h - T_c^{-1}$
σ	electrical conductivity, $[\Omega m]^{-1}$

PREFACE

Food preservation has existed since the early times. Food is not only important for relief from hunger, but food with nutrition is also the demand of today's millennium. Food with convenience, good sensorial attributes, longer shelf life is today's challenge. The preservation in the current scenario is done by different methods like heating, cooling, drying, concentration, freezing, and cooling. All these techniques have their own pros and cons. Therefore, now a challenge to save the food from deteriorating its quality in terms of loss of nutrients, texture, sensorial characteristics with longer shelf life has been taken by the food scientists and industrialist. Several new food processing, preservation, and quality assessment technologies have been investigated by researchers to preserve the quality and design of specific nutrient-rich food.

Handbook of Research on Food Processing and Preservation Technologies is a bouquet of various emerging techniques in the food processing sector and quite relevant for the food industry and academic community. Nonthermal techniques (such as high-pressure processing [HPP], pulsed electric field [PEF], pulsed light [PL], ultraviolet [UV], microwave, ohmic heating, electrospinning, nano, and microencapsulation) are a few novel processing techniques that are being investigated thoroughly. The role and application of minimal processing techniques (such as ozone treatment, vacuum drying, osmotic dehydration, dense phase carbon dioxide [DPCD] treatment, and high pressure-assisted freezing) have also been included with a wide range of applications. Literature has reported successful applications on juices, meat, fish, fruits and vegetable slices, food surfaces, purees, milk, and milk products, extraction, drying enhancement, and encapsulation of micro-macro nutrients.

Furthermore, this handbook also covers some computer-aided techniques emerging in the food processing sector, such as robotics, radio frequency identification (RFID), three-dimensional food printing, artificial intelligence, etc. Enough emphasis has also been given on nondestructive quality evaluation techniques (such as image processing, terahertz spectroscopy imaging technique, near-infrared [NIR], Fourier transform infrared [FTIR] spectroscopy technique, etc., for food quality and safety evaluation. A significant role of food properties in the design of specific food and edible films have been elucidated. Thus, this handbook would have significant scope in food processing, preservation, safety, and quality evaluation. The handbook is organized under five volumes.

Volume 2: Nonthermal Food Preservation and Processing Novel Strategies illustrates various applications of novel nonthermal food processing techniques. This book will serve many professionals working in the area of food science, technology and engineering around the world. The book will also act as a reference book for researchers, students, scholars, industries, universities, and research centers. Each chapter covers major aspects pertaining to the principles, design, and applications of various novel and advanced food processing techniques. This book is divided into two parts: Part I: Novel Nonthermal Food Processing Technologies, which discusses various techniques viz. HPP, ultrasonication (US) of foods, microwave vacuum dehydration (MVD), thermoelectric refrigeration technology for preservation of fruits and vegetables, electro-spinnability of food-grade biopolymers; Part II: Advances in Food Processing and Preservation Techniques emphasizes technologies for shelf life enhancement of herbs and leafy vegetables, advanced methods of encapsulation, ozonation: potential application in oilseeds storage, and the role of vacuum technology in food preservation. Part III: Food Processing Techniques for Product Formulation elucidates encapsulation of probiotics by electrospinning for food applications, advance encapsulation methodologies for herbal food products, and recent advances in design and application of mechanical expellers for dairy, food, and agricultural processing.

This book volume will serve as a valuable resource of information and excellent reference for researchers, scientists, students, growers, traders, processors, industries, and others for food preservation, processing, and quality aspects of food products.

This book has taken its present shape due to the excellent contribution by the contributing authors, who have been the soul of this compendium. We have mentioned their names in each chapter and also in the list of contributors. We are indeed indebted to them for their knowledge, dedication, and enthusiasm. We expect this book to prove a helpful resource for all the food processing and engineering academicians, food processors, and students.

We also extend our sincere thanks to Apple Academic Press, Inc., for their immense facilitation and suggestions right through this assignment.

We take this opportunity to thank: (1) our families for their motivation, moral support, and blessings in counteracting every obstacle coming in

the way; (2) our spouses for their understanding, patience, and encourage-
ment throughout this project; (3) our institutes: PAU, Ludhiana, Punjab;
ICAR-NDRI, Karnal; Southern Regional Station, ICAR-NDRI, Bangalore;
and UPRM, Mayaguez, for their support during the compilation of this
publication.

—Editors

PART I

Novel Non-Thermal Food Processing Technologies

CHAPTER 1

HIGH-PRESSURE PROCESSING AND PRESERVATION OF FOODS

ROBINA RAI, SWASTIKA DAS, and SHYAM KUMAR SINGH

ABSTRACT

High-pressure isostatic processing is a nonthermal minimum processing method to satisfy both scientific and consumer demands in the food industry. In addition to inactivating microorganisms and enzymes, it also ensures maximum retention of nutrients and sensory characteristics, thereby enabling the production of high-quality shelf-stable products. This chapter discusses the efficacy of high-pressure treatment in food processing and its prospective applications. Its subsequent effects on the microbial, enzymatic, and qualitative attributes of the food product are also reviewed. This chapter also explores the usage of high-pressure in conjunction with other techniques for the enhanced outcome. A brief legislative and commercial comprehension of food products vulnerable to high-pressure is also provided considering its increasing marketability in the current scenario.

1.1 INTRODUCTION

The food industry has seen a paradigm shift of consumer demand towards more "natural" and "fresh" products. Mild processing technologies have restricted the ability to guarantee safe consumption after prolonged storage. Thus, there is a necessity for processing techniques to enhance the shelf life of a product without compromising its quality. Several "emerging techniques" have been developed and studied over the past decades to meet these demands. The impact of pressure on food products was discovered about a century ago but owing to its high cost and technical complexity, the preservation capability of high-pressure processing (HPP) was not actively researched until the 1980s [62]. By 1991, Japan had semi-continuous high-pressure setups and was merchandizing high-pressure processed products (such as fruit sauces, salad dressings, yogurts, and fruit jellies) [166]. HPP operates on the following two principles [27, 85, 172]:

- **Isostatic Principle:** Instant and uniform distribution of pressure throughout a sample regardless of its shape or size.
- **Le Chatelier's Principle:** Pressure favors any reaction that is followed by volume reduction and hinders any reaction followed by volume rise.

Food is a complex system comprising of many interdependent biological and physicochemical fractions. Because of the extensive research in the past several years, a large amount of information is now available to the scientific community, and such data is being utilized for further study and applications [47, 64, 85].

In this chapter, the process of HPP in food is discussed, along with its applications and possible combinations.

1.2 EFFECTS OF HIGH-PRESSURE PROCESSING (HPP) ON FOODS

The marketability of food and food products depends primarily on safety for consumption and consumer acceptance of food. Many biological and chemical factors affect the safety of the product, while its acceptance depends on its functional and organoleptic characteristics. HPP can be used to counter and control these parameters to develop a desirable and safe food product.

1.2.1 MICROORGANISMS

High-pressure affects microorganisms in a manner identical to high tempera-ture. Unlike thermal operation, HPP is uniform and almost instantaneous. Pressure influences the system volume, which then interferes with the equilibrium of the microorganism. The pressure-reaction rate dependence can be explained with the help of transition state theory, and at a constant temperature, this relation is entirely governed by the activation volume [153, 159] as follows:

$$\left(\frac{\partial \ln k}{\partial P}\right)_T = -\left(\frac{\Delta V}{RT}\right) \tag{1}$$

where; k, P, R, and T denote reaction velocity constant, pressure, universal gas constant (8.314 cm³ MPa K⁻¹ mol⁻¹), and temperature (K), respectively.

These elements in Eqn. (1) direct microbial growth and survival and can be utilized for lengthening the product's shelf-life. Food preservation can be achieved by arresting the growth of and inactivating microorganisms [53]. While refrigeration, pH regulation, additives, modified atmosphere packaging (MAP) are examples of the former, the latter encompasses more severe treatments, such as thermal sterilization, high-pressure, ultrasonication (US), etc. Thermal technologies are the most employed methods of preservation owing to their simplicity and ease of operation. However, a general concern regarding the "freshness" of the product has prompted the use of nonthermal techniques. Additionally, combined technologies are also increasingly being implemented to protect the food against microbial attacks while keeping the harshness of the process in check.

HPP can be utilized in the pasteurization of food without significant temperature elevation, thereby ensuring a safe product with retained nutritional and sensory properties. The inactivation of pathogens depends on their sensitivity and the pressure employed. Pressure treatment is less effective on spores when compared to their vegetative counterparts, and it is more threatening to Gram-negative bacteria than Gram-positive bacteria [106].

1.2.1.1 VEGETATIVE CELLS

The degree of microbial inactivation accomplished at a specific pressure treatment relies on various elements, which interact with each other. While most bacteria can grow in a range of 20–30 MPa pressure, some microorganisms have displayed growth at pressure >50 MPa (barophiles) while some (*Escherichia coli, Saccharomyces cerevisiae*) could survive at a pressure >200 MPa or more (baroduric or barotolerant) [181].

The greater sensibility of Gram-negative bacteria (*Vibrio parahaemolyticus*) against pressure can be attributed to their complex membrane, which is more vulnerable against the changes in the surroundings [143]. Furthermore, HPP was more lethal in the exponential stage of the life cycle of the organism [38, 98, 110, 181]. The duration of treatment had also positive effect on its lethality for a specific species until a certain pressure, beyond which the treatment time displayed no significant influence [23]. Temperature can also be employed in conjunction with HPP to enhance the overall effectiveness of the process and to reduce its process time. Moderate high (40°C) or sub-zero (0 to –20°C) temperatures may aid inactivation of *S. cerevisiae* [57]. Mild heat augments the inactivation process, thereby lessening the required pressure, while low temperature modifies the structure and nature of the microbes, making them more sensitive to the applied pressure [23, 24].

1.2.1.2   SPORES

Spores exhibit greater resistance to pressure inactivation in comparison to their vegetative counterparts due to their structure and coat thickness. Spores develop in the limited supply of nutrition and moisture in harsh environmental conditions, enabling prolonged survival until favorable conditions have been re-established when they finally break dormancy through germination [45]. Owing to their high resistance, higher-pressure level for an extended period is suggested for spore inactivation [63]. HPP at ambient temperature has often been found to be inadequate for the purpose [136, 167]. Interestingly, some studies reported that spores were more vulnerable at moderate to low pressure than at higher pressure levels and accordingly a two-step procedure to inactivate spores was suggested [167]: The first stage stimulates germination of the spores followed by the second stage of inactivation of the germinated spore [5]. The application of temperature significantly amplifies the effectiveness of pressure treatment, especially in the germination step [100, 128]. Temperature and pressure can react synergistically, resulting in enhanced spore inactivation [43, 44]. Thermal assisted pressure processing has resulted in superior products with greater quality and consumer acceptance [99]. Bacterial spores are the most resistant species among the microbial groups, including yeast and mold spores. Spore resistance is also dependent on the nature of food, and hence the most resistant spore in a specific food is used as the target/reference for successful inactivation; and accordingly treatment is decided. For example:

- For low acid foods (pH > 4.6) like milk, meat, and vegetables, *C. botulinum* proteolytic type B spores are the target organisms and a minimum temperature-pressure treatment of >900 MPa, >110°C for >5 minutes is suggested [44].
- In case of stronger acidic foods (pH < 4.6) like juices and purees, depending on the critical organism viz., *A. acidoterrestris* spore or *B. Nivea*, ≥621 MPa, ≥90°C for >5 minutes or ≥600 MPa, ≥90°C for 15 minutes is recommended [44].

1.2.1.3 YEASTS AND MOLDS

In general, yeasts are not connected with foodborne diseases. However, owing to their ability to thrive in unfavorable conditions like low water activity and high organic acid concentration, they can significantly damage the food product. Yeasts and molds are relatively more vulnerable to pressure treatment, except in their spore form [51]. Some varieties of molds have toxic effects and hence must be killed or inactivated for preservation. A pressure of 400 MPa is usually sufficient to inactivate yeasts and molds [145].

Varela-Santos et al. [104] reported that fruit juice treated at <400 MPa for 5 minutes or 350 MPa for 30 minutes resulted in a 5-log reduction of nine species of yeasts and molds. The inactivation behavior in mold spores was different from that observed in bacterial spores. Molds were not appreciably affected at pressures below 300 MPa, above which a rapid decline in mold population was observed [169].

An important example of the toxigenicity of molds is the production of mycotoxins, which can grow in a variety of crops and food and pose a serious threat to humans and animals. One such mycotoxin is the Patulin (Figure 1.1) produced by over 60 fungi species (*P. expansum*, *P. claviforme*, *P. urticae*, *P. patulum*, *A. clavatus*, etc.), spanning over 30 genera [14, 36, 84]. A 56.24% reduction in patulin content in fresh apple juice containing 100 ppb of mycotoxin was obtained with HPP treatments of 30 to 500 MPa at 30 to 50°C [9]. A patulin reduction of 31% was reported in fruit and vegetable juices processed at 600 MPa for 5 min [56]. The extent of patulin degradation is dependent on the pressure intensity and exposure time. For apple-spinach juice, the most prominent patulin decline (43 ppb and 0.28 μM) was observed at a pressure of 600 MPa for 5 min [39]. Another example of mycotoxin is the citrinin in black table olive oils, which underwent 64–100% deterioration under 250 MPa at the end of 5 minutes [156]. Complete reduction of the mycotoxins deoxynivalenol (DON), zearalenone (ZEA) was attained in maize grains subjected to 550 MPa and 45°C for 20 minutes [73].

1.2.1.4 VIRUSES

The application of HPP for inactivation of viruses is a relatively new. Viruses are comparatively less sensitive to pressure treatment. Tobacco mosaic virus (TMV) was completely inactivated under 800 MPa pressure for 45 minutes [15]. Pressure up to 600 MPa significantly inactivated viruses (5 log reduction) in various non-enveloped food- and waterborne viruses [22, 54,

94]. The mechanisms of virus inactivation may be explained by the genetic material damage, structural, and functional damage or disruption of the viral envelope during pressure treatment [95].

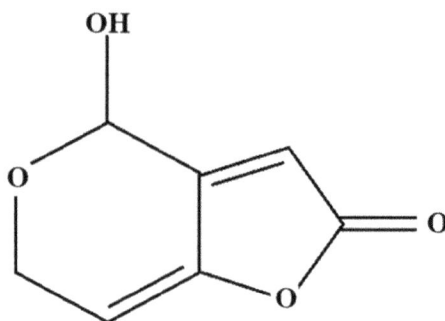

FIGURE 1.1   Chemical structure of patulin ($C_7H_6O_4$).

Lou et al. [94] found that applying 350 MPa pressure for 5 minutes on a norovirus (MNV-1) completely deformed the viral capsid protein's tertiary and quaternary structures with no significant reaction on its primary and secondary proteins. Tang et al. [152] observed that the MNV-1 capsid lost its ability to bind with receptors after HPP treatment. Regardless, HPP had no impact on the viral genomic RNA, which can be attributed to the fact that the pressure levels employed in high-pressure treatment of food is insufficient to affect covalent bonds [94, 152]. The degree of virus inactivation via HPP depends on several extrinsic (pressure intensity, holding time, process temperature) and intrinsic (pH, food matrix, presence of salt, etc.), parameters [95].

In general, enveloped viruses are more sensitive to environmental and biological distress than non-enveloped viruses, thus rendering them more vulnerable to pressure treatment. However, this trend is inconsistent as VSV (vesicular stomatitis virus), an enveloped virus, exhibited greater stability than most non-enveloped viruses (FCV, MNV-1, etc.), under the effect of pressure. The susceptibilities of enveloped viruses also widely differ based on their nature, shape, and size, thermodynamic behavior, receptor binding ability, composition, protein structure and isoelectric point [95]. HPP can be effectively used in surface and internal viral inactivation, with no adverse impact on the sensory characteristics of food, especially in fresh produce [35, 123, 165]. HPP efficiently reduced the population of MNV-1 in fresh lettuce, strawberry puree, and fresh-cut strawberries, HAV (hepatitis A virus), and FCV in salsa, HAV in strawberry puree and sliced green onion, MNV-1,

HAV, and adenovirus type 41 in green onions and MNV-1 in cabbage *Kimchi* [61, 79, 94, 108].

1.2.2 ENZYMES

High-pressure also influences the enzymatic activity of food products. Appropriate pressure treatment can be applied to activate or inactivate enzymes. The pH, substrate concentration, enzyme structure, and pressurization temperature have substantial effect on the process [64]. Enzyme inactivation in the range of 100–300 MPa can be reversed, while enzymes subjected to pressure treatment above 300 MPa cannot be reactivated [96, 151]. High-pressure enzyme inactivation can be theorized to resemble protein denaturation [96]. Pressure stimulates various reversible and irreversible modifications and partial or complete unfolding of the enzyme structure [97]. Since pressure has no influence on covalent bonds, the enzyme retained its primary structure, while its tertiary and quaternary structures underwent alterations under the impact of HPP [59, 101].

Owing to their unpredictable character, the impact of HPP on enzymes cannot be predetermined. Oxidative enzymes like polyphenol oxidase (PPO) and peroxidase (POD) are usually inactivated with pressure-temperature combinations. In the pressure inactivation of PPO (450 MPa, 60 min, 50°C) in apple juice, pressure, and temperature acted synergistically at pressure >400 MPa and temperature >40°C with a maximum inactivation of 91% [18]. Pure pressure treatment without thermal assistance is inadequate for complete inactivation of oxidative enzymes [25]. However, the degree of enzyme inactivation, in addition to the process factors, also depends upon the nature and source of the sample and higher inactivation has also been reported at ambient temperatures.

The highest inactivation values obtained by Liu et al. [93] for PPO and for POD were 88% and 42%, respectively under 600 MPa pressure for 60 minutes at 25C. Several pectic (PME, PG, β-galactosidase, and α-arabinofuranosidas) enzymes in tomato puree and non-pectic (actinidin, amylase, LOX (lipoxygenases), and HPL) enzymes in kiwi fruit juice, apple juice, and tomato juice have been inactivated with the help of HPP [65, 75, 82, 112, 126, 129]. Furthermore, high-pressure can also enhance enzymatic activity due to enzyme conformation or active-site reconfiguration, modification of substrate or media, favoring enzyme release, and activation of latent enzymes [39, 58, 154].

1.2.3 QUALITY ATTRIBUTES

The quality of a food product depends on nutrient content, color, taste, texture, etc. Many of these quality parameters undergo degradation during thermal preservation because of the high temperature involved. Since HPP does not engage extreme temperatures, heat damage can be avoided, but the application of intense pressure may affect the quality of the product in a different manner. The influence of HPP on a food quality can be roughly classified into three groups as described in this section.

1.2.3.1 FUNCTIONAL CHANGES

High-pressure supports hydrogen bond development while it disrupts weaker bonds in protein [111]. The pressure range usually employed in HPP of food is inadequate to break low energy covalent bonds that are found in the primary structure of protein or fatty acid molecule [32, 125]. The pressure rise followed by a decrease in volume distorts the protein structure due to altered ionic bonds and hydrophobic interactions [10, 32, 125]. The decrease in volume also supports hydrogen bond formation due to reduced inter-atomic distance, which is responsible for the stability of α-helical and β-pleated protein structures [111]. The diminished particle size and lowered inter-particle interaction lead to lower viscosity and higher bulk phase motility [135].

Most food systems undergo adiabatic heating due to compression and the temperature rise is directly affected by the degree of compression, further amplifying flow properties. Pressure systems can be developed to manufacture food products with desired rheological properties. The intra and intermolecular interactions and/or conformational changes due to HPP affect the functional property of many food products favorable or undesirable, which is determined by the food's nature and the consumer demand. The combined action of high-pressure homogenization and trehalose addition on probiotic enriched mandarin juice resulted in increased hydrophobicity of the incubated probiotic species [16, 20].

High-pressure homogenization also enhances functional characteristics (solubility, emulsifying, and foaming properties) of a product due to partial unfolding of protein structure. Pressure >100 MPa leads to increased unfolding followed by protein aggregation and inferior functional properties [2, 135, 174]. While starch integrity suffered no major effect due to pressure treatment, however surface marks and fractures proportional to treatment severity were reported [140]. Depletion of starch crystallinity was also noticed at the end of high-pressure homogenization, resulting in modified starches with stronger gel-network and smoother gels [74, 140, 163]. The pressure strength also induces an escalated degree of fat crystallization and improved stability of fat systems, which find their potential application in the dairy industry [84–86].

1.2.3.2   SENSORY CHANGES

High-pressure technology has little effect on the covalent bonds responsible for the organoleptic and nutritional characteristics of a food product. This serves as a major advantage when compared to conventional preservation techniques. However, sensory behaviors depending upon the functional attributes may be altered due to the pressure treatment. Unlike traditional methods, high-pressure treatment has the ability to maintain the original flavor and color of a product [164, 176]. A similar observation was reported in the high-pressure treatment of partially cooked steaks producing lighter and more yellow product [150].

Sun et al. [149] studied the effect of *sous vide* (SV) cooking, HPP treatment and their combination on beefsteaks and discovered similar sensory values but lower juiciness, tenderness, and overall acceptability. Nevertheless, due to shorter cooking time and greater convenience, combination cooking has probable application in the meat processing industry. Cooking loss in meat, generally observed during thermal operation, was significantly reduced in HPP of horse mackerel muscle with greater lightness value [157]. Serra et al. [139] conducted a study to determine the effect of HPP on dry curing of frozen ham and reported lower crumbliness, higher fibrousness, and overall improved texture. Delgado et al. [34] subjected *Torta de Casar* cheese to 600 MPa pressure before maturation to reduce enzymatic and microbial activity since it cannot be thermally pasteurized owing to its Protected Designation of Origin. This prevents over-ripening of the cheese during storage and the off-flavor and excessive bitterness associated with it. This effect, characterized by the release of volatile compounds, was more pronounced under higher pressure [7, 8].

1.2.3.3   NUTRITIONAL CHANGES

Pressure has minimal influence on the antioxidant activity of fruits juices when compared to thermal processes. Fruit smoothies processed at 450 MPa showed antioxidant levels similar to fresh smoothies, while reduced values were observed at the end of 600 MPa treatments, suggesting a detrimental effect of pressure intensity on the food antioxidants during HPP [76]. This result was further affirmed during the pressurization of red sweet peppers at 100–500 MPa for 15 minutes and purple waxy corn at 250–550 MPa for 30–45 minutes [60, 132]. However, the highest antioxidant activity in pressure treated waxy corn was observed at 700 MPa, documenting a loss of 5.41% with respect to untreated kernels [132]. A similar trend was noticed in strawberry purees processed at 400 and 600 MPa for 15 minutes [109], a smoothie at 450 and 600 MPa for 3 minutes [6], and fermented pomegranate beverage at 500, 550, and 600 MPa for 5–10 minutes [127]. This difference in behavior can be explained by the selection of time-pressure combination and the nature of the food product.

Processing temperature also has an impact on the antioxidant activity of the treated product, with temperature generally displaying a destructive effect on antioxidants [101]. The total phenol content (TPC) of a sample also incurred a loss under high-pressure as demonstrated by Jez et al. [69] in the HPP treatment of tomato puree, where the lowest phenol levels were observed for 400 and 550 MPa treated samples. This can be ascribed to the leaching of the phytochemicals along with the exudate. Yet a rise in TPC was observed when the same puree was processed at 700 MPa. Comparable results were reported in the pressure processing of smoothies [6], pumpkin [180], and pomegranate juice [160], due to structural modification and cell membrane disruption leading to an overall increase in bioavailability. The effect of HPP on the tocopherols and Vitamin E in a food product depends on its cellular matrix. Contradictory results have been recorded where HPP has a positive effect on the tocopherol and Vitamin E content in orange, milk, and juice [33], negative effect on soy-based fruit beverage [13] and neutral effect on açaí juice [31].

Sharabi et al. [141] analyzed the potential of ultra high-pressure homogenization in milk and remarked a significant fall in degradation rate of riboflavin because of reduced transparency of milk particles and an inconsiderable impact on vitamin C.

1.3 COMBINATION TREATMENTS

While high-pressure treatment is a competent food processing technique owing to the final product's superior quality, commercial pressure sterilization or pasteurization is often inefficacious to ensure the product safety in the given range. Thus, researchers are now exploring the prospects of combination pressure treatments to counter such issues and strike a balance between product quality and safety. Furthermore, the implementation of other techniques along with HPP can decrease the required pressure intensity and thus make the process more economical.

1.3.1 TEMPERATURE-PRESSURE

The processing time of HPP can be greatly reduced with the assistance of heat. While HPP at ambient condition was adequate to reduce the microbial population of a fresh pineapple slice to a desired level, enzyme inactivation required thermal aid [25]. Raj et al. [69] examined the integrated impact of pressure (200–500 MPa) and temperature (30–60°C) on Indian gooseberry juice for duration of 1 second to 20 minutes. The thermal assisted pressure processed juice manifested superior quality over thermal sterilization in terms of color, TPC, antioxidant activity and ascorbic acid retention. The processing of pineapple puree at 200–600 MPa and 50–70°C for 10–20 minutes exhibited similar results [26]. The pressure-heat combination has also been employed for achieving efficient gelation. Pressure-heat combination (400 MPa, 25 and 75°C for 30 minutes) on chicken myofibrillar proteins resulted in reduced denaturation and increased porosity of the gel [179]. A pressure-temperature combination of 450 MPa and 30°C for 10 min was adequate for the partial gelatinization of paddy during parboiling and thereby reducing soaking time [122]. Li et al. [88] investigated the influence of low and moderate temperature assisted pressure treatment (300–450 MPa, 10°C, 40–70°C, 10–40 minutes) on fresh lychee juice and observed improved color (higher L* values) at moderate temperature processing.

1.3.2 ULTRASOUND-PRESSURE

Studies related to the combined application of ultrasound and pressure is scarce. The simultaneous operation of the two processes is impossible since the rarefaction regions necessary for the cavitation phenomenon in the US is unobtainable under such high-pressure. Pyatkovskyy et al. [114] analyzed the impact of HPP and sonication in sequence. The pressure was varied from 200–400 MPa and the on-off cycle was maintained at 10 seconds. Pressure and sonication manifested synergistic effect in the HPP-sonication sequence and additive effect in the sonication-HPP sequence.

1.3.3 *PULSED ELECTRIC FIELD (PEF)-PRESSURE*

The combined effect of HPP, US, and pulsed electric field (PEF) was analyzed by Pyatkovskyy et al. [114] in different permutations for the inactivation of *Listeria innocua* cells. In sequential application, PEF, and HPP displayed additive effects while synergistic effect was observed in the case of simultaneous application. Simultaneous PEF-HPP processing resulted in the maximum leaching of intracellular non-protein components in the processing medium. However, further research and comprehensive analyzes are essential before HPP-PEF combination can be employed for the commercial processing of food.

1.4   ADVANTAGES OF HIGH-PRESSURE PROCESSING (HPP)

Several advantages give HPP an edge over conventional technologies. Since the process is isostatic, the applied pressure and therefore its subsequent effect is uniform in the entire product and is not dependent on its size and shape. The instantaneous distribution of pressure throughout the system also ensures uniform treatment such that no part of the product is under-or over-processed. Moreover, thermal processes are often accompanied with quality degradation due to the loss of bioactive compounds, nutritional and sensory properties. Higher retention of phenols, antioxidants, vitamins, and other phytochemicals has been reported in HPP treated products as compared to traditional methods [6, 33, 69, 141].

Studies have also displayed greater acceptability of some HPP products, where heat treatment impaired the sensory characteristics of the food. Pressure has a mild effect on the product's color, texture, and flavor resulting in an appreciable improvement in its overall quality [139, 150, 157]. This technology can be implemented to manufacture additive-free products. Additionally, with the help of high-pressure, the functional properties of the product can be modified, leading to development of novel food products [21, 74, 140]. HPP can be used to sterilize or pasteurize food with adequate pressure and can be combined with other hurdles like temperature and acidification [130]. Besides, food products are subjected to pressure treatment after packaging, and thus any contamination after processing can be avoided. It is important to note that HPP does not generate by-products and the pressure medium can be reused and recycled, making it a sustainable technology for the future.

1.5 LIMITATIONS OF HIGH-PRESSURE PROCESSING (HPP)

HPP comes with a series of challenges that are yet to be surmounted. The HPP setup demands a significant capital investment as a result of which HPP products are priced higher than traditionally processed food. The economy of operation also suffers a setback because a continuous HPP system is yet to be developed. Existing equipments are batch-type or pseudo-continuous, which increase the treatment time and consequently has an adverse effect on its overall throughput capacity. The products to be processed also require an adequate packaging material, which should be flexible and impermeable in nature. Food material in tins, glass, or other hard containers cannot be subjected to this treatment.

The equipment also demands regular cleaning and proper sanitation to avoid contamination and instrument damage. Besides, porous food may undergo irreversible structural changes under pressure, while products with low moisture content is inefficient in transferring the applied pressure necessary for inactivating microorganisms. This limits the variety of products that can be processed using this technology.

1.6 COMMERCIAL APPLICATIONS OF HPP IN FOOD INDUSTRY

1.6.1 STERILIZATION AND PASTEURIZATION

Pressure is high as 1000 MPa is being used to sterilize and pasteurize various food products [10]. Most of the focus is on fresh juice, milk, and other organic beverages. However, recently this focus has extended to decontamination of meat in ready-to-eat (RTE) food products [137]. A pressure range of 400–600 MPa has been sufficient to ensure the microbial safety of such products [50]. The shelf-life of cured pork was significantly increased by HPP treatment (400 and 600 MPa, −35 and −15°C) with an appreciable drop in growth rate of lactic acid bacteria (LAB) during storage [124]. Vacuum packaged dry-cured beef processed under 500 MPa at 18°C for 5 minutes and stored at 6°C was safe for consumption up to 210 days [131]. The Food Safety and Inspection Service (FSIS) have approved HPP for the inactivation of *Listeria monocytogenes* in the processed meat [131].

1.6.2   BLANCHING

Enzymatic activity exerts an influence on the quality of the food at the time of storage. Thermal blanching aids in the prevention of browning by enzyme inactivation, although degradation in sensory and nutritional properties have also been observed [142]. Gentle pressure processing can be used to disable vegetative cells, molds, and yeasts that are vulnerable to pressure. Temperature or pH-assisted pressure treatment has significantly reduced the pressure resistance of PPO. Peach slices blanched in citric acid solution (1–1.2%) at 25°C under pressure >300 MPa was effective in inactivating PPO [79]. However, Shayanfar et al. [142] reported the inefficiency of HPP in blanching of cut apples at pressures <600 MPa. The HPP blanched products were softer and the incorporation of texture hardening agents like calcium ion was suggested.

1.6.3   FREEZING AND THAWING

Depending on the path of crystallization and the type of ice crystal formed, high-pressure freezing can be classified into: pressure-assisted freezing, pressure shift freezing and pressure-induced freezing [30, 146]. Faster freezing rates have been observed at elevated pressure [81, 138] due to reduced latent heat of crystallization and other physico-thermal changes. However, higher pressure decreases the freezing point of the product, and therefore, in order to maintain sufficient temperature difference between the product and the cooling medium, the temperature of the medium must be lowered further, and this type of freezing is called pressure-assisted freezing.

Pressure cooling can also be utilized for sub-zero cooling without ice crystal formation, thus preventing freezing damage while protecting against microbial action. In addition to this, the pressurized sub-zero sample may be subjected to sudden release of pressure resulting in supercooling and instant uniform freezing. This kind of freezing resulting from pressure release is called pressure shift freezing. A comparison between pressure-assisted freezing and pressure shift freezing showed that the latter resulted in smaller and more number of crystals under identical pressure [48]. The uniform crystallization can be ascribed to uniform temperature prior to pressure release which is almost instantaneous (Isostatic principle). The short freezing time results in the development of small-sized crystals which are less deleterious to the quality of the frozen food [81, 86, 87].

The combined effect of pressure and low temperature resulted in denaturation of protein. There were significant modifications in the secondary structure of whey protein while no effect on its tertiary structure was observed when it was subjected to 500 MPa at –15 and –35°C [11]. The 300 MPa pressure treatments at –16°C induced 3% denaturation in milk protein as a result of which organoleptic changes were observed [162]. The pressure denaturation can be explained by the volume decrease due to an increase in pressure (Le Chatelier's principle) causing the unfolding of protein [11].

Thawing is the process of allowing the ice in the frozen sample to melt at ambient condition. Frozen products must be thawed before further processing or consumption or distribution for retail. Thawing can be detrimental to the structure or quality of the food because of the physical and chemical changes causing damage. While thawing at higher temperature is faster, it also results in greater tissue damage. Though thawing at sub-zero temperatures is said to prevent quality loss, the loss in temperature difference slows down the process of thawing [105, 151]. The application of appropriate temperature at low temperature can be used instead. This is because under the influence of pressure, the melting point temperature of ice is significantly lower. Pressure also reduces the latent heat of fusion of ice, resulting in a greater rate of thawing. The thawing rate increases appreciably due to higher temperature difference between the thawing medium and the frozen product.

When the thawing process is complete, the temperature must be raised above 0°C before releasing pressure in order to avoid ice formation due to adiabatic cooling [28].

1.6.4 DRYING AND OSMOTIC DEHYDRATION

The application of pressure on food samples with the primary purpose of the drying is not observed in general. Nevertheless, HPP has been observed to hasten the drying process. This can be ascribed to the fact that under the effect of pressure, the permeability of cell walls and membranes increases, as a result of which the tissue structure is significantly altered [120, 161]. A comparative analysis of green beans, carrot, and potato pressurized at 600 MPa and 70°C for 15 minutes revealed that upon fluidized bed drying, the drying rate of potato was increased considerably while no such effect was seen in the case of green bean and carrot [41]. Thus, the extent of impact of pressure treatment on the drying rates varies according to the original permeability of the product [120]. Moreover, process conditions (pressure, time, temperature) and matrix of the product also influence the mass transfer rate during the ensuing drying process [103]. A 400 MPa pretreatment at 25°C for 10 minutes resulted in drying rates similar to samples subjected to chemical pretreatment and thus can be employed to replace the latter and to avoid the risk of chemical residues [1]. Due to the increase in mass transfer rates, high-pressure pretreated products exhibited shorter drying time. Strawberry slices subjected to 100 MPa for 5 minutes reduced vacuum drying time by 4 hours and the final moisture content (wet basis) was 5.69% as opposed to 7.7% of the control sample [177].

The osmotic dehydration is a moisture removal process from a lower concentration region to a higher concentration region via a semi-permeable membrane. The concentration gradient across the membrane is the driving force behind the process as a result of which it is generally slow. High-pressure increases the permeability of the cell membrane, which can be utilized to augment the mass transfer during osmosis. HP pretreatments (100–800 MPa) enhanced water loss and solid gain in the sample [121]. Moisture diffusivity and solute diffusivity values were reported to be about four times and twice that of the control sample, respectively.

A significant percentage of water removal can be accredited to the cell wall fracture under the impact of the pressure [121]. This phenomenon has been utilized to facilitate infusion of solute (sucrose) into potato cylinders (200–800 MPa, 20–60°C) reporting diffusivity values as high as eight times compared to those at ambient conditions [147]. However, pressure levels higher than 400 MPa induced starch gelatinization and consequently retarded diffusion and thus gelatinization pressure must be considered for the food product to be pressure treated prior to drying.

1.6.5   REHYDRATION

Rehydration is the process of restoring lost water in dried products and is usually done before consumption of dehydrated foods. However, it is always accompanied by seepage of nutrients and solid loss. Rehydration is a measure of damage undertaken by the product during its prior drying or pre-drying process [168]. High-pressure permeabilizes the cell membrane of treated food, which lowers the rehydration capacity of the product. Consequently, significantly lower values of water and solute diffusivity values were reported during the rehydration of pressure-pretreated samples [118, 121]. Hydration time for common beans (*Phaseolus vulgaris* L.) was considerably reduced after HPP treatment thereby minimizing cooking time and leaching losses [18].

The utilization of high-pressure at the time of soaking appreciably improved the imbibition properties of the product. Pressure treatment (≤600 MPa) in conjunction with soaking was found to enhance water uptake in rice and minimize the soaking time [3]. The swelling capacity of the rice grain was directly governed by process parameters like pressure, temperature, and time.

A study of soybeans soaked in water under 300–500 MPa at 20°C for 0–380 minutes revealed that the seed coat and helium underwent irreversible microstructural changes [175]. This resulted in enhanced water absorption ability. Furthermore, the water distribution was more uniform in the HPP sample compared to the control sample due to restricted water mobility. Nonetheless, proteins in HPP soaked samples underwent partial denaturation because of the pressure. In a comparative study of hydration and cooking of chickpeas, with, and without a pressure pretreatment (275–690 MPa), cooking time was inversely proportional to the soaking period. The 5-minute pressure treatment resulted in cooking time analogous to that obtained in samples soaked for 90 minutes. Higher water intake and lower solute loss was observed in high-pressure hydration of navy beans under moderate pressure (33 MPa, 3 minutes, 55°C) [117]. In pressure soaking of common beans, high-pressure-short time and low pressure-long time favor hydration, resulting in softer grains and reduced cooking time [18].

1.6.6 TEXTURIZATION

The structural and morphological changes in food under the influence of the pressure have been utilized to develop food products with novel texture or improve the texture of ready-to-eat products. A drop in hardness was observed in green beans processed under 500 MPa at 20°C for 5 minutes [83]. A similar observation was reported in the HP processing (400–900 MPa, 60°C, 5–10 minutes) of green peas [115]. This reduction in hardness has been explained by the β-elimination and enzymatic demethylation of pectin in the vegetable matrix and the cell wall damage under the effect of pressure [83, 144, 148]. The denaturation of protein is also a major factor in the texturization of the product.

Saricaoglu et al. [135] studied the impact of HPP (0–150 MPa) on hazelnut meal proteins and reported enhanced rheological and functional characteristics of the meal at HPP up to 100 MPa. It is interesting to note that textural modifications in carrot were dependent on pressure only up to a threshold level of 400 MPa, beyond which changing the pressure (400–1200 MPa) had no significant effect [150]. However, this aspect of HP application requires further research before appropriate application.

1.6.7   EXTRACTION

Conventional extraction achieved with the help of soaking, solvents, maceration, etc., are not favored due to slower extraction, lower yield, inferior extraction efficiency, etc., [178]. HPP has been employed to overcome these drawbacks. High-pressure extraction operates at pressure levels varying from 100 to 500 MPa and temperatures ranging from 20 to 50°C and has been declared eco-friendly by the United States Food and Drug Administration (USDA) [70, 170]. The structural collapse and cell wall rupture due to pressure accelerate mass transfer, which in turn enhances extraction resulting in a decline in process duration and solvent consumption [29, 71, 170]. A 200 MPa treatment on ginseng roots at 60°C for 5 minutes revealed the inception of hollow openings and small particles. Additionally, modification of cellulose and cell wall formation was also observed [29].

Knorr [80] was the first to report caffeine extraction from coffee with the help of HPP. Higher carotenoid content was achieved in tomato puree extracted at 100–400 MPa [134]. The magnitude of pressure, holding time, process temperature, solvent concentration and type, etc., favored the extraction process [91, 170].

The solvent selection is made after a careful consideration of the nature, strength, and polarity of target compounds followed by liquid: solid determination [4, 55, 70, 72, 178]. The solvent must be adequate to submerge the entire solid. Ethanol is the most common solvent used in bioactive extraction from plant sources. In pressure extraction (100–600 MPa, 1–10 minutes) of polyphenols from green tea, pressure level was found to directly affect the extraction yield [170], while the temperature was reported to have a positive effect on the extraction of phenols from longan fruit pericarp [113]. A comparative analysis by Guo et al. [55] between pressure extraction, heating reflux extraction, and microwave-assisted extraction of pectin from navel orange peel revealed that pressure extraction was the most efficient method out of the three methods with higher extraction yield and shorter process time. Furthermore, extracts obtained with the help of pressure displayed superior antioxidant activity [170]; and low processing temperature was seen to favor the extraction of heat sensible compounds [70, 170].

1.6.8 GELATINIZATION

Pressure can be applied in the gelatinization of carbohydrates and proteins [12, 66, 89, 171]. HPP can gelatinize starch granules at ambient conditions. The degree of gelatinization depends on process parameters (pressure, temperature, time) and samples characteristics (type, moisture content) [17, 89]. In a study conducted by Famelart et al. [46], milk processed at 200–400 MPa for 10–30 minutes did not exhibit any gel formation. Gel formation was observed in ultra-filtered and micro-filtered milk concentrate, albeit the gel firmness varied with the concentrations of citrate and protein. While protein concentration had a positive impact on the firmness of the gel, softer gels were obtained at higher concentration of citrate, displaying the highest firmness values at a pH of 5.9. Whey concentrate gel was obtained at a pH of 9.0. Unlike milk concentrate gels, the gel firmness was decided by the magnitude of pressure instead of protein concentration [46]. A stable whey protein gel was obtained with the help of 600 MPa pressure at 30°C for 0–30 min [77].

With the increase of processing time, the amount of alpha-lactalbumin and beta-lactoglobulin A and B declined while the intermolecular disulphide bonds rose in number resulting in stronger and more elastic gels. Pressure treatment of whey proteins (400–600 MPa for 20 min) exhibited lower gel strength and Young's modulus values than thermally induced gels [102]. Moreover, up to 150 MPa, pressure had a direct effect on the gelatinization temperature, while the opposite trend was observed in the range 250–450 MPa [172]. The combined gelation of polysaccharides and proteins affects the final gel network due to intermolecular interaction [40]. Li et al. [90] evaluated the characteristics of pressure-induced β-Lg/κ-car mixed gels at varying pH (3, 5, 7) and reported improved compactness and uniformity of the mixed gel network at higher pressure levels.

1.7 COMMERCIAL AND LEGISLATIVE COMPREHENSION

Despite its numerous advantages, HPP is still not actively employed in the food industry. Several technical limitations hinder its rapid commercialization, a major example of which is the high initial capital investment. The relative unfamiliarity also alienates potential investors from this technology since it comes with its own set of regulations. All industries dealing with food processing or packaging must abide by certain standards.

Any business planning to utilize HPP in any step of their commercial production must report and execute strategies complying with HACCP standards. The processing conditions used, viz. pressure level, treatment time, temperature, etc., must be documented. The chosen HPP parameters must be justified for the product being processed. The safety of any manufactured food product is to be approved by the food safety and standards authority of the country. The HPP has been approved by the USDA to supersede conventional pasteurization methods in the food industry.

According to the United States Department of Agriculture, 5-log cycle reduction of *E. coli* O157:H7 is an adequate microbial safety indicator for the food products [158]. An inspection program personnel (IPP) is required to verify the industry's activity and its compliance with 9 CFR 417.2(a) (1) and 417.2(a)(2). Any business using HPP as a part of their preservation or quality enhancement (whether on-site or outsourced) must present all documents explaining the process flaw and hazard analysis. European companies, likewise, require a single safety assessment and compliance with Regulation (EC) 258/97 [42]. Any major change to the food product needs additional approval [49]. The USA, Japan, and Europe have several functioning companies successfully manufacturing various high-pressure processed products [67].

1.8   SUMMARY AND FUTURE PROSPECTS

The ability of HPP in extending shelf-life while retaining nutritional and sensory properties of food products has been established through a number of studies. This chapter documents and summarizes the researches exploring high-pressure technology, particularly its effect and commercial implementation. HPP has been found to successfully inactivate many pathogenic microorganisms, enzymes, and viruses. The combination of pressure with other preservation technologies can add hurdles to microbial action and thus reduce the severity of a single technique. It has also been reported to enhance several other processes like freezing, gelation, etc. Unlike thermal pasteurization, because of the avoidance of extreme temperatures, HPP has no adverse effect on the organoleptic properties of the product. Temperature-sensitive nutrients are also retained. It helps in eliminating the need for chemical additives [52]. The label of "clean technology" has made it more appealing to the health-conscious generation, enabling industries to compete in the "natural and organic products" market. The development of larger setups, automation, and pseudo-continuous systems will possibly lower the price of the products making it more affordable for the consumers. Further, the reduction of consumed time and energy in high-pressure treatment aligns it with the goals of sustainable technology [107, 155]. HPP has the potential to replace and surpass conventional technologies.

Currently, the commercial application of this technology is restricted to a few developed countries like Japan, the USA, and Canada. This lack of commercialization is due to the high capital investments, technical complexities and the need for more industrial research. The current study on HPP must be expanded to the industrial level and an appropriate scale up must be designed. With the rise in research on pressure processing, the production cost will likely decline over the next decade and pressure-processed food will be easily accessible to everyone.

KEYWORDS

- high-pressure processing
- HPP applications
- HPP affects
- inspection program personnel
- minimal processing
- pulsed electric field

REFERENCES

1. Ade-Omowaye, B. I., Rastogi, N. K., Angersbach, A., & Knorr, D., (2001). Effect of high-pressure or high electrical field pulse pretreatment on dehydration characteristics of paprika. *Innovative Food Science and Emerging Technologies, 2*(1), 1–7.
2. Ahmed, J., Al-Ruwaih, N., Mulla, M., & Rahman, M. H., (2018). Effect of high-pressure treatment on functional, rheological and structural properties of kidney bean protein isolate. *LWT-Food Science and Technology, 91*, 191–197.
3. Ahromrit, A., Ledward, D. A., & Niranjan, K., (2004). High-pressure induced water uptake characteristics of Thai glutinous rice. *Journal of Food Engineering, 72*(3), 225–233.

4. Altuner, E. M., Işlek, C., Çeter, T., & Alpas, H., (2012). High hydrostatic pressure extraction of phenolic compounds from *Maclura pomifera* fruits. *African Journal of Biotechnology, 11*(4), 930–937.

5. Ananta, E., Heinz, V., Schlüter, O., & Knorr, D., (2001). Kinetic studies on high-pressure inactivation of *Bacillus stearothermophilus* spores suspended in food matrices. *Innovative Food Science and Emerging Technologies, 2*(4), 261–272.

6. Andrés, V., Villanueva, M. J., & Tenorio, M. D., (2016). The effect of high-pressure processing on color, bioactive compounds, and antioxidant activity in smoothies during refrigerated storage. *Food Chemistry, 192*, 328–335.

7. Ávila, M., Gómez-Torres, N., Delgado, D., Gaya, P., & Garde, S., (2016). Application of high-pressure processing for controlling *Clostridium tyrobutyricum* and late blowing defect on semi-hard cheese. *Food Microbiology, 60*, 165–173.

8. Ávila, M., Gómez-Torres, N., Delgado, D., Gaya, P., & Garde, S., (2017). Effect of high-pressure treatments on proteolysis, volatile compounds, texture, color, and sensory characteristics of semi-hard raw ewe milk cheese. *Food Research International, 100*, 595–602.

9. Avsaroglu, M., Bozoglu, F., Alpas, H., Largeteau, A., & Demazeau, G., (2015). use of pulsed-high hydrostatic pressure treatment to decrease patulin in apple juice. *High-Pressure Research, 35*(2), 214–222.

10. Aymerich, T., Picouet, P. A., & Monfort, J. M., (2008). Decontamination technologies for meat products. *Meat Science, 78*(1/2), 114–129.

11. Baier, D., Purschke, B., Schmitt, C., Rawel, H. M., & Knorr, D., (2015). Effect of high-pressure-low temperature treatments on structural characteristics of whey proteins and micellar caseins. *Food Chemistry, 187*, 354–363.

12. Balny, C., Masson, P., & Heremans, K., (2002). High-pressure effects on biological macromolecules: From structural changes to alteration of cellular processes. *Biochimica et Biophysica Acta (BBA)-Protein Structure and Molecular Enzymology, 1595*(1/2), 3–10.

13. Barba, F. J., Esteve, M. J., & Frigola, A., (2012). Impact of high-pressure processing on vitamin E (α-, γ-, and δ-tocopherol), vitamin D (cholecalciferol and ergocalciferol), and fatty acid profiles in liquid foods. *Journal of Agricultural and Food Chemistry, 60*(14), 3763–3768.

14. Barug, D., Bhattnagar, D., & Van, E. H. P., (2006). *The Mycotoxin Factbook* (p. 384). Wageningen; Netherlands: Wageningen Academic Publishers.

15. Basset, J., Gratia, A., Macheboeuf, M., & Manil, P., (1938). Action of high-pressures on plant viruses. *Proceedings of the Society for Experimental Biology and Medicine, 38*(2), 248–251.

16. Basson, M. D., Craig, D. H., & Zhang, J., (2007). Cytoskeletal signaling by way of aactinin-1 mediates ERK1/2 activation by repetitive deformation in human $CaCO_2$ intestinal epithelial cells. *The American Journal of Surgery, 194*(5), 618–622.

17. Bauer, B. A., & Knorr, D., (2005). The impact of pressure, temperature and treatment time on starches: Pressure-induced starch gelatinization as pressure time-temperature indicator for high hydrostatic pressure processing. *Journal of Food Engineering, 68*(3), 329–334.

18. Bayindirli, A., Alpas, H., Bozoglu, F., & Hizal, M., (2006). Efficiency of high-pressure treatment on inactivation of pathogenic microorganisms and enzymes in apple, orange, apricot, and sour cherry juices. *Food Control, 17*(1), 52–58.

19. Belmiro, R. H., Tribst, A. A. L., & Cristianini, M., (2018). Impact of high-pressure processing in hydration and drying curves of common beans (*Phaseolus vulgaris L.*). *Innovative Food Science and Emerging Technologies, 47*, 279–285.

20. Betoret, E., Calabuig-Jiménez, L. A. U. R. A., Patrignani, F., Lanciotti, R., & Dalla, R. M., (2017). Effect of high-pressure processing and trehalose addition on functional properties of mandarin juice enriched with probiotic microorganisms. *LWT-Food Science and Technology, 85*(1), 418–422.

21. Buchheim, W., & El-Nour, A. M. A., (1992). Induction of milkfat crystallization in the emulsified state by high hydrostatic pressure. *Lipid/Fett., 94*(10), 369–373.

22. Buckow, R., Isbarn, S., Knorr, D., Heinz, V., & Lehmacher, A., (2008). Predictive model for inactivation of feline calicivirus, a norovirus surrogate, by heat and high hydrostatic pressure. *Applied and Environmental Microbiology, 74*(4), 1030–1038.

23. Butz, P., Heinisch, O., & Tauscher, B., (1994). Hydrostatic high-pressure applied to food sterilization III: Application to spices and spice mixtures. *International Journal of High-Pressure Research, 12*(4–6), 239–243.

24. Carlez, A., Rosec, J. P., Richard, N., & Cheftel, J. C., (1993). High-pressure inactivation of *Citrobacter freundii, Pseudomonas fluorescens*, and *Listeria innocua* in inoculated minced beef muscle. *LWT-Food Science and Technology, 26*(4), 357–363.

25. Chakraborty, S., Kaushik, N., Rao, P. S., & Mishra, H. N., (2014). High-pressure inactivation of enzymes: A review on its recent applications on fruit purees and juices. *Comprehensive Reviews in Food Science and Food Safety, 13*(4), 578–596.

26. Chakraborty, S., Rao, P. S., & Mishra, H. N., (2015). Effect of combined high-pressure-temperature treatments on color and nutritional quality attributes of pineapple (*Ananas comosus* L.) puree. *Innovative Food Science and Emerging Technologies, 28*, 10–21.

27. Cheftel, J. C., (1995). High-pressure, microbial inactivation and food preservation. *Food Science and Technology International, 1*, 75–90.

28. Cheftel, J. C., Lévy, J., & Dumay, E., (2000). Pressure-assisted freezing and thawing: Principles and potential applications. *Food Reviews International, 16*(4), 453–483.

29. Chen, R., Meng, F., Zhang, S., & Liu, Z., (2009). Effects of ultrahigh-pressure extraction conditions on yields and antioxidant activity of ginsenoside from ginseng. *Separation and Purification Technology, 66*(2), 340–346.

30. Choukroun, M., & Grasset, O., (2007). Thermodynamic model for water and high-pressure ices up to 2.2 GPa and down to the metastable domain. *The Journal of Chemical Physics, 127*(12), 124506.

31. Cilla, A., Alegría, A., & De Ancos, B., (2012). Bio-accessibility of tocopherols, carotenoids, and ascorbic acid from milk- and soy-based fruit beverages: Influence of food matrix and processing. *Journal of Agricultural and Food Chemistry, 60*(29), 7282–7290.

32. Considine, K. M., Kelly, A. L., Fitzgerald, G. F., Hill, C., & Sleator, R. D., (2008). High-pressure processing effects on microbial food safety and food quality. *FEMS Microbiology Letters, 281*(1), 1–9.

33. Da Silveira, T. F. F., Cristianini, M., & Kuhnle, G. G., (2019). Anthocyanins, non-anthocyanin phenolics, tocopherols and antioxidant capacity of açaí juice (*Euterpe oleracea*) as affected by high-pressure processing and thermal pasteurization. *Innovative Food Science and Emerging Technologies, 55*, 88–96.

34. Delgado-Martínez, F. J., Carrapiso, A. I., Contador, R., & Ramírez, M. R., (2019). Volatile compounds and sensory changes after high-pressure processing of mature "torta

del Casar" (Raw Ewe's Milk Cheese) during refrigerated storage. *Innovative Food Science and Emerging Technologies, 52,* 34–41.

35. DiCaprio, E., Ma, Y., Purgianto, A., Hughes, J., & Li, J., (2012). Internalization and dissemination of human norovirus and animal caliciviruses in hydroponically grown romaine lettuce. *Applied and Environmental Microbiology, 78*(17), 6143–6152.

36. Drusch, S., & Ragab, W., (2003). Mycotoxins in fruits, fruit juices, and dried fruits. *Journal of Food Protection, 66,* 1514–1527.

37. Dumay, E., Lambert, C., Funtenberger, S., & Cheftel, J. C., (1996). Effects of high-pressure on the Physico-chemical characteristics of dairy creams and model oil/water emulsions. *LWT-Food Science and Technology, 29*(7), 606–625.

38. Earnshaw, R. G., (1995). Kinetics of high-pressure inactivation of microorganisms. In: Ledward, D. A., Johnston, D. E., Earnshaw, R. G., & Hasting, A. P. M., (eds.), *High-Pressure Processing of Foods* (pp. 37–46). Nottingham, England: Nottingham University Press.

39. Eisenmenger, M. J., & Reyes-De-Corcuera, J. I., (2009). High-pressure enhancement of enzymes: A review. *Enzyme and Microbial Technology, 45*(5), 331–347.

40. Ersch, C., Meinders, M. B., & Bouwman, W. G., (2016). Microstructure and rheology of globular protein gels in the presence of gelatin. *Food Hydrocolloids, 55,* 34–46.

41. Eshtiaghi, M. N., Stute, R., & Knorr, D., (1994). High-pressure and freezing pretreatment effects on drying, rehydration, texture, and color of green beans, carrots, and potatoes. *Journal of Food Science, 59*(6), 1168–1170.

42. European Commission (EC), (2004). *Evaluation Report on Novel Food Regulation 258/97 Concerning Novel Foods and Novel Food Ingredient.* DOI: https://ec.europa.eu/food/sites/food/files/safety/docs/novel-food_applications-status_en.pdf (accessed on 11 March 2021).

43. Evelyn, Kim, H. J., & Silva, F. V. M., (2016). Modeling the inactivation of *Neosartorya fischeri* ascospores in apple juice by high-pressure, power ultrasound and thermal processing. *Food Control, 59,* 530–537.

44. Evelyn, Milani, E., & Silva, F. V., (2017). Comparing high-pressure thermal processing and thermosonication with thermal processing for the inactivation of bacteria, molds, and yeasts spores in foods. *Journal of Food Engineering, 214,* 90–96.

45. Evelyn, & Silva, F. V. M., (2019). Heat assisted HPP for the inactivation of bacteria, molds, and yeasts spores in foods: Log reductions and mathematical models. *Trends in Food Science and Technology, 88,* 143–156.

46. Famelart, M. H., Chapron, L., Piot, M., Brulé, G., & Durier, C., (1998). High-pressure-induced gel formation of milk and whey concentrates. *Journal of Food Engineering, 36*(2), 149–164.

47. Farr, D., (1990). High-pressure technology in the food industry. *Trends in Food Science and Technology, 1,* 14–16.

48. Fernández, P. P., Otero, L., Guignon, B., & Sanz, P. D., (2006). High-pressure shift freezing versus high-pressure assisted freezing: Effects on the microstructure of a food model. *Food Hydrocolloids, 20*(4), 510–522.

49. Food Standards Agency (FSA), (2001). *Advisory Committee on Novel Foods and Process* (p. 124). Annual report by FSA.

50. Fulladosa, E., Sala, X., Gou, P., Garriga, M., & Arnau, J., (2012). K-lactate and high-pressure effects on the safety and quality of restructured hams. *Meat Science, 91*(1), 56–61.

51. Georget, E., Sevenich, R., & Reineke, K., (2015). Inactivation of microorganisms by high isostatic pressure processing in complex matrices: A review. *Innovative Food Science and Emerging Technologies, 27*, 1–14.

52. Glass, K. A., McDonnell, L. M., Rassel, R. C., & Zierke, K. L., (2007). Controlling *Listeria monocytogenes* on sliced ham and turkey products using benzoate, propionate, and sorbate. *Journal of Food Protection, 70*(10), 2306–2312.

53. Gould, G. W., & Sale, A. J. H., (1970). Initiation of germination of bacterial spores by hydrostatic pressure. *Journal of General Microbiology, 60*(3), 335–346.

54. Grove, S. F., Lee, A., Lewis, T., Stewart, C. M., Chen, H. Q., & Hoover, D. G., (2006). Inactivation of foodborne viruses of significance by high-pressure and other processes. *Journal of Food Protection, 69*, 957–968.

55. Guo, X., Han, D., Xi, H., Rao, L., Liao, X., Hu, X., & Wu, J., (2012). Extraction of pectin from navel orange peel assisted by ultra-high-pressure, microwave, or traditional heating: A comparison. *Carbohydrate Polymers, 88*(2), 441–448.

56. Hao, H., Zhou, T., Koutchma, T., Wu, F., & Warriner, K., (2016). High hydrostatic pressure assisted degradation of patulin in fruit and vegetable juice blends. *Food Control, 62*, 237–242.

57. Hashizume, C., Kimura, K., & Hayashi, R., (1995). Kinetic analysis of yeast inactivation by high-pressure treatment at low temperatures. *Bioscience, Biotechnology and Biochemistry, 59*(8), 1455–1458.

58. Hendrickx, M., Ludikhuyze, L., Van, D. B. I., & Weemaes, C., (1998). Effects of high-pressure on enzymes related to food quality. *Trends in Food Science and Technology, 9*(5), 197–203.

59. Heremans, K., (1993). The behavior of proteins under pressure. In: Winter, R., & Jonas, J., (eds.), *High-Pressure Chemistry, Biochemistry and Materials Science* (pp. 443–469). Dordrecht, Netherlands: Springer.

60. Hernández-Carrión, M., Hernando, I., & Quiles, A., (2014). High hydrostatic pressure treatment as an alternative to pasteurization to maintain bioactive compound content and texture in red sweet pepper. *Innovative Food Science and Emerging Technologies, 26*, 76–85.

61. Hirneisen, K. A., & Kniel, K. E., (2013). Inactivation of internalized and surface contaminated enteric viruses in green onions. *International Journal of Food Microbiology, 166*(2), 201–206.

62. Hite, B. H., (1899). The effect of pressure in the preservation of milk: A preliminary report. *West Virginia Agricultural Experiment Station, 58*, 15–35.

63. Hoover, D. G., (1993). Pressure effects on biological systems. *Food Technology, 47*(6), 150–155.

64. Hoover, D. G., Metrick, C., Papineau, A. M., Farkas, D. F., & Knorr, D., (1989). Biological effects of high hydrostatic pressure on food microorganisms. *Food Technology (Chicago), 43*(3), 99–107.

65. Houben, K., Kermani, Z. J., & Van, B. S., (2013). Thermal and high-pressure stability of pectin methylesterase, polygalacturonase, β-galactosidase and α-arabinofuranosidase in a tomato matrix: Towards the creation of specific endogenous enzyme populations through processing. *Food and Bioprocess Technology, 6*(12), 3368–3380.

66. Hu, X., Xu, X., Jin, Z., Tian, Y., Bai, Y., & Xie, Z., (2011). Retrogradation properties of rice starch gelatinized by heat and high hydrostatic pressure (HHP). *Journal of Food Engineering, 106*(3), 262–266.

67. Huang, H. W., Wu, S. J., Lu, J. K., Shyu, Y. T., & Wang, C. Y., (2017). Current status and future trends of high-pressure processing in food industry. *Food Control, 72*, 1–8.
68. Jaenicke, R., (1981). Enzymes under extremes of physical conditions. *Annual Review of Biophysics and Bioengineering, 10*(1), 1–67.
69. Jeż, M., Wiczkowski, W., Zielińska, D., Białobrzewski, I., & Błaszczak, W., (2018). The impact of high-pressure processing on the phenolic profile, hydrophilic antioxidant and reducing capacity of purée obtained from commercial tomato varieties. *Food Chemistry, 261*, 201–209.
70. Joo, C. G., Lee, K. H., Park, C., Joo, I. W., Choe, T. B., & Lee, B. C., (2012). Correlation of increased antioxidation with the phenolic compound and amino acids contents of *Camellia sinensis* leaf extracts following ultra high-pressure extraction. *Journal of Industrial and Engineering Chemistry, 18*(2), 623–628.
71. Jun, X., Deji, S., Ye, L., & Rui, Z., (2011). Micro mechanism of ultra high-pressure extraction of active ingredients from green tea leaves. *Food Control, 22*(8), 1473–1476.
72. Jun, X., Shuo, Z., Bingbing, L., Rui, Z., Ye, L., Deji, S., & Guofeng, Z., (2010). Separation of major catechins from green tea by ultrahigh-pressure extraction. *International Journal of Pharmaceutics, 386*(1/2), 229–231.
73. Kalagatur, N. K., Kamasani, J. R., Mudili, V., & Krishna, K., (2018). Effect of high-pressure processing on growth and mycotoxin production of *Fusarium graminearum* in Maize. *Food Bioscience, 21*, 53–59.
74. Kasemwong, K., Ruktanonchai, U. R., & Srinuanchai, W., (2011). Effect of high-pressure micro fluidization on the structure of cassava starch granule. *Starch/ Stärke, 63*(3), 160–170.
75. Katsaros, G. I., Tsevdou, M., Panagiotou, T., & Taoukis, P. S., (2010). Kinetic study of high-pressure microbial and enzyme inactivation and selection of pasteurization conditions for Valencia orange juice. *International Journal of Food Science and Technology, 45*(6), 1119–1129.
76. Keenan, D. F., Rößle, C., Gormley, R., Butler, F., & Brunton, N. P., (2012). Effect of high hydrostatic pressure and thermal processing on the nutritional quality and enzyme activity of fruit smoothies. *LWT-Food Science and Technology, 45*(1), 50–57.
77. Keim, S., & Hinrichs, J., (2004). Influence of stabilizing bonds on the texture properties of high-pressure-induced whey protein gels. *International Dairy Journal, 14*(4), 355–363.
78. Kingsley, D. H., Guan, D., & Hoover, D. G., (2005). Pressure inactivation of hepatitis a virus in strawberry puree and sliced green onions. *Journal of Food Protection, 68*(8), 1748–1751.
79. Kingsly, A. R. P., Balasubramaniam, V. M., & Rastogi, N. K., (2009). Influence of high-pressure blanching on polyphenol oxidase activity of peach fruits and its drying behavior. *International Journal of Food Properties, 12*(3), 671–680.
80. Knorr, D., (1999). Process assessment of high-pressure processing of foods: An overview. In: Oliveira, J. C., (ed.), *Processing Foods: Quality Optimization and Process Assessment* (pp. 249–267). USA: CRC Press.
81. Knorr, D., Schlüter, O., & Heinz, V., (1998). Impact of high hydrostatic pressure on phase transitions of foods. *Food Technology, 52*(9), 42–45.
82. Krebbers, B., Matser, A. M., & Hoogerwerf, S. W., (2003). Combined high-pressure and thermal treatments for processing of tomato puree: Evaluation of microbial inactivation

and quality parameters. *Innovative Food Science and Emerging Technologies, 4*(4), 377–385.

83. Krebbers, B., Matser, A. M., Koets, M., & Van, D. B. R. W., (2002). Quality and storage-stability of high-pressure preserved green beans. *Journal of Food Engineering, 54*(1), 27–33.

84. Lai, C. L., Fuh, Y. M., & Shih, D. Y. C., (2000). Detection of mycotoxin patulin in apple juice. *Journal of Food and Drug Analysis, 8*(2), 85–96.

85. Ledward, D. A., (1995). High-pressure processing-the potential. In: Ledward, D. A., Johnston, D. E., Earnshaw, R. G., & Hasting, A. P. M., (eds.), *High-Pressure Processing of Foods* (pp. 1–5). Nottingham, England: Nottingham University Press.

86. Lévy, J., Dumay, E., Kolodjziejczyk, E., & Cheftel, J. C., (1999). Freezing kinetics of a model oil-in-water emulsion under high-pressure or by pressure release. Impact on ice crystals and oil droplets. *LWT-Food Science and Technology, 32*, 396–405.

87. Lévy, J., Dumay, E., Kolodziejczyk, E., & Cheftel, J. C., (2000). Kinetics of freezing of an oil-in-water emulsion by pressure release. Impact on ice crystals and oil droplets. *International Journal of High-Pressure Research, 19*(1–6), 183–189.

88. Li, B. S., Zhu, Y. F., Zhang, W., & Ruan, Z., (2017). Effect of high-pressure processing (HPP) combined with low and moderate temperature treatments on the color and enzyme inactivation in freshly squeezed lychee juice. *Modern Food Science and Technology, 33*(7), 151–156.

89. Li, W., Bai, Y., Mousaa, S. A., Zhang, Q., & Shen, Q., (2012). Effect of high hydrostatic pressure on physicochemical and structural properties of rice starch. *Food and Bioprocess Technology, 5*(6), 2233–2241.

90. Li, X., He, X., Mao, L., Gao, Y., & Yuan, F., (2020). Modification of the structural and rheological properties of β-lactoglobulin/κ-carrageenan Mixed gels induced by high-pressure processing. *Journal of Food Engineering, 274*, 109851.

91. Liu, F., Wang, D., Liu, W., Wang, X., Bai, A., & Huang, L., (2013). Ionic liquid-based ultrahigh-pressure extraction of five tanshinones from *Salvia miltiorrhiza* Bunge. *Separation and Purification Technology, 110*, 86–92.

92. Liu, S., Chen, Y., Gu, L., Li, Y., & Wang, B., (2013). Effects of ultrahigh-pressure extraction conditions on yields of berberine and palmatine from cortex *Phellodendri amurensis*. *Analytical Methods, 5*(17), 4506–4512.

93. Liu, Y., Zhao, X. Y., Zou, L., & Hu, X. S., (2013). Effect of high hydrostatic pressure on overall quality parameters of watermelon juice. *Food Science and Technology International, 19*(3), 197–207.

94. Lou, F., Neetoo, H., Chen, H., & Li, J., (2011). Inactivation of a human norovirus surrogate by high-pressure processing: Effectiveness, mechanism, and potential application in the fresh produce industry. *Applied and Environmental Microbiology, 77*, 1862–1871.

95. Lou, F., Neetoo, H., Chen, H., & Li, J., (2015). High hydrostatic pressure processing: A promising nonthermal technology to inactivate viruses in high-risk foods. *Annual Review of Food Science and Technology, 6*(1), 389–409.

96. Ludikhuyze, L., Van, L. A., Denys, I. S., & Hendrickx, M. E., (2001). Effects of high-pressure on enzymes related to food quality. In: Hendrickx, M. E. G., Knorr, D., Ludikhuyze, L., Van, L. A., & Heinz, V., (eds.), *Ultra High-Pressure Treatments of Foods* (pp. 115–166). Boston, USA: Springer.

97. Ludikhuyze, L., Van, L. A., Indrawati, Smout, C., & Hendrickx, M., (2003). Effects of combined pressure and temperature on enzymes related to quality of fruits and vegetables: From kinetic information to process engineering aspects. *Critical Reviews in Food Science and Nutrition, 43*(5), 527–586.

98. Mackey, B. M., Forestiere, K., & Isaacs, N., (1995). Factors affecting the resistance of *Listeria monocytogenes* to high hydrostatic pressure. *Food Biotechnology, 9*(1/2), 1–11.

99. Matser, A. M., Krebbers, B., Van, D. B. R. W., & Bartels, P. V., (2004). Advantages of high-pressure sterilization on quality of food products. *Trends in Food Science and Technology, 15*(2), 79–85.

100. Mills, G., Earnshaw, R., & Patterson, M. F., (1998). Effects of high hydrostatic pressure on *Clostridium sporogenes* spores. *Letters in Applied Microbiology, 26*(3), 227–230.

101. Mozhaev, V. V., Heremans, K., Frank, J., Masson, P., & Balny, C., (1994). Exploiting the effects of high hydrostatic pressure in biotechnological applications. *Trends in Biotechnology, 12*(12), 493–501.

102. Ngarize, S., Adams, A., & Howell, N., (2005). A comparative study of heat and high-pressure induced gels of whey and egg albumen proteins and their binary mixtures. *Food Hydrocolloids, 19*(6), 984–996.

103. Oey, I., Van, D. P. I., Van, L. A., & Hendrickx, M., (2008). Does high-pressure processing influence nutritional aspects of plant based food systems? *Trends in Food Science and Technology, 19*(6), 300–308.

104. Ogawa, H., Fukuhisa, K., Kubo, Y., & Fukumoto, H., (1990). Pressure inactivation of yeasts, molds, and pectinesterase in Satsuma mandarin juice: Effects of juice concentration, pH, and organic acids, and comparison with heat sanitation. *Agricultural and Biological Chemistry, 54*(5), 1219–1225.

105. Okamoto, M., Kawamura, Y., & Hayashi, R., (1990). Application of high-pressure to food processing: Textural comparison of pressure and heat-induced gels of food proteins. *Agriculture and Biological Chemistry, 54*(1), 185–189.

106. Palou, E., Lopez-Malo, A., Barbosa-Canovas, G. V., & Swanson, B. G., (2007). High-pressure treatment in food preservation. In: Rahman, M. S., (ed.), *Handbook of Food Preservation* (pp. 833–872). New York, USA: CRC Press.

107. Pardo, G., & Zufía, J., (2012). Life cycle assessment of food-preservation technologies. *Journal of Cleaner Production, 28*, 198–207.

108. Park, S. Y., Ha, J. H., Kim, S. H., & Ha, S. D., (2017). Effects of high hydrostatic pressure on the inactivation of norovirus and quality of cabbage *kimchi. Food Control, 81*, 40–45.

109. Patras, A., Brunton, N. P., Da Pieve, S., & Butler, F., (2009). Impact of high-pressure processing on total antioxidant activity, phenolic, ascorbic acid, anthocyanin content, and color of strawberry and blackberry purées. *Innovative Food Science and Emerging Technologies, 10*(3), 308–313.

110. Patterson, M. F., Ledward, D. A., & Rogers, N., (2006). High-pressure processing: Chapter 6. In: Brennan, J. G., (ed.), *Food Processing Handbook* (pp. 173–197). Weinheim, Germany: Wiley-VCH.

111. Patterson, M. F., Quinn, M., Simpson, R., & Gilmour, A., (1995). Effects of high-pressure on vegetative pathogens. In: Ledward, D. A., Johnston, D. E., Earnshaw, R. G., & Hasting, A. P. M., (eds.), *High-Pressure Processing of Foods* (pp. 47–63). Nottingham, England: Nottingham University Press.

112. Plaza, L., Muñoz, M., De Ancos, B., & Cano, M. P., (2003). Effect of combined treatments of high-pressure, citric acid and sodium chloride on quality parameters of tomato puree. *European Food Research and Technology, 216*(6), 514–519.

113. Prasad, K. N., Yang, B., Zhao, M., Wei, X., Jiang, Y., & Chen, F., (2009). High-pressure extraction of Corilagin from Longan (*Dimocarpus longan Lour.*) fruit pericarp. *Separation and Purification Technology, 70*(1), 41–45.

114. Pyatkovskyy, T. I., Shynkaryk, M. V., Mohamed, H. M., Yousef, A. E., & Sastry, S. K., (2018). Effects of combined high-pressure (HPP), pulsed electric field (PEF) and sonication treatments on inactivation of *Listeria innocua*. *Journal of Food Engineering, 233*, 49–56.

115. Quaglia, G. B., Gravina, R., Paperi, R., & Paoletti, F., (1996). Effect of high-pressure treatments on peroxidase activity, ascorbic acid content, and texture in green peas. *LWT-Food Science and Technology, 29*(5/6), 552–555.

116. Raj, A. S., Chakraborty, S., & Rao, P. S., (2019). Thermal assisted high-pressure processing of Indian gooseberry (*Embilica officinalis L.*) juice-impact on color and nutritional attributes. *LWT-Food Science and Technology, 99*, 119–127.

117. Ramaswamy, R., Balasubramaniam, V. M., & Sastry, S. K., (2005). Effect of high-pressure and irradiation treatments on hydration characteristics of navy beans. *International Journal of Food Engineering, 1*(4) 1–17.

118. Rastogi, N. K., Angersbach, A., & Knorr, D., (2000). Synergistic effect of high hydrostatic pressure pretreatment and osmotic stress on mass transfer during osmotic dehydration. *Journal of Food Engineering, 45*(1), 25–31.

119. 119.Rastogi, N. K., & Raghavarao, K. S. M. S., (2007). Opportunities and challenges in high-pressure processing of foods. *Critical Reviews in Food Science and Nutrition, 47*(1), 69–112.

120. Rastogi, N. K., Raghavarao, K. S. M. S., & Niranjan, K., (2005). Developments in osmotic dehydration. In: Sun, D. W., (ed.), *Emerging Technologies for Food Processing* (pp. 221–250). London, UK: Elsevier Academic Press.

121. Rastogi, N. K., & Niranjan, K., (1998). Enhanced mass transfer during osmotic dehydration of high-pressure treated pineapple. *Journal of Food Science, 63*(3), 508–511.

122. Ravichandran, C., Purohit, S. R., & Rao, P. S., (2018). High-pressure induced water absorption and gelatinization kinetics of paddy. *Innovative Food Science and Emerging Technologies, 47*, 146–152.

123. Rawsthorne, H., Phister, T. G., & Jaykus, L. A., (2009). Development of a fluorescent *in situ* method for visualization of enteric viruses. *Applied and Environmental Microbiology, 75*(24), 7822–7827.

124. Realini, C. E., Guàrdia, M. D., Garriga, M., Pérez-Juan, M., & Arnau, J., (2011). High-pressure and freezing temperature effect on quality and microbial inactivation of cured pork carpaccio. *Meat Science, 88*(3), 542–547.

125. Rendueles, E., & Omer, M. K., (2011). Microbiological food safety assessment of high hydrostatic pressure processing: A review. *LWT-Food Science and Technology, 44*(5), 1251–1260.

126. Riahi, E., & Ramaswamy, H. S., (2003). High-pressure processing of apple juice: Kinetics of pectin methylesterase inactivation. *Biotechnology Progress, 19*(3), 908–914.

127. Rios-Corripio, G., Welti-Chanes, J., Rodríguez-Martínez, V., & Guerrero-Beltrán, J. Á., (2020). Influence of high hydrostatic pressure processing on physicochemical

characteristics of a fermented pomegranate (*Punica granatum L.*) beverage. *Innovative Food Science and Emerging Technologies, 59*, 102249.

128. Roberts, C. M., & Hoover, D. G., (1996). Sensitivity of *Bacillus coagulans* spores to combinations of high hydrostatic pressure, heat, acidity, and nisin. *Journal of Applied Bacteriology, 81*(4), 363–368.

129. Rodrigo, D., Jolie, R., Van, L. A., & Hendrickx, M., (2007). Thermal and high-pressure stability of tomato lipoxygenase and hydroperoxide lyase. *Journal of Food Engineering, 79*(2), 423–429.

130. Rosenthal, A., Pokhrel, P. R., & Da Rocha, F. E. H., (2018). High-pressure processing of fruit products: Chapter 13. In: Rosenthal, A., Deliza, R., Welti-Chanes, J., & Barbosa-Cánovas, G., (eds.), *Fruit Preservation. Food Engineering Series* (pp. 351–398). Springer, New York, USA: Springer.

131. Rubio, B., Martinez, B., Garcia-Cachan, M. D., Rovira, J., & Jaime, I., (2007). Effect of high-pressure preservation on the quality of dry-cured beef "Cecina de Leon". *Innovative Food Science and Emerging Technologies, 8*(1), 102–110.

132. Saikaew, K., Lertrat, K., Meenune, M., & Tangwongchai, R., (2018). Effect of high-pressure processing on color, phytochemical contents, and antioxidant activities of purple waxy corn (*Zea mays L. var. ceratina*) kernels. *Food Chemistry, 243*, 328–337.

133. Sale, A. J. H., Gould, G. W., & Hamilton, W. A., (1970). Inactivation of bacterial spores by hydrostatic pressure. *Microbiology, 60*(3), 323–334.

134. Sánchez-Moreno, C., Plaza, L., De Ancos, B., & Cano, M. P., (2004). Effect of combined treatments of high-pressure and natural additives on carotenoid extractability and antioxidant activity of tomato puree (*Lycopersicum esculentum Mill.*). *European Food Research and Technology, 219*(2), 151–160.

135. Saricaoglu, F. T., Gul, O., Besir, A., & Atalar, I., (2018). Effect of high-pressure homogenization (HPH) on functional and rheological properties of hazelnut meal proteins obtained from hazelnut oil industry by-products. *Journal of Food Engineering, 233*, 98–108.

136. Sarker, M. R., Akhtar, S., Torres, J. A., & Paredes-Sabja, D., (2015). High hydrostatic pressure-induced inactivation of bacterial spores. *Critical Reviews in Microbiology, 41*(1), 18–26.

137. Scheinberg, J. A., Svoboda, A. L., & Cutter, C. N., (2014). High-pressure processing and boiling water treatments for reducing *Listeria monocytogenes*, *Escherichia coli* O157: H7, *Salmonella spp.*, and *Staphylococcus aureus* during beef jerky processing. *Food Control, 39*, 105–110.

138. Schlüter, O., George, S., Heinz, V., & Knorr, D., (1999). Pressure-assisted thawing of potato cylinders. In: Ludwig, H., (ed.), *Advances in High-Pressure Bioscience and Biotechnology* (pp. 475–480). Verlag, Berlin: Springer.

139. Serra, X., Grèbol, N., Guàrdia, M. D., & Guerrero, L., (2007). High-pressure applied to frozen ham at different process stages. Effect on the sensory attributes and on the color characteristics of dry-cured ham. *Meat Science, 75*(1), 21–28.

140. Shahbazi, M., Majzoobi, M., & Farahnaky, A., (2018). Physical modification of starch by high-pressure homogenization for improving functional properties of κ-carrageenan/starch blend film. *Food Hydrocolloids, 85*, 204–214.

141. Sharabi, S., Okun, Z., & Shpigelman, A., (2018). Changes in the shelf-life stability of riboflavin, vitamin C and antioxidant properties of milk after (Ultra) high-pressure

homogenization: Direct and indirect effects. *Innovative Food Science and Emerging Technologies, 47*, 161–169.

142. Shayanfar, S., Chauhan, O. P., Toepfl, S., & Heinz, V., (2014). Effect of nonthermal hurdles in extending shelf-life of cut apples. *Journal of Food Science and Technology, 51*(12), 4033–4039.

143. Shigehisa, T., Ohmori, T., Saito, A., Taji, S., & Hayashi, R., (1991). Effects of high hydrostatic pressure on characteristics of pork slurries and inactivation of microorganisms associated with meat and meat products. *International Journal of Food Microbiology, 12*(2/3), 207–215.

144. Sila, D. N., Duvetter, T., De Roeck, A., & Verlent, I., (2008). Texture changes of processed fruits and vegetables: Potential use of high-pressure processing. *Trends in Food Science and Technology, 19*(6), 309–319.

145. Smelt, J., (1998). Recent advances in the microbiology of high-pressure processing. *Trends in Food Science and Technology, 9*(4), 152–158.

146. Smith, N. A. S., Burlakov, V. M., & Ramos, A. M., (2013). Mathematical modeling of the growth and coarsening of ice particles in the context of high-pressure shift freezing processes. *The Journal of Physical Chemistry B, 117*(29), 8887–8895.

147. Sopanangkul, A., Ledward, D. A., & Niranjan, K., (2002). Mass transfer during sucrose infusion into potatoes under high-pressure. *Journal of Food Science, 67*(6), 2217–2220.

148. Stute, R., Klingler, R. W., & Boguslawski, S., (1996). Effects of high-pressure treatment on starches. *Starch Stärke, 48*(11/12), 399–408.

149. Sun, S., Rasmussen, F. D., Cavender, G. A., & Sullivan, G. A., (2019). Texture, color, and sensory evaluation of sous-vide cooked beef steaks processed using high-pressure processing as a method of microbial control. *LWT-Food Science and Technology, 103*, 169–177.

150. Sun, Y., Kang, X., Chen, F., Liao, X., & Hu, X., (2019). Mechanisms of carrot texture alteration induced by pure effect of high-pressure processing. *Innovative Food Science and Emerging Technologies, 54*, 260–269.

151. Suzuki, C., & Suzuki, K., (1963). The gelation of ovalbumin solutions by high-pressures. *Archives of Biochemistry and Biophysics, 102*(3), 367–372.

152. Tang, Q., Li, D., Xu, J., Wang, J., Zhao, Y., Li, Z., & Xue, C., (2010). Mechanism of inactivation of murine norovirus-1 by high-pressure processing. *International Journal of Food Microbiology, 137*: 186–189.

153. Tauscher, B., (1995). Pasteurization of food by hydrostatic high-pressure: Chemical aspects. *European Food Research and Technology, 200*(1), 3–13.

154. Terefe, N. S., Yang, Y. H., Knoerzer, K., Buckow, R., & Versteeg, C., (2010). High-pressure and thermal inactivation kinetics of polyphenol oxidase and peroxidase in strawberry puree. *Innovative Food Science and Emerging Technologies, 11*(1), 52–60.

155. Toepfl, S., Mathys, A., Heinz, V., & Knorr, D., (2006). Potential of high hydrostatic pressure and pulsed electric fields for energy-efficient and environmentally friendly food processing. *Food Reviews International, 22*(4), 405–423.

156. Tokuşoğlu, Ö., Alpas, H., & Bozoğlu, F., (2010). High hydrostatic pressure effects on mold flora, citrinin mycotoxin, hydroxytyrosol, oleuropein phenolics and antioxidant activity of black table olives. *Innovative Food Science and Emerging Technologies, 11*(2), 250–258.

157. Torres, J. A., Saraiva, J. A., & Guerra-Rodríguez, E., (2014). Effect of combining high-pressure processing and frozen storage on the functional and sensory properties of horse

mackerel (*Trachurus trachurus*). *Innovative Food Science and Emerging Technologies, 21*, 2–11.

158. United States Department of Agriculture (USDA), (2012). *High-Pressure Processing (HPP) and Inspection Program Personnel (IPP) Verification Responsibilities* (p. 37). FSIS Directive 6120.1; USDA.

159. Van, E. R., Asano, T., & Le Noble, W. J., (1989). Activation and reaction volumes in solution. *Chemical Reviews, 89*, 549–688.

160. Varela-Santos, E., & Ochoa-Martinez, A., (2012). Effect of high hydrostatic pressure (HHP) processing on physicochemical properties, bioactive compounds and shelf-life of pomegranate juice. *Innovative Food Science and Emerging Technologies, 13*, 13–22.

161. Vega-Gálvez, A., Uribe, E., & Perez, M., (2011). Effect of high hydrostatic pressure pretreatment on drying kinetics, antioxidant activity, firmness, and microstructure of aloe vera (*Aloe barbadensis* Miller) gel. *LWT-Food Science and Technology, 44*(2), 384–391.

162. Volkert, M., Puaud, M., Wille, H. J., & Knorr, D., (2012). Effects of high-pressure-low temperature treatment on freezing behavior, sensorial properties, and air cell distribution in sugar-rich dairy-based frozen food foam and emulsions. *Innovative Food Science and Emerging Technologies, 13*, 75–85.

163. Wang, B., Li, D., Wang, L. J., Liu, Y. H., & Adhikari, B., (2012). Effect of high-pressure homogenization on microstructure and rheological properties of alkali-treated high-amylose maize starch. *Journal of Food Engineering, 113*(1), 61–68.

164. Wang, C. Y., Huang, H. W., Hsu, C. P., & Yang, B. B., (2016). Recent advances in food processing using high hydrostatic pressure technology. *Critical Reviews in Food Science and Nutrition, 56*(4), 527–540.

165. Wei, J., Jin, Y., Sims, T., & Kniel, K. E., (2010). Manure-and biosolids-resident murine norovirus 1 attachment to and internalization by romaine lettuce. *Applied and Environmental Microbiology, 76*(2), 578–583.

166. Williams, A., (1994). New technologies in food preservation and processing: Part II. *Nutrition and Food Science, 94*(1), 20–23.

167. Wilson, D. R., Dabrowski, L., Stringer, S., Moezelaar, R., & Brocklehurst, T. F., (2008). High-pressure in combination with elevated temperature as a method for the sterilization of food. *Trends in Food Science and Technology, 19*(6), 289–299.

168. Witrowa-Rajchert, D., & Lewicki, P. P., (2006). Rehydration properties of dried plant tissues. *International Journal of Food Science and Technology, 41*(9), 1040–1046.

169. Woldemariam, H. W., & Emire, S. A., (2019). High-pressure processing of foods for microbial and mycotoxins control: Current trends and future prospects. *Cogent Food and Agriculture, 5*(1), 1622184.

170. Xi, J., & Wang, B., (2013). Optimization of ultrahigh-pressure extraction of polyphenolic antioxidants from green tea by response surface methodology. *Food and Bioprocess Technology, 6*(9), 2538–2546.

171. Yang, Z., Gu, Q., & Hemar, Y., (2013). In situ study of maize starch gelatinization under ultra-high hydrostatic pressure using x-ray diffraction. *Carbohydrate Polymers, 97*(1), 235–238.

172. Ye, H. Y., Yang, S. L., & Ye, T. H., (2000). Effects of high-pressure on the properties of starch gelatinization. *Journal of the Chinese Cereals and Oils Association, 15*(1), 10–13.

173. Yordanov, D. G., & Angelova, G. V., (2010). High-pressure processing for foods preserving. *Biotechnology and Biotechnological Equipment, 24*(3), 1940–1945.

174. Yu, C., Wu, F., Cha, Y., Zou, H., Bao, J., Xu, R., & Du, M., (2018). Effects of high-pressure homogenization on functional properties and structure of mussel (*Mytilus edulis*) myofibrillar proteins. *International Journal of Biological Macromolecules, 118*, 741–746.

175. Zhang, H., Ishida, N., & Isobe, S., (2004). High-pressure hydration treatment for soybean processing. *Transactions of the ASAE, 47*(4), 1151–1158.

176. Zhang, L., Dai, S., & Brannan, R. G., (2017). Effect of high-pressure processing, browning treatments, and refrigerated storage on sensory analysis, color, and polyphenol oxidase activity in pawpaw (*Asimina triloba L.*) pulp. *LWT-Food Science and Technology, 86*, 49–54.

177. Zhang, L., Liao, L., Qiao, Y., Wang, C., Shi, D., An, K., & Hu, J., (2020). Effects of ultrahigh-pressure and ultrasound pretreatments on properties of strawberry chips prepared by vacuum-freeze drying. *Food Chemistry, 303*, 125386.

178. Zhang, S., Wang, Z., Cheng, C. H., Wang, T., Ni, Z., & Lin, J., (2011). Optimization for ultra high-pressure extraction of *Scutellaria barbata* by central composite design response surface methodology. *African Journal of Biotechnology, 10*(65), 14637–14343.

179. Zheng, H., Han, M., Bai, Y., Xu, X., & Zhou, G., (2019). Combination of high-pressure and heat on the gelation of chicken myofibrillar proteins. *Innovative Food Science and Emerging Technologies, 52*, 122–130.

180. Zhou, C. L., Liu, W., Zhao, J., Yuan, C., Song, Y., Chen, D., Ni, Y. Y., & Li, Q. H., (2014). The effect of high hydrostatic pressure on the microbiological quality and physical-chemical characteristics of pumpkin (*Cucurbita maxima Duch.*) during refrigerated storage. *Innovative Food Science and Emerging Technologies, 21*, 24–34.

181. ZoBell, C. E., (1970). Pressure effects on morphology and life processes of bacteria. In: Zimmerman, A. M., (ed.), *High-Pressure Effects on Cellular Processes* (pp. 85–130). New York, USA: Academic Press.

182. Zulkurnain, M., Maleky, F., & Balasubramaniam, V. M., (2016). High-pressure crystallization of binary fat blend: A feasibility study. *Innovative Food Science and Emerging Technologies, 38*, 302–311.

CHAPTER 2

ULTRASONICATION OF FOODS

BRINDHA DEIVENDRAN, MADHUMITHA MARAN, and
S. SUBHASHINI

ABSTRACT

Ultrasonic is one of the upcoming technologies, which saves time, has
non-destructive applications. Sonication is a non-thermal technology with a
frequency of sound waves >18 kHz. In many fields of food technology, the
ultrasound sound is used in crystallization, freezing, bleaching, degassing,
etc. Ultrasound is also for food processing to increase the shelf-life of foods
and to make the food taste better. It offers advantages, such as productivity,
yield, better quality product, less time, economical, and eco-friendly.

2.1 INTRODUCTION

Major applications of ultrasound that has attracted the interest of Research
and Development (R&D) in the past two decades are classified into: food
processing and food analysis. Ultrasound has emerged as a novel perspective
method in food science and packaging industry. Ultrasound can be defined
as the sound waves of sounds with a frequency range higher than the
hearing capacity of human ears (~20 kHz) [44]. It ensures the quality and
safety of food products [33]. When the ultrasound energy is channeled into
food materials, it induces compression and decompression (also called
rarefactions) in particles of the medium, because of which, a higher amount
of energy is produced. In ultrasound, the sound ranges will be differentiated
into major two categories:

- **Low Energy:** High frequency ultrasound used for diagnosis in the
  range of megahertz; and

- **Superior Energy:** The frequency that will be lesser and kilohertz will be the range of power ultrasound.

Ultrasound may be used to monitor the composition of meat products, poultry, raw fish, and the fermented stages of increased energy (intensity that is high, high power: 2500 kHz with intensities >one Wcm^2). It is also used for controlling microstructures and modification, emulsification, defoaming of food products, etc., [43]. Apart from the food industry, ultrasound is also used in various fields that are listed in Table 2.1. This chapter focuses on the applications of ultrasonic technology in foods.

TABLE 2.1   Frequency Ranges and its Applications in Medicine

Frequency	Application	Device Used	Description
30–150 kHz	Teeth related issues	Micro probes	Taking away of debris
0.1–1 MHz	To remove stone formation in kidneys	Lithotripter	Stones formed in the kidney broken by ultrasound
0.5–1.5 MHz	The surgical removal of body tissue	High intensity focused ultrasound (HIFU)	Cell death due to heating of tissues; temperature increase to 42°C
0.7–3 MHz	Therapies DID in humans.	Normal probes	Removal of malignant tumor-forming cells by heating certain points of internal organs.
1–20 MHz	Photocopying the human organs.	Ultrasonic imaging	Image capturing of organs.

2.2   GENERATION OF POWER ULTRASOUND

Power ultrasound uses the frequency ranging from 20 to 100 kHz with an intensity of sound varying from 10 to 1000 W/cm^2. The ultrasonic transducers can change electrical energy into sound energy, which consists of two types: *Normal and common usage:* piezoelectric and magnetostrictive, where piezoelectric considered to be common. Magnetostrictive transducers are not that efficient based on usage of electric power, but have strong construction for use in heavy-duty acoustic cavitation in sonication. Cavitation is an issue because liquid moves quickly due to erosion crossing across the surface of metals (e.g., impellors and pipes). When the sound produced by ultrasound passes through the liquid medium, then the cavitation bubbles are generated.

Compression and rarefaction are the series in which the sound waves travel to affect the molecules in liquid. When the cycle of rarefaction goes beyond the forces that may attract between the liquid molecules, there must be a formation of a void or a cavity. The void is designed to take a subtle amount of water vapor in the structure from the given solution during the compressed period the bubble will not collapse in a total state, but rather will start to get bigger than its usual size for the upcoming process cycles to form a bubble of acoustic cavitation. There may be the countless number of bubbles in a liquid medium, where certain bubbles will not have much reaction but whereas others will grow bigger and will undergo a violent collapse due to instability in size and will generate a temperature of about 5,000 K and a pressure of about 50 MPa. The microbial cell disruption causes the pressure to change from these implosions [9, 46, 56]. The bubbles formed by cavitation can be categorized as follows:

- **Stable Cavitation:** These are bubbles that moving in to- and fro-motion are often non-linear, towards some equilibrium size of acoustic pressure during the process cycles. Large bubble clods are formed since they stay for a longer time and will gather around them.
- **Transient Cavitation:** It is also called inertial cavitation, the bubbles will lead to violent collapse, and these are for short duration [1, 38, 65].

Due to a venturi effect (the forcing of a liquid through a small opening or orifice at high pressure), cavitation can also be formed. After coming back from this small orifice, liquid flows in the direction of a pressure gradient, and cavitation of bubbles is produced by decompression. This might be the driving force for the technology that has been followed by the food industries for mixing and emulsification [47].

2.2.1 MAGNETO STRICTIVE TRANSDUCER

This system is of extremely strong construction with a very strong driving force. There will be a loss in electrical energy as heat for about 40%, and therefore, something external is required for cooling with an operating frequency to maximum cut-off of 100 kHz [55].

2.2.2 PIEZOELECTRIC TRANSDUCER

It has high electromechanical conversion and over 95% are electrically efficient. In addition, it can easily operate at the overall ultrasonic range [55].

2.3 FUNDAMENTALS OF ULTRASOUND METHOD

2.3.1 DISINTEGRATION OF CELL STRUCTURES USING ULTRASOUND

It is most probably used in the method of extraction of intracellular materials, where extracting starch from the cell wall is an example. The disintegration of ultrasonics can be tested easily in any method. It can be used at laboratory scale for about one ml to 5 L. At the upper limit, it can be used at 0.1 to 20 L/minute. The production scale starts at 20 L/minute [55].

2.3.2 EXTRACTION OF PROTEIN AND ENZYME

Proteins and enzymes, which are extracted and stored in cells and sub-cellular particles, are also processed by ultrasound [55].

2.3.3 LIPID AND PROTEIN EXTRACTION

The residual oil or fat, which comes out during the pressing of cake, is facilitated by the destruction of the cell wall during pressing. This technique can be used to analyze fruits (citrus oil), oil extracted from mustard (grounded), etc. [55].

2.3.4 SILENCING OF MICROBES AND ENZYMES

Microbes and enzyme inactivation of fruit juices and sauces are also carried out. Undesirable alterations of sensory attributes can be caused by thermal treatments that are texture, flavor, color, smell, and other nutritional aspects like proteins, vitamins, etc. Ultrasound can be a minimal processing efficient technique and is non-thermal processing [55].

2.3.5 ULTRASONIC DEAGGLOMERATION AND DISPERSION

Ultrasonic cavitation can break the particles, because it generates high shear energy and into single dispersed particles by agglomeration [55].

2.3.6 TEMPERATURE AND PRESSURE SYNERGIES OF ULTRASOUND

The effectiveness of ultrasonication (US) is higher when the anti-microbial method uses: thermo-sonication, which involves heat and ultrasound; mano-sonication, which involves pressure and ultrasound and mano-thermo-sonication pressure, which involves a combination of heat and ultrasound [55].

2.3.7 SONICATION FOR LEAK DETECTION

Ultrasound is widely used in the leak detection of bottle and cans in the filling process. This is done by detecting leakage of carbon dioxide during its release then the leakage [55].

2.3.8 LIBERATION OF PHENOLIC COMPOUNDS AND ANTHOCYANINS

The processing steps followed that is thawing, mashing, and enzyme incubation are used to extract juice from bilberries species and black currants species. The matrix of grape and berry is used to obtain phenolic compounds and anthocyanins [55].

2.4 MEASUREMENT OF ULTRASONIC ENERGY

2.4.1 DOSIMETRY

The equipment for ultrasonic method can change or differentiate the supply of power that is being produced. The absolute power for the liquid processing will not be affected because this parameter is not responsible for the measurement in general terms. Usually, the measurement of acoustic power in a liquid is done by a hydrophone; however, in food processing applications, the intensity in use is quite high, and this equipment can prove to be fragile. It is improbable to generate a precise amount of power because it reflects the sound wave that is used for the production of a cavity called cavitation. If there is no loss of heat in a reactor, the calorimetry can provide the correct estimation. However, most of the reactors do have some losses; therefore, it is only used as a temporary tool.

The amount of energy being transferred to a food medium is called an acoustic energy density (Wcm^3 or W/ml) or ultrasound intensity (W/cm^2). Each type of equation needs a power measurement and the intensity can be achieved by dividing the density and total volume of the liquid or area of the transducers for the emitting face. Energy density is generally the most useful among these. If one is interested in the chemical effects of sonication, then a different measurement technique might be required. In a scenario like this, it will be a mandatory to determine the level of radicals produced rather than the heating effect [17].

2.4.2 CHEMICAL DOSIMETRY

The energy produced during the collapse of cavitation is enough to break-down the water chemical bonds for yielding OH and H radicals. They are extremely reactive with short-term life-span and the hydrogen peroxide (H_2O_2) is produced as a byproduct. Iodine dosimetry method works on the principle that H_2O_2 reacts quite fast with I_2 in the medium of the solution to give out iodine in a molecular form that will absorb at 355 nm in the spectrum produced by the visible light. Therefore, the iodine yield can be determined by shooting up the absorbance range to 355 nm on ultraviolet (UV) or spectrophotometer that is changed to H_2O_2 concentration by using Beer and Lamberts Law and stoichiometry [4]. However, estimation for H_2O_2 production in a food system during an ultrasound treatment is complicated because there will be a presence of food components, such as different charged ions and particles. Till now, there has been no novel trustworthy method to measure the activity of cavitation in a food system. Therefore, amplitude vibration measurement of the transducer serves as an indicator to the cavitation of ultrasonic and might be a dependable way to determine the power of ultrasound [56, 68].

2.5 TYPES OF ULTRASOUND

2.5.1 DOPPLER ULTRASOUND

These are used to determine the speed and direction of the cells in blood that propagate along veins, arteries, and capillaries. The imaging helps the physician to detect any malfunction within the flow of blood and problems in the congenital parts. It is commonly used in obstetrics to assess and monitor the growth of a baby in the womb and also to detect the heart rate in the fetus.

2.5.2 OBSTETRIC ULTRASOUND

This is a non-invasive ultrasound method to scan the fallopian tubes of the uterus, ovaries, and baby in the womb. They are also used to confirm early pregnancies and also monitor fetal size, age, and weight. In addition, they can be used to identify and diagnose the problems in the pregnancy.

2.5.3 3D AND 4D ULTRASOUND TECHNOLOGIES

The 3D-ultrasound technology produces an invariable 3D image of the unborn child or anything else to be examined, whereas the 4D-ultrasound can additionally track the movement of a 3D image of an unborn child or any other part of the body. The 4D-ultrasound permits to observe 3D images in movement, which permits to see the baby's movements before its birth.

2.5.4 ECHOCARDIOGRAM (ECG)

The ECG monitors the heartbeat, valves functions, and the blood flow. It can determine the amount of blood pumped by the heart and can monitor the motion of heart walls.

2.5.5 CAROTID ULTRASOUND

This method helps to monitor and to view the arteries that are located along the neck. These arteries are responsible for the blood supply to the brain.

2.6 TYPES OF SONICATION

Preservation of food uses ultrasound in combination with other methods to increase the effectiveness of ultrasound technique. Many studies have come up with a combination of ultrasound with either oppression, temperature or oppression and temperature.

2.6.1   ULTRASONICATION (US)

Ultrasonication (US) is a process of how ultrasound can be applied with the lowest temperature. Therefore, this method is used for temperature-sensitive products concerning about the loss of nutrients, such as vitamin-C, denaturation of protein, non-enzymatic browning, etc. However, it needs exposure for a long period to kill/deactivate stable enzymes and/or microorganisms causing a need for excessive energy. During application of ultrasound, there are chances for temperature to rise up depending on the power of ultrasonication and exposure time [55].

2.6.2   THERMO SONICATION (TS)

Thermo sonication (TS) is a combination of thermal energy and waves produced by ultrasound. In this technique, the material is exposed to ultrasound in combination with subtle heat. As a result of additional heat, ultrasound produces a high amount of cavitation, giving positive results on inactivation of microbes than the heat alone. Therefore, combining low-frequency ultrasound with mild heat helps in reduction of the processing time by 55% and processing temperature by 16% and this helps in the sensory quality of the product [55].

2.6.3   MANOSONICATION (MS)

Manosonication (MS) is the combination of ultrasound with pressure. In this method, enzymes or microbes are inactivated, when ultrasound is combined with tolerable pressure with the temperature at the lowest stage. At the same temperature, the efficiency of inactivation will be higher than the sound waves produced by ultrasound [55].

2.6.4   MANOTHERMOSONICATION (MTS)

Manothermosonication (MTS) technology combines heat, ultrasound, and pressure. The applied pressure and temperature maximizes the cavitation giving higher efficiency for inactivation of enzymes and microorganisms. At the same temperature, MTS inactivates several enzymes in a shorter time than the thermal treatments. MTS can inactivate microorganisms having high thermo-tolerance. MTS can also inactivate thermo-resistant enzymes, such as lipoxygenase, peroxidase (POD), and polyphenol oxidase (PPO) [55].

2.7   EFFECTS OF ULTRASOUND

When the ultrasound has high intensity to propagate across a medium, the effects are produced, and the need for this technique is determined by the attributes of the source. Ideally, this sound wave creates an oscillating compression and decompression of the media. However, in a liquid medium, the rarefaction wave motion can often exceed the attractive forces as ultrasonic power reaches a threshold value, thus creating cavitation bubbles from the existing gas nuclei [67]. These bubbles further give rise to a cavitation phase, which is stable, resulting in the micro agitation of the liquid by maintaining a stable increase and decrease in size. However, the bubbles can also expand and collapse, which in turn can generate an extreme temperature (4726°C) and pressure (15871.6 psi). This leads to the formation of a shear wave of increased energy and disturbance within the cavitation zone. This formed effect is called temporary cavitation [40]. It generates a micro-jet that hits a solid surface if the breakdowns of bubbles are uniformly produced and so on [44]. Hence, this is one of the important effects when ultrasound works on high power for sanitizing operations. The mechanism is schematized in Figure 2.1. Moreover, an injection of fluid to the inner side of the solid can take place when these microjets hit its surface [45].

The magnitude of this phenomenon and its consequences are influenced by the attributes of the liquid, such as viscosity, and process parameters (such as: ultrasonic pressure or frequency and ultrasonic intensity). However, if the medium is gas during application of ultrasound, it is very difficult to achieve an effective propagation of sound waves at increased frequencies. This is because of a mismatch of acoustic impedance that exists between transducers and the gas. Nevertheless, application of ultrasound in an efficient manner can result in varying effects within the interface, like micro stirring or variations of pressure, thus minimizing the width or thickness of the matrix boundary by having an influence on mass transfer phenomena [11].

Ultrasonic waves generate alternating compressions and expansions and these effects are similar to the effects seen on subsequent squeezing and releasing of a sponge [18]. This effect leads to the release of fluid from a particle's inner side and it is called sponge effect. This movement directs to the solid surface and the liquid enters from the external phase. Various microscopic passages are created and the matter interchange takes place easily as water molecules are maintained within the capillaries. This happens because, in the above mechanism, the forces involved exceed the force due to surface tension. Other repercussions can include solid structure degradation or deformation and the variation in surface tension or viscosity. Often, the phenomenon of heat and mass transfer is influenced by ultrasound effects. When the ultrasound waves are allowed to solid matrix immersed in a medium, the internal transport is facilitated easily by the entry of liquid into the solid matrix and/or exit in an easy manner. The exchange between the solid interface and surrounding fluid also tends to be easy. Hence, when ultrasound is applied effectively, it creates many applications that involve reducing internal resistance and external resistance to transport or heat and mass transport [52].

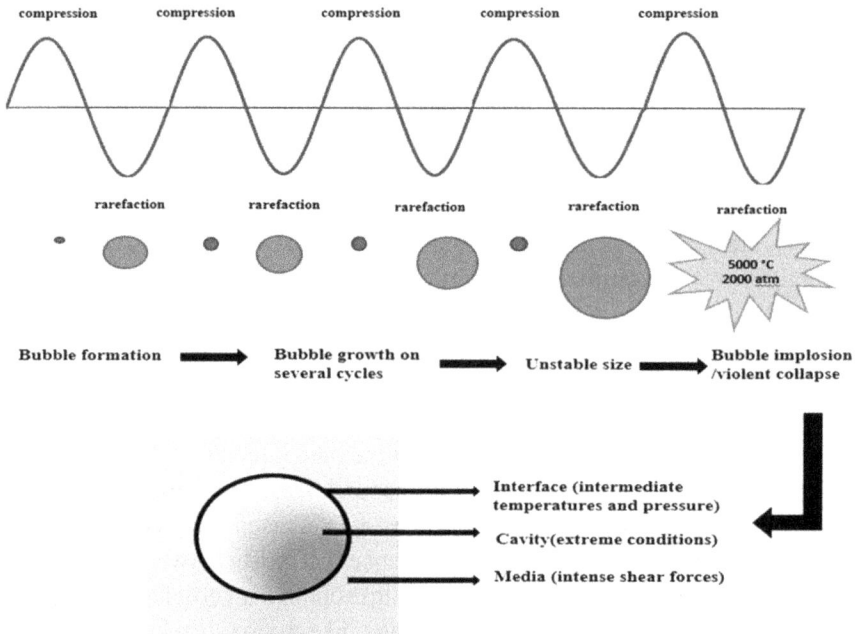

FIGURE 2.1   Effects of ultrasound on bubble cavitation.

2.8   ULTRASONIC SENSORS

Transducers are used for sensing the medium in ultrasound, which is categorized in various arrangements. Ultrasound involves Pulse-echo method, which is the most frequent configuration. On the exterior area of piping, there is a single transducer transmitting an ultrasonic pulse; the echoed signals are received by the respective transducer. The reflections and echoes are produced when ultrasonic pulse joins interfaces or boundaries connecting divergent voids or matter that are present inside the material. Possibly, this method offers a reliable way to estimate the depth of a fouling film reason being the limits that are present in the middle of the pipe, the film, and pipe wall and also in between the film and the outcome. Measurement of time is necessary for producing the data for concluding the depth of several materials, and data is collected from the time taken by the reflection to come back to the transducer. However, the velocity of sound in the matter should be well studied for application of this technique. The film area and, in turn, the margin between film and outcome is usually irregular; as a result, ultrasound pulse dissipates leaving a small fraction of the signal [70].

2.9   ULTRASONICS IN EMULSIFICATION

Among both flaxseed oil and fish oil, the former has an additional advantage of satisfying odor. The creamier taste and greater stability of emulsion come from small droplet sizes. Moreover, oil droplet <100 nm can produce an emulsion that is translucent and this emulsion can be introduced in the beverages and food-gels without losing its clarity. Numerous techniques are available for the fabrication of such emulsions.

Valve homogenization is the conventional mechanism used in the food industry. This mechanism is energy-intensive due to the effectiveness of a small part of the applied energy. The principle of this technique is the particle size reduction by the collision of particles through the architecture of micro-fluidic channel compared to a straight shear field. However, major work so far has been the concentration on the composition of synthetic emulsions. Emulsification of ultrasound exists because of techniques, such as First technique being the emission of oil stage in the water channel in the arrangement of small droplets reason being when the acoustic field is applied, it constructs unstable interfacial waves. Secondly, when ultrasound of low frequency is used, it causes acoustic cavitation [34].

2.10   INACTIVATION OF MICROBES

The eradication of microorganisms is an issue for customers. In food processing, constructive destruction of microbes is crucial; one report on adulterated or spoilt food is a threat to the position, name, and success of the manufacturer. For minimizing the bacterial load of a product, initial contamination should be reduced, microorganisms should be inactivated, and procedures should be implemented for preventing the subsequent growth of inactivated microbial population [60]. Irradiations of ultrasound can be utilized for the inactivated population of bacteria. Traditional techniques for inactivated bacteria generally require heat processing and the processing forms the unpleasant flavors and result in the nutrient loss [55].

2.11   APPLICATIONS

2.11.1   DRYING

Acoustic drying shows great potential and has commercial importance [15]. The oldest method of dehydrating edible products has been done by high temperature air. Although the method might sound economical, yet there is a problem of retention of interior moisture. In addition, high temperature air can cause damage to the food, by affecting the taste, color, and nutritional properties [25]. On the other hand, ultrasonic osmotic dehydration technology utilizes lower solution temperature and obtains higher solute grain rates and water loss [66]. Therefore, color, flavor, and nutritional properties are not unaffected. With this treatment, a reduction in rehydration properties and subsequent conventional freeze-drying time has been observed [25].

2.11.2   FILTRATION TECHNIQUES

The filtration method is a separation of unwanted materials in the food industry and this process isolates solids from its mother liquid or to produce solids-free liquid. In this case, two specific effects are involved: (1) fine particles agglomeration in the nodes of the acoustic waves; and (2) sufficient vibrational energy generation. Acoustic filtration is also called ultrasonically assisted filtration that is an efficient method to increase the vacuum filtration of difficult mixtures to separate [64]. Surface accumulation of particles on membrane utilized for filtration can be one of the problems observed during filtration. Ultrasound shooting up the flux by disrupting the polarization quantity and deposits are found on the membrane surfaces and thus having no impact on membrane permeability. Extraction of fruit juice and drinks from the pulp is the main applications of this method [58].

2.11.3 DEPOLYMERIZATION

Depolymerization employs ultrasound involving polymer deterioration [62] through two possible mechanisms: collapsed cavitation bubble mechanical degradation of the polymer and chemical degradation. In the chemical method, there is a reaction between the high-energy molecules and the polymer. Cavitation phenomenon produces high-energy molecules like hydroxyl radicals [27]. Ultrasound plays an active role in the food industry with its ability to depolymerize starch. Ultrasound has the potential to convert raw materials like carbohydrates, which are polymeric in nature, to use lightweight molecules or its simpler components [72].

2.11.4 DEFOAMING

Foam is a gas in liquid dispersion colloidal system. They are thermodynamically unstable and have density approaching that of the gas. Though foam has various applications in industrial processes such as cosmetics, intensive foaming or persistent foams are undesirable in various processes, yet it may lead to problems like loss of products, decrease in productivity, etc. In the past, the foam has been controlled by the addition of chemical antifoams, use of mechanical breakers and lowering the temperature [50]. In the defoaming system for ultrasound in a rotation system, a focused airborne emitter is placed that is controlled by electricity. A complex movement is created when the transducer rotates and it covers a large area of defoaming with various rotational velocities, and almost all bubbles collapse immediately under ultrasound beams.

2.11.5 EXTRUSION AND DEMOLDING

In industrial cooking of foods, the cooked product is difficult to remove from the mold because of the adhesion of the product to the mold during cooking. To counteract this issue, the surface of the molds is coated with white grease, PTFE (polytetrafluoroethylene) or a thin layer of silicone. However, replacing them over a span of time is expensive and not completely successful. A solution to this problem is achieved by connecting the vessel to a matrix, where ultrasound is the source to release food products [63]. Extrusion is also a similar property of ultrasound in which the material is released from a matrix, by lowering the force. The food is extruded by providing ultrasound. This process reduces the drag resistance, thereby improving the fluidity of highly viscous or sticky materials through tube [51].

2.11.6   DEAERATION/DEGASSING

Deaeration in an ultrasonic field is done when waves formed by the acoustic because gas bubbles to undergo rapid vibration, due to which the adjacent bubbles move within the liquid and unite together. To displace or remove the air from the liquid surface is the main role in the manufacture of carbonated drinks. This prevents damages caused by bacteria and oxygen. This technique is mainly used in deaeration of aerated drinks like beer prior to bottling [7]. In viscous liquids like melted chocolate, it is difficult to remove gas by ultrasonically assisted degassing technique though it is useful in the case of aqueous systems.

2.11.7   THAWING/DEFROSTING

Freezing is a commonly used technique for preservation of various food products [59]. In processing industry, acoustic thawing has become a promising technology but optimum acoustic power and frequencies have to be chosen its success. Since thawing is a slow process, it involves damaging foodstuffs by the contamination of microorganisms through chemical and physical changes with time. Therefore, quick thawing at low temperature becomes important for retaining the food quality and to prevent excessive dehydration of foods. Several research studies have been conducted on the relaxation phenomenon based on the fact that frozen foods can absorb more sound energy within the normal frequency range pertaining to crystals of ice [37]. In addition, the mechanism of thawing is faster in this relaxation frequency than the process using only conductive heating. The acoustic thawing shortens the thawing time and thereby improving the product quality [8].

2.11.8   FREEZING AND CRYSTALLIZATION

In conventional freezing, the problems like non-uniform crystal development arise, leading to the destruction of cell structure, texture, and drip loss while thawing because of the continuous formation and growth of these ice crystals [53]. Ultrasound influences in the crystal formation also. The range of ultrasound from 20–100 kHz is useful in crystallization process [61]. When ultrasound is exposed towards the medium, it produces a number of nucleation sites, thereby enhancing both nucleation and crystal growth rates [53]. The dwell time is reduced because even traditional cooling is more rapid when ultrasound is applied [2].

2.11.9   BRINING, PICKLING, AND MARINATING

The meat and vegetables use pickling and marinating; and commonly used bringing of salt or pickling fermentation mainly has three limitations:

- For brining, a high quantity of sodium chloride is used, which requires the desalting process for the use of food products;
- During fermentation, the process is difficult to control; and
- By soaking, the products may soften, bloated or structural damage can occur.

Alternative technologies are required to eradicate these limitations. For food products especially with crunchy texture, ultrasound helps in reducing the pickling time [36]. There is no need for desalting of low sodium content products. Uniformly salted product is the result of this process. Two main mass transfer processes are involved in brining: water gets migrated from the flesh to the liquid, and the solute gets migrated from the liquid to the flesh. Ultrasound energy should be used with a combination of methods like brining or marinating raw foodstuffs by submerging it in the brine or marinade [30]. In sonicated samples, water, and brine content has been found at elevated levels than samples treated with no sonication.

2.11.10   CUTTING

Since the 1950s, the US has been used for the accurate cutting of brittle materials like ceramics, glass, fibers, etc., due to improved food processing by providing an improvised way to slice or cut. Using this method, maintenance costs, and product waste can be minimized. In this technique, an ultrasonic source is linked to a shaft, which is attached to a knife-type blade [57].

2.11.11   COOKING

In conventional cooking method may sometimes reduce the quality of food. Therefore, ultrasound technology has been utilized in cooking with several benefits [31]. Cooking by ultrasound leads to greater cooking speed. It provides an energy-efficient and quick method to improve the texture of food. The after-cooking moisture in the food is also preserved. Thus, the properties of water-binding of meat are increased by ultrasound power [48].

2.11.12   MEAT TENDERIZATION

Mechanical pounding is the traditional method of meat tenderization, but this method leads to poor quality of meat. Power ultrasound is based on two methods: breakage of integrity of muscular cells and increasing the rate of enzymatic reactions by using biochemical effects [6]. Ultrasound is used for producing processed meats. Meat products are present in recombined forms, such as beef rolls. During processing, myofibrillar proteins are released by which a protein gel is formed and it holds the meat pieces together [48]. Adding salt or tumbling the meat pieces by sonication helps to tenderize the meat. Therefore, these samples treated with ultrasound are of superior quality.

2.11.13   PASTEURIZATION/STERILIZATION

Pasteurization and sterilization are commonly used for inactivating enzymes and microorganisms in food products. However, these methods may lead to loss of nutrients, development of unpleasant flavor and degradation of the quality of food products. By using ultrasound, these processes can be improved on the basis of the cavitation effects. A great high acoustic power inputs, it breaks cells, but at lower intensity, a cell can be inactivated [54]. Ultrasound is effective in the dairy industry for pasteurization [10]. Investigation on the efficiency of ultrasound shows enzymes inactivation like peroxidise PPOs and pectin methylesterase to cause degradation of vegetable and fruit juices and several enzymes pertaining to the quality of milk. Therefore, thermosonication (TS) and MTS can inactivate various enzymes. Ultrasound in combination with heat has the potential to elevate the sterilization rate, as it reduces damage and the intensity of thermal treatments.

2.11.14   EMULSIFICATION/HOMOGENIZATION

Emulsification/homogenization is a technique of incorporating the hydrophobic bioactive compounds into different food products. Acoustic emulsifications have a unique improvement over the conventional methods. This technique has the ability to produce emulsion, which has sub-micron distribution with higher stability and utilizing less energy than conventional methods, without the addition of surfactants [5]. It has its application in the food industry for various products like fruit juices, mayonnaise, ketchup, etc. [13].

2.11.15   SPORE INACTIVATION

Spores produced by microorganisms can withstand extreme environments, such as variable pH, high temperature, mechanical shock, and osmotic pressure. Due to the survivability of these bacterial spores to heat treatment, these might reduce the shelf-life of thermally processed foods. Bacterial spores are resistant and metabolically inert to agents like desiccation, heat, toxic chemicals, and radiation [16, 39]. Among the non-heating technologies, ultrasound has been studied for bacteria inactivation, thus showing a potential in reducing food contamination by microbes [3, 19, 42]. Since sonication alone is not enough to inactivate the spores (due to its high resistivity), studies are done for improving the efficiency for inactivation of spores by merging several techniques, such as heat in combination with US [3, 21–24, 28, 35, 49].

2.11.16   ENZYME INACTIVATION

Ultrasound using power effectively disrupts the cells to release its components. It has been used for the inhibition of enzymes and their activity. Pepsin in its purest form has been inactivated by ultrasound application due to cavitation [12, 32].

2.11.17   EXTRACTION

The US takes less extraction time to extract natural compounds than other conventional techniques. The mechanical effect of US facilitates penetration of the solvent into the cells to a greater extent, thereby upgrading conveying and cavitation repercussion. This originates the wall of the cell to shatter, resulting in release of their components in the channel [8, 14, 20, 26, 31, 69].

2.11.18   MICROORGANISM INACTIVATION

Ultrasound has emerged as one of the greener technologies that wipe out the microorganisms. Power ultrasound method can destroy the microorganisms due to cavitation. However, the free radicals that are produced are short-lived and therefore there is no harmful effect on human beings consuming these ultrasound treated foods [71].

2.11.19   ANTI-FOULING

One of the stubborn problems in cooling systems due to the usage of unused water from various water sources is biofouling. By application of US before chlorination, the mussel extinct time can be reduced up to 12%. Therefore, this treatment method can effectively decrease the chlorine dose and exposure time [29].

2.12   SUMMARY

Ultrasonic technology is a non-destructive method applied to the food industry to ensure maximum quality with minimum processing so that the nutritional quality is not affected. It is also an innovative preservation tool for many food products. Thus, it offers better quality and higher productivity of safe foods.

KEYWORDS

- cavitation
- echocardiogram
- high intensity focused ultrasound
- mano-Thermo sonication
- preservation
- ultrasonication

REFERENCES

1. Abramov, V. O., (1998). *High-Intensity Ultrasonics: Theory and Industrial Applications* (p. 369). Amsterdam-Netherlands: Gordon and Breach Science Publishers.
2. Acton, E., & Morris, G. J., (1992). *Method and Apparatus for the Control of Solidification in Liquids.* Worldwide Patent WO, 99/20420.
3. Ansari, J. A., Ismail, M., & Farid, M., (2017). Investigation of the use of ultrasonication followed by heat for spore inactivation. *Food and Bioproducts Processing, 104,* 32–39.
4. Ashok, K. M., Sunartio, D., & Kentish, S., (2008). Modification of food ingredients by ultrasound to improve functionality: A preliminary study on a model system. *Innovative Food Science and Emerging Technologies, 9*(2), 155–160.
5. Behrend, O., & Schubert, H., (2001). Influence of hydrostatic pressure and gas content on continuous ultrasound emulsification. *Ultrasonics Sonochemistry, 8*(3), 271–276.
6. Boistier-Marquis, E., Lagsir-Oulahal, N., & Callard, M., (1999). Applications of power ultrasound applications in food industries. *Food and Agricultural Industries, 116*(3), 23–31.
7. Brown, B., & Goodman, J. E., (1965). *High-Intensity Ultrasonics: Industrial Applications* (p. 310). London, U K: Iliffe Books.
8. Brown, T., James, S. J., & Purnell, G. L., (2005). Cutting forces in foods: Experimental measurements. *Journal of Food Engineering, 70*(2), 165–170.
9. Butz, P., & Tauscher, B., (2002). Emerging technologies: Chemical aspects. *Food Research International, 35*(2/3), 279–284.
10. Cameron, M., McMaster, L. D., & Britz, T. J., (2009). Impact of ultrasound on dairy spoilage microbes and milk components. *Dairy Science Technology, 89*(1), 83–98.
11. Cárcel, J. A., García-Pérez, J. V., Riera, E., & Mulet, A., (2007). Influence of high intensity ultrasound on drying kinetics of persimmon. *Drying Technology, 25*(1), 185–193.
12. Chandrapala, J., Oliver, C., Kentish, S., & Ashokkumar, M., (2012). Ultrasonics in food processing. *Ultrasonics Sonochemistry, 19*(5), 975–983.
13. Chemat, F., Grondin, I., Sing, A. S. C., & Smadja, J., (2004). Deterioration of edible oils during food processing by ultrasound. *Ultrasonics Sonochemistry, 11*(1), 13–15.
14. Clark, J. P., (2008). An update on ultrasonics. *Food Technology, 62*(7), 75–77.
15. Cohen, J. S., & Yang, T. C. S., (1995). Progress in food dehydration. *Trends in Food Science and Technology, 6*(1), 20–25.
16. Coleman, W. H., Zhang, P., Li, Y. Q., & Setlow, P., (2010). Mechanism of killing of spores of *Bacillus cereus* and *Bacillus megaterium* by wet heat. *Letters in Applied Microbiology, 50*(5), 507–514.
17. De La Fuente, S., Riera, E., Acosta, V. M., Blanco, A., & Gallego-Juárez, J. A., (2006). Food drying process by power ultrasound. *Ultrasonics, 44,* e523–e527.
18. Ding, T., Xuan, X. T., Li, J., Chen, S. G., & Liu, D. H., (2016). Disinfection efficacy and mechanism of slightly acidic electrolyzed water on *Staphylococcus aureus* in pure culture. *Food Control, 60*(2), 505–510.
19. Dolatowski, Z. J., Stadnik, J., & Stasiak, D., (2007). Applications of ultrasound in food technology. *Acta Scientiarum Polonorum Technologia Alimentaria, 6*(3), 89–99.
20. Evelyn, & Silva, F. V. M., (2018). Differences in the resistance of microbial spores to thermosonication, high pressure thermal processing, and thermal treatment alone. *Journal of Food Engineering, 222,* 292–297.

21. Evelyn, & Silva, F. V. M., (2015a). Thermosonication versus Thermal processing of skim milk and beef slurry: Modeling the inactivation kinetics of psychrotrophic *Bacillus cereus* spores. *Food Research International, 67*, 67–74.

22. Evelyn, & Silva, F. V. M., (2015b). Use of power ultrasound to enhance the thermal inactivation of *Clostridium perfringens* spores in beef slurry. *International Journal of Food Microbiology, 206*, 17–23.

23. Fan, L., Hou, F., Muhammad, A. I., & Ruiling, L. V., (2019). Synergistic inactivation and mechanism of thermal and ultrasound treatments against *Bacillus subtilis* spores. *Food Research International, 116*, 1094–1102.

24. Fernandes, F. A. N., Linhares, Jr. F. E., & Rodrigues, S., (2008). Ultrasound as pre-treatment for drying of pineapple. *Ultrasonics Sonochemistry, 15*, 1049–1054.

25. Gallo, M., Ferrara, L., & Naviglio, D., (2018). Application of ultrasound in food science and technology: A perspective. *Foods, 7*(10), 164.

26. Grönroos, A., Pirkonen, P., & Ruppert, O., (2004). Ultrasonic depolymerization of aqueous carboxymethylcellulose. *Ultrasonics Sonochemistry, 11*(1), 9–12.

27. Ha, J. H., Kim, H. J., & Ha, S. D., (2012). Effect of combined radiation and NaOCl/ ultrasonication on reduction of *Bacillus cereus* spores in rice. *Radiation Physics and Chemistry, 81*(8), 1177–1180.

28. Haque, M. N., & Kwon, S., (2018). Effect of ultra-sonication and its use with sodium hypochlorite as antifouling method against *Mytilus edulis* larvae and mussel. *Environmental Geochemistry and Health, 40*(1), 209–215.

29. Hatloe, J., (1995). *Methods for Pickling and/or Marinating Non-Vegetable Foodstuff Raw Material.* International Patent, WO 9518537.

30. Hausgerate, B. S., (1978). *Process and Device for Treating Foods Using Ultrasonic Frequency Energy.* Germany Patent, DE 2950-384.

31. Islam, M. N., Zhang, M., & Adhikari, B., (2014). The inactivation of enzymes by ultrasound: A review of potential mechanisms. *Food Reviews International, 30*(1), 1–21.

32. Jambrak, A. R., (2013). Application of high-power ultrasound and microwave in food processing: Extraction. *Food Processing and Technology, 117*(4), 437–442.

33. Kentish, S., Wooster, T. J., Ashokkumar, M., Balachandran, S., Mawson, R., & Simons, L., (2008). The use of ultrasonics for nanoemulsion preparation. *Innovative Food Science and Emerging Technologies, 9*(2), 170–175.

34. Khanal, S. N., Anand, S., & Muthukumarappan, K., (2014). Inactivation of thermoduric aerobic spore formers in milk by ultrasonication. *Food Control, 37*(1), 232–239.

35. Kingsley, I. S., & Farkas, P., (1990). Pickling process and product. *International Polymers for Advanced Technologies.* WO 1990/ 005458.

36. Kissam, A. D., Nelson, R. W., Ngao, J., & Hunter, P., (1981). Water thawing of fish using low frequency acoustics. *Journal of Food Science, 47*, 71–75.

37. Laborde, J. L., Bouyer, C., Caltagirone, J. P., & Ge´rard, A., (1998). Acoustic bubble cavitation at low frequencies. *Ultrasonics, 36*(1–5), 589–594.

38. Larysa, P., (2005). Application of ultrasound: Chapter 15. In: Da-Wen, S., (ed.), *Emerging Technologies for Food Processing* (pp. 271–292). Cambridge MA: Academic Press.

39. Lei, R., Xu, Z., Wang, Y., Zhao, F., Hu, X., & Liao, X., (2015). Inactivation of *Bacillus subtilis* spores by high pressure CO_2 with high temperature. *International Journal of Food Microbiology, 205*, 73–80.

40. Leighton, T. G., (1998). The principles of cavitation: Chapter 9. In: Povey, M. J. W., & Mason, T. J., (eds.), *Ultrasound in Food Processing* (Vol. 59, pp. 151–182). UK: Blackie Academic & Professional.

41. Li, B., & Sun, D. W., (2002). Novel methods for rapid freezing and thawing of foods: A review. *Journal of Food Engineering, 54,* 175–182.

42. Li, J., Suo, Y., Liao, X., Ahn, J., Liu, D., Chen, S., Ye, X., & Ding, T., (2017). Analysis of *Staphylococcus aureus* cell viability, sublethal injury, and death induced by synergistic combination of ultrasound and mild heat. *Ultrasonics Sonochemistry, 39,* 101–110.

43. Mason, T. J., Chemat, F., & Vinatoru, M., (2011). The extraction of natural products using ultrasound or microwaves. *Current Organic Chemistry, 15*(2), 237–247.

44. Mason, T. J., & Cordemans, E. D., (1996). Ultrasonic intensification of chemical processing and related operations: A review. *Transactions of the Institution of Chemical Engineers, 74,* 511–516.

45. Mason, T. J., (1998). Power ultrasound in food processing-the way forward: Chapter 6. In: Povey, M. J. W., & Mason, T. J., (eds.), *Ultrasound in Food Processing* (Vol. 59, pp. 105–126). UK: Blackie Academic & Professional.

46. Mason, T. J., & Peters, D., (2003). *Practical Sonochemistry* (p. 166). Cambridge MA: Woodhead Publishing.

47. McClements, D. J., (1995). Advances in the application of ultrasound in food analysis and processing. *Trend Food Science and Technology, 6,* 293–299.

48. Milani, E. A., & Silva, F. V. M., (2017). Ultrasound-assisted thermal pasteurization of beers with different alcohol levels: Inactivation of *Saccharomyces cerevisiae* ascospores. *Journal of Food Engineering, 198*(C), 45–53.

49. Morey, M. D., Deshpande, N. S., & Barigou, M., (1999). Foam destabilization by mechanical and ultrasonic vibrations. *Journal of Colloid and Interface* Science, *219*(1), 90–98.

50. Mousavi, S. A. A. A., Feizi, H., & Madoliat, R., (2007). Investigations on the effects of ultrasonic vibrations in the extrusion process. *Journal of Material Processing and Technology, 187,* 657–661.

51. Muralidhara, H. S., Ensminger, D., & Putnam, A., (1985). Acoustic dewatering and drying (low and high frequency): State of the art review. *Drying Technology, 3*(4), 529–566.

52. Norton, T., Delgado, A., Hogan, E., Grace, P., & Sun, D. W., (2009). Simulation of high-pressure freezing processes by enthalpy method. *Journal of Food Engineering, 91,* 260–268.

53. Piyasena, P., Mohareb, E., & McKellar, R. C., (2003). Inactivation of microbes using ultrasound: A review. *International Journal of Food Microbiology, 87*(3), 207–216.

54. Rana, A., (2017). Ultrasonic processing and its use in food industry: A review. *International Journal of Chemical Studies, 5*(6), 1961–1968.

55. Raviyan, P., Zhang, Z., & Feng, H., (2005). Ultrasonication for tomato pectin methylesterase inactivation: Effect of cavitation intensity and temperature on inactivation. *Journal of Food Engineering, 70*(2), 189–196.

56. Rawson, F. F., (1998). An introduction to ultrasonic food cutting: Chapter 14. In: Povey, M. J. W., & Mason, T. J., (eds.), *Ultrasound in Food Processing* (Vol. 59, pp. 254–270). UK: Blackie Academic & Professional.

57. Riera, E., & Gallego-Juarez, J. A., (2000). Application of high-power ultrasound to enhance fluid/solid particle separation processes. *Ultrasonics, 38*(1–8), 642–646.

58. Rouillé, J., Lebail, A., Ramaswamy, H. S., & Leclerc, L., (2002). High pressure thawing of fish and shellfish. *Journal of Food Engineering, 53*(1), 83–88.

59. Sala, F. J., Burgos, J., Condon, S., Lopez, P., & Raso, J., (1995). Effect of heat and ultrasound on microorganisms and enzymes: Chapter 9. In: Gould, G. W., (ed.), *New*

Methods of Food Preservation (pp. 176–204). London, UK: Springer Science + Business Media Dordrecht.

60. Sanz, P. D., Otero, L., De Elvira, C., & Carrasco, J. A., (1997). Freezing processes in high-pressure domains. *International Journal of Refrigeration, 20*(5), 301–307.

61. Schmid, G., & Rommel, O., (1939). Rupture of macromolecules with ultrasound. *Zeitschrift für Physikalische Chemie, 185*, 97–39.

62. Scotto, A., (1988). Device for demolding industrial food products. French patent. *Food Research, 2*(604), 63–70.

63. Senapati, N., (1991). Ultrasound in chemical processing. In: Mason, T. J., (ed.), *Advances in Sonochemistry* (Vol. 2, pp. 187–210). London: JAI Press.

64. Show, K. Y., Mao, T. H., & Lee, D. J., (2007). Optimization of sludge disruption by sonication. *Water Research, 41*(20), 4741–4747.

65. Simal, S. D., Mirabo, F. B., Deya, E., & Rossello, C. A., (1997). Simple model to predict the mass transfers in osmotic dehydration. *Lebensm. Untersuchung. Forsch, 204*(3), 210–214.

66. Soria, A. C., & Villamiel, M., (2010). Effect of ultrasound on the technological properties and bioactivity of food: A review. *Trends in Food Science and Technology, 21*(7), 323–331.

67. Tiwari, B. K., & Mason, T. J., (2011). Ultrasound processing of fluid foods: Chapter 6. In: Cullen, P. J., Tiwari, B. K., & Valdramidis, V., (eds.), *Novel Thermal and Non-Thermal Technologies for Fluid Foods* (Vol. 21, pp. 135–165). New York, USA: Academic Press.

68. Tsukamoto, I., Yim, B., & Stavarache, C. E., (2004). Inactivation of *Saccharomyces cerevisiae* by ultrasonic irradiation. *Ultrasonics Sonochemistry, 11*(2), 61–65.

69. Vinatoru, M., (2001). Overview of the ultrasonically assisted extraction of bioactive principles from herbs. *Ultrasonics Sonochemistry, 8*(3), 303–313.

70. Withers, P. M., (1996). Ultrasonic, acoustic, and optical techniques for the non-invasive detection of fouling in food processing equipment. *Trends in Food Science and Technology, 7*(9), 293–298.

71. Yusaf, T., & Al-Juboori, R. A., (2014). Alternative methods of microorganism disruption for agricultural applications. *Applied Energy, 114*, 909–923.

72. Zuo, J. Y., Knoerzer, K., Mawson, R., Kentish, S., & Ashokkumar, M., (2009). The pasting properties of sonicated waxy rice starch suspensions. *Ultrasonics Sonochemistry, 16*(4), 462–468.

CHAPTER 3

MICROWAVE VACUUM DEHYDRATION TECHNOLOGY IN FOOD PROCESSING

RONIT MANDAL and ANUBHAV PRATAP SINGH

ABSTRACT

Drying is the process of removal of moisture from foods for extending its shelf-life and preserving it. Microwave vacuum drying uses microwave radiation energy under a vacuum environment to dry the food products. Microwave energy enables faster drying and products do not experience higher drying temperatures. The dried products retain their characteristics as a whole. This chapter deals with current advances in the microwave vacuum drying technology (MVDT) discussing principles and applications in recent decades.

3.1 INTRODUCTION

Food products, especially plant materials like fruits and vegetables are good sources for several health contributing nutrients. However, they have high water activity owing to a higher moisture content, which leads to the degradation of their quality. This could be due to the growth of microorganism, enzymatic, and oxidative reactions [25]. There are many preservation technologies to tackle this problem. Such technologies include, but are not limited to thermal preservation, freezing, cold storage, drying, etc. Reducing the moisture content by drying is one of the solutions to this problem. Drying is thus one of the oldest methods which help in not only controlling and preservation of food by checking the microbial growth, enzymatic, and oxidative deterioration, but also reduces the bulk weight of the food materials. This eases the transportation of bulk goods, reducing storage and transportation costs.

Drying is the process of removal of moisture from food materials by simultaneous heat and mass transfer operations. This is carried out by one or the other forms of drying. Traditional methods of drying include sun drying, where products are kept in the open under the sun for a prolonged period. This technique has been improved by the introduction of solar driers. Other conventional drying methods like convective air-drying, fluidized bed drying, vacuum drying, and freeze-drying are prevalent in the food industry. However, these require longer drying times and energy [24, 25]. These increase overall drying cost as well as hamper the organoleptic quality of dehydrated products in some cases.

To overcome such drawbacks, novel drying methods have been introduced which use radiant energy to dry the food materials. The infrared, radiofrequency, and microwave radiation energy have been recently employed in the industry for drying of several agricultural commodities. These help in producing high quality and commercial dehydrated fruit and vegetable products and are very promising to overcome the limitations of conventional drying techniques. Microwave energy-assisted drying techniques have high-energy efficiency and show great potential in food processing [15].

Microwave can penetrate the food product and heat them to the required temperature for drying. Vacuum drying processes, on the other hand, have been used for a long time. These processes involve lowering the drying temperature by reducing the boiling temperature of water in food products. In addition, as the surrounding pressure decreases, there is a greater pressure gradient between the product and the surrounding environment. This enhances the mass transfer rate to a great extent. Microwave vacuum drying (MVD) is the application of microwave-assisted drying under vacuum with rapid volumetric heating and drying and gives superior product quality. With its first inception in the late 1990s, the process has gained tremendous exploration. This can be seen by the steady rise of a number of research publications over the last 25 years (Figure 3.1).

This chapter explores the potential and recent advances of MVD technology for the food processing industry. The chapter starts with brief information on drying and its augmentation with vacuum and microwave energy to inform the readers about the peculiarities. The principles, salient features, and some applications of MVD in the food industries are also discussed.

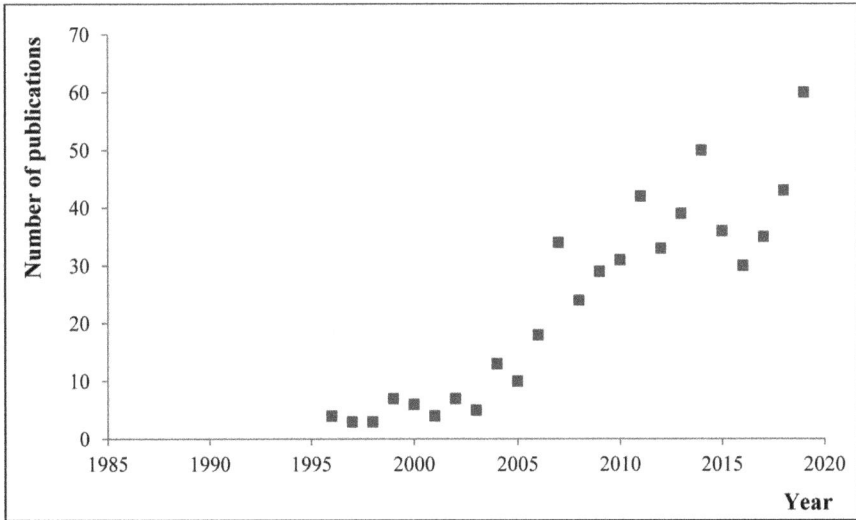

FIGURE 3.1 Total number of articles containing the terms "vacuum microwave dehydration" or "vacuum microwave drying" or "microwave vacuum drying" that were published between 1995 and 2020.

3.2 PRINCIPLES OF DRYING AND MICROWAVE VACUUM DEHYDRATION (MVD)

The drying process is used in the food industry to produce food products with low moisture content and enhanced shelf life. Drying is a process wherein simultaneous heat and mass transfer are occurring. The heating of food occurs, and the moisture is removed simultaneously. The moisture in food is removed by evaporation as the vapor pressure of food exceeds the surrounding partial pressure of water vapor around the food. Food and biological materials are characterized by the presence of free water in their inner layers and crack and crevices or pores, which are easily removed by a physical process. Free water is available for microbial growth and enzymatic and chemical reactions [28]. Drying removes the free water of the product, thus enhancing its shelf life. During drying, the water is removed initially at a constant rate up to a certain moisture content called critical moisture content. Water is removed from the surface layers of the food. The surface of the food remains wet and its temperature remains around the wet-bulb temperature of the drying air surrounding the food. This is the reason the food surface remains moist during the initial stages of drying.

As the drying progresses, the rate of drying ceases to be constant and falls with time. During this falling rate period, water is drawn from the inner pores and the process is driven by vapor and capillary diffusion. The surface of food dries up and demands more moisture from inside due to surrounding pressure difference. The temperature of the food rises, and it also suffers considerable shrinkage during this stage due to pore collapse on water removal. This leads to a significant deterioration of food quality parameters like color and flavor. The prolonged drying times enhance these deteriorative chemical reactions. Thereafter, the nutrients and functional components of food are also destroyed under higher drying temperatures.

Vacuum applications in drying have been shown to be advantageous as compared to the conventional atmospheric drying. Superior product quality is obtained by vacuum drying. In addition, due to no oxygen surrounding the product, the oxygen-prone products are protected against deterioration [28]. It is known that water evaporates when the vapor pressure is equal to the surrounding atmospheric pressure. At 101.325 kPa at sea level, water boils at 100°C. The evaporation temperature is related to the pressure by the famous Clausius-Clapeyron equation (Eqn. (1)). Water boils at a lower temperature at reduced pressure. Thus, under vacuum, the drying at a lower temperature is achieved since the vapor pressure of water equates to the lower partial pressure of water vapor surrounding the product.

$$\ln\left(\frac{P_2}{P_1}\right) = \frac{\Delta H_v}{R}\left(\frac{1}{T_1} - \frac{1}{T_2}\right) \tag{1}$$

where; P_1 and P_2 = surrounding pressures (kPa); T_1 and T_2 = Boiling temperatures (°C); DH_v = Latent heat of vaporization (kJ/mol); R = universal gas constant (kJ/mol-K).

Due to the lower evaporation temperature, the product quality remains higher than the conventional hot air-drying processes. It is also supplemented by a higher drying rate due to a greater mass transfer because of higher-pressure gradient [28]. In a study by Kardum et al. [17] for drying of chlorpropamide, maximum drying rate of 3.22×10^{-2} kg/m^2-min was achieved for vacuum drying at 0.1 atm and 60°C against the rate of 2.39×10^{-2} kg/m^2-min for conventional hot air-drying at 60°C. Typical vacuum drying processes include drying in vacuum oven, similar to a hot air oven drier and freeze-drying, drying under vacuum and low temperatures. Even better product quality is obtained by freeze-drying.

Microwave processing was first employed in the food industry back in the 1960s mainly as a concluding step in the drying of potato chips, processed meat products like bacon, poultry, pasta, and thawing of frozen products [29]. Microwave is an electromagnetic wave that has a frequency ranging from 300 MHz to 300 GHz. Commercial microwave equipment usually operates at 915 (industrially) and 2450 MHz (domestically). The dipole compounds like water effectively absorb microwave and oscillates/rotates 915 (or 2450) times according to the applied microwave frequency. Thus, the water molecules gain energy and lead to frictional heat generation in the product. Consequently, the product temperature rises quickly and without any thermal gradients. The microwave has the tendency to penetrate deeper into the product leading to more or less uniform heating. Overall the product is heated rapidly in a volumetric way. Thus, when microwave energy is utilized in the drying process and the moisture in the product quickly reaches the boiling temperature and gets evaporated. Higher drying rates, lower drying times and better product quality are obtained as a result of microwave-assisted drying [28].

The drying rates in microwave drying are even higher than vacuum drying rates. In the same study mentioned earlier, Kardum et al. [17] obtained even drying rates up to 3.76×10^{-2} kg/m^2-min for microwave drying as compared to 3.22×10^{-2} kg/m^2-min for vacuum drying. The applications of microwave in the drying process are limited as there still exist some complexities. The dielectric properties like relative permittivity ε, dielectric constant ε' and dielectric loss constant ε'' are important in relation to microwave processing of food and vary as per the food composition. These properties are affected by the presence of free and bound water, ions, electrolytes, major food components like carbohydrates, fats, moisture. These properties determine the microwave energy absorption by food. As the chemical composition of food varies, the absorbed energy by food product varies which lead to non-uniform heating in several cases as well as overheating. As a matter of fact, frozen state of food can significantly affect the energy absorption by food as these properties get affected. These properties are related by the following equation [28].

$$\varepsilon = \varepsilon' - \varepsilon'' \qquad (2)$$

Microwave vacuum dehydration (MVD) is the dehydration process where drying is assisted by microwave energy under vacuum application to enhance drying rates and obtain rapid volumetric heating. MVD process first appeared during the end of the 20th century. Drouzas and Schubert [9] were the first researchers to formally investigate the influence of vacuum level and microwave power level on the drying rates and quality characteristics of banana. During MVD, the moisture quickly migrates to the surface and boiling point is lowered. The drying rates are increased, and drying times are significantly decreased. With decrease in microwave power and increase in vacuum level lowered drying times have been observed. Nahimana and Zhang [26] used MVD for drying of carrots and obtained lowered drying times from 42 to 24 min with increasing power level.

Abano et al. [1] obtained decreased drying time from 84 to 14 min on drying tomato slices. It can be seen that more than 50% reductions in drying times have been observed by MVD. The equipment for MVD involves a drying chamber with a motorized turntable, on which the product is kept. The magnetron generates the microwave energy and waveguide channels the microwave on the turntable. Vacuum is maintained inside the chamber with the help of a vacuum pump. Typical schematic diagram of MVD equipment is shown in Figure 3.2.

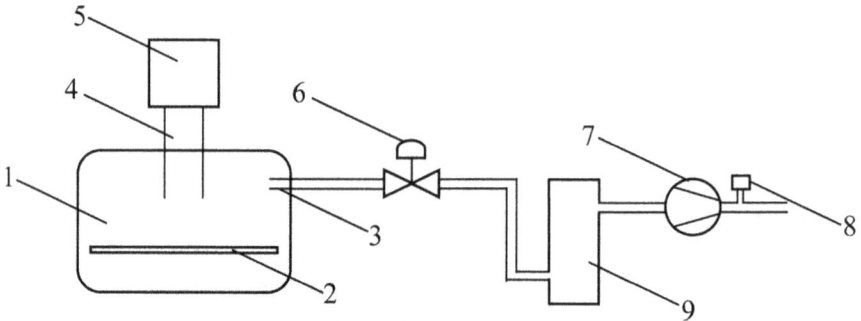

FIGURE 3.2 MVD equipment schematic diagram. (1) Drying chamber; (2) turntable; (3) air exhaust pipe; (4) waveguide; (5) magnetron; (6) valve; (7) vacuum pump; (8) pressure gauge; (9) air compensation tank.

3.3 DRYING KINETIC MODELS OF MICROWAVE VACUUM DRYING

Mathematical models and simulations help in a deeper understanding of the drying processes. It becomes easier to understand the effect of different parameters on drying rates and times. Food dehydration experts and researchers have developed several intrinsic parametric models for predicting the drying behavior of food materials. The drying kinetics of various fruits and vegetables undergoing different drying processes has been studied by researchers in line with the kinetic studies for conventional drying methods. Hence, the MVD drying kinetics has also been implemented by several researchers. Although numerous models have been used in the researches for modeling drying kinetics, the applicable model depends on the process variables, product dielectric characteristics, and pretreatments. Initially, theoretical models based on process variables and conservation of energy was introduced by the researcher.

Kiranoudis [18] for the first time studied drying kinetics of fruits, namely apple, pear, and kiwi. Three levels of microwave power (425 kW. 595 kW and 850 kW) and vacuum pressure (20 MBar, 40 MBar and 67 MBar) for periods up to 240 s were used for experimentation. The researchers modeled the drying process as a three-parameter empirical mass transfer model (Eqn. (3)) influenced by process variables.

$$X_s = X_{sO}\exp[-k_m t] \tag{3}$$

$$k_m = k_0 Q^{k_1} P^{k_2} \tag{4}$$

where; P = pressure level (MBar); Q = power level (kW); t = time (s); k_m, k_0, k_1, k_2 = drying constants; X_s = moisture content (kg/kg dry solids); X_{s0} = initial moisture content. At a particular vacuum level, the drying rate was increased with higher power level, but at a particular power level, the drying rate was slightly affected by vacuum level.

Later on, the drying kinetics of vegetables like carrot slices were studied by Cui et al. [7], using theoretical drying model based on energy conservation in the dryer. They used three power levels (336.5, 267.5, and 162.8 W) and pressure levels (30, 51, and 71 MBar) with a drying time up to 25 min. The model they developed was as follows.

$$X_s : \left\{ X_w = X_0 - \frac{Q_{abs} t}{m_0 \lambda_p} (X_w \geq 2) \right\} \tag{5}$$

$$or \ \{\varphi X_w = \varphi(X_0 - \frac{Q_{abs} t}{m_0 \lambda_p})\varphi = 1.3662 X_w^{-0.5741} \ (0 < X_w < 2) \}$$

where; X_w = sample moisture content by theoretical calculation (kg/kg dry solids); Q_{abs} = heat absorbed by samples (kW); λ_p = latent heat of vaporization (at vacuum pressure P); m_0 = mass of sample (kg); φ = correction factor.

The researchers observed that the drying rate varied linearly with the microwave power and inversely proportional to latent heat of evaporation at vacuum pressure P. Similar to Kirandous [18], the effect of microwave power was more than that of vacuum level on drying rates.

Giri and Prasad [14] carried out the drying kinetics of MVD treatment of button mushrooms (*Agaricus bisporus*) using central composite design (CCD). They varied the power level from 115 to 285 W, vacuum pressure from 6.5 to 23.5 kPa and product thickness from 5.8 to 14.2 mm. They evaluated the typical conventional drying models like exponential and empirical Page's model to predict the drying behavior.

$$MR = exp\ (-Kt) \tag{6}$$

$$MR = exp\ (-Kt^n) \tag{7}$$

where; MR = moisture ratio; K = drying rate constant (min^{-1}); k, n = parameter of Page's model.

The model parameters and drying rate constant and the effects of power, vacuum pressure, and product thickness were calculated using regression analysis. Page's model adequately described the drying process.

Alginate-starch hydrocolloid gels drying was investigated by Jaya and Durance [16] using MVD at different power levels of 300–1100 kW of 200 W increments under 25 mm Hg vacuum. The moisture content of the sample during drying was modeled with respect to microwave power as per the following equation.

$$MR = a\ exp\ (-bP) \tag{8}$$

Not only they observed higher drying rates at higher power levels, at any point in drying, lower moisture content was measured in samples treated under higher power.

Figiel [13] studied the drying behavior of garlic cloves (whole, halved, and sliced). The garlics were dried under power levels of 240, 480, and 720 W, as 30 s microwave pulses at 4–6 kPa. The drying was described as a 4-parameter sigmoid curve for moisture content.

$$X_s = a + \frac{b}{1 + e^{\frac{t-c}{d}}} \tag{9}$$

where; a, b, c, and d = model parameters. They observed that increase in microwave power increased the drying rate.

Song et al. [30] used MVD for drying of potato slices and used three power levels (140, 240 and 340 W) and vacuum pressure levels (–0.04, –0.06, and –0.08 MPa). Drying kinetic models were established using a modified page, and Henderson and Pabis models.

$$MR = exp\ [-KtP^{nP}] \tag{10}$$

$$MR = a\ exp\ [-kt] \tag{11}$$

where; K, n = Page model parameters; a, k = Henderson and Pabis model parameters. The model parameters were obtained by regression analysis and depended on process variables.

In another study by Figiel [12], beetroot cubes were dried using MVD under pressure levels of 240, 360 or 480 W at 65 Pa pressure. The exponential model as Eqn. (11) was used to fit the drying data.

Processed starch products extruded, cornstarch-based pellets have been dried and their drying kinetics haven introduced by Kraus et al. [19] They carried out drying by MVD process taking different amounts of samples (100, 200, and 300 g), microwave power (400, 600, and 800 W) and vacuum pressure (20, 50 and 100 MBar). However, the drying behavior was different from natural products. They fitted the drying data to thirteen theoretical models and found that Balbin and Sahin model (Eqn. (12)) best described the drying behavior.

$$MR = (1 - a)\ exp\ [-kt^n] + b \tag{12}$$

where; a, b, k, n = model parameters. It was observed by the researchers that the increase in the microwave power and decrease in initial sample amount reduced the drying time.

Calín-Sánchez [3] dried black chokeberry (*Aronia melanocarpa*) using MVD at power levels of 240 and 360 W at 4–6 kPa. They used a modified Page's model to describe the drying behavior of berries.

$$MR = a\ exp\ [-kt^n] \tag{13}$$

Blueberries were dried by Zielinska [33] using MVD at the specific power level of 1.3 W/g at 4–6 kPa. The drying process was described by the exponential model (Eqn. (11)).

In a combined convective-MVD drying process, Lech et al. [22] dehydrated pear cubes using MVD at power levels of 240, 480 and 720 W, as a finishing drying method to convective process. The drying was described by modified Page's model as per Eqn. (13). The MVD treatment increased drying rates with higher microwave powers.

Granular products, like green soybean, were dried by Cao et al. [2] MVD treatment at 800 W and 0.09 MPa. They observed faster drying rates as compared to another microwave-assisted drying (microwave freeze-drying). However, no kinetic models were implemented by them.

MVD has been combined with osmotic dehydration as well. In a study, Lech et al. [21] studied the drying kinetics of pumpkin slices in solution. Power levels of 120–480 W were used, and a logarithmic model was used to describe the drying process.

$$MR = a\ exp\ [-kt] + b \tag{14}$$

Overall, it seems that most of the preexisting empirical drying models fit the MVD drying processes for biomaterials. Although depending on the types of food matrices, the models may vary amongst these. Though, modified forms of exponential and Page's model could better approximate these processes.

The research group of this chapter was carried at Food Process Engineering Lab, the University of British Columbia, for drying of brewers' spent grains (unpublished data) under MVD, convective drying, and freeze-drying methods. The drying was undertaken at three different temperatures of 60, 65, and 70°C for conventional drying and 250 W under MVD. Drying data were adequately fitted to Page's model (Eqn. (7)). The drying rate was faster for the MVD process than convective drying and took lesser drying time than the latter. Figure 3.3 shows the MR vs. drying time for different drying conditions.

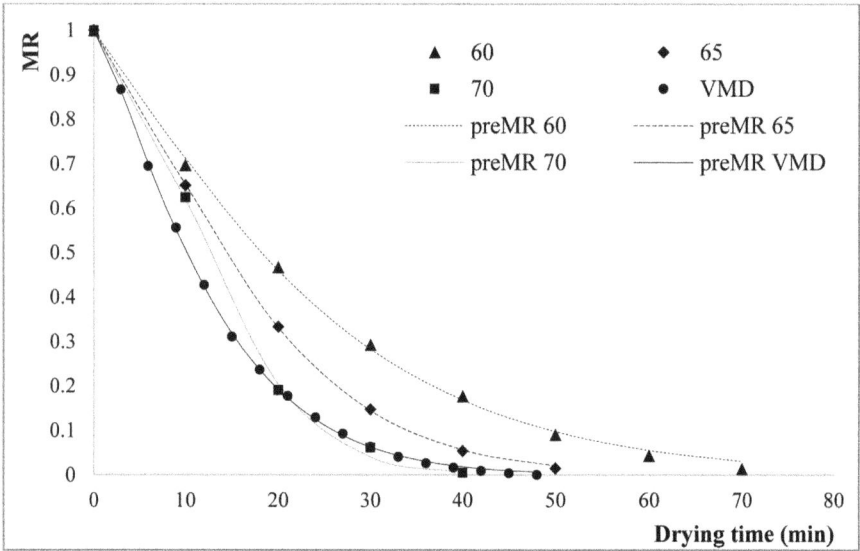

FIGURE 3.3   MR vs. drying time for different conditions (▲) 60°C, (♦) 65°C, (■) 70°C and (●) MVD at 250 W for drying brewers' spent grains. The lines represent the predicted MR for different conditions.

3.4 MICROWAVE VACUUM DEHYDRATION (MVD): FOCUS ON FOOD QUALITY PARAMETERS

Numerous researches have been conducted in the last two decades regarding the MVD drying of several fruits and vegetables. This section discusses some of these studies that have been done on the MVD process in the past years. A favorable dehydration technology would be rapid and brings about minimum changes in product characteristics. For being a viable and effective food dehydration technology, it is important to study how the food quality parameters are affected by MVD. Researchers have shown that the drying process is favored by MVD with minimum changes in product quality and good rehydration characteristics.

Drouzas and Schubert [9] firstly pioneered the use of MVD where they used it for drying of bananas. They varied the microwave power from 150 to 850 W and chamber pressure from 15 to 350 MBar in a pulsed mode for 20 s treatments, with pauses for 10 s in between. They observed higher drying rates with increased power levels. However, there was significant product burning at higher power levels. The researchers found optimized process parameters to be 25 MBar and 150 W. They obtained a product closer to freeze-dried one, with a final water activity of 0.35 having a bright yellow color for banana, no shrinkage, good aroma and taste and also a good rehydration characteristic of the finished bananas.

Drouzas et al. [10] dried model fruit gel simulating orange juice concentrate using convective tunnel drying and MVD process. The vacuum level was varied between 640 and 710 W at pressure levels of 30–50 MBar. They measured the effect of sample position and pressure on the drying rates and color changes. They also studied the sorption isotherm of the dried gel. The shape of the sorption isotherm resembled that for a high-sugar material. Higher the pressure used more was the browning of the color.

Other fruits like apples and strawberries were studied by Erle and Schubert [11], where they investigated the effect of osmotic treatment on the MVD process while drying. Osmotic treatment conditions were: temperature 20–70°C, osmotic solution of 54 wt% sucrose with 6 wt% calcium chloride or 60 wt% without calcium chloride. MVD parameters were: 195 and 390 W, the pressure of 5 kPa for treatment between 13 and 39 min. The sugar uptake of the product was 25–30% on dry basis. Vitamin C content remained 60% after MVD as compared to 100% after osmotic treatment. Product volume was also retained up to 20–50% for strawberries and up to 20–60% in apples due to the formation of steam inside the tissues. Calcium chloride addition led to a gel-like product with a rigid and brittle texture. Electron microscopy revealed that the osmotic pretreatment preserved the dried sample more than MVD processed ones.

Clary et al. [5] evaluated the effects of MVD treatment on grapes quality. The drying process comprised two different types of drying modes: fixed power from 0.5 to 1.5 kW and decreasing power level from 3 to 0.5 kW under a pressure of 2.7 kPa. They observed that fixed power led to over-heating of product at higher power levels and a moist product at lower power levels. Optimum specific energy of 0.92 Wh/g (based on applied power to grapes) when used, produced grapes of low moisture content and puffed/crispy characters.

Another study was done by Clary et al. [4] on grapes for the production of dried products using MVD experiments. They varied the power level applied (0–3 kW) during the process by sensing and controlling the product surface temperature rise during drying. The pressure of 2.7 kPa was maintained. A product surface temperature of 70°C was found optimum. The process also led to higher retention of nutrients than sun-dried grapes. Dried grapes were also free from any burning. This shows MVD as a potential technology for raisins production.

Carrot slices were dried by Cui et al. [8], where they studied the variation in product temperature during MVD drying. Temperature changes occurred in stages with initial warming-up, water vaporization, and heating up. The total temperature was also observed to be uniform throughout the slices.

Cui et al. [6] investigated the MVD under varied processing conditions, namely pressure levels of 30–50 MBar and power levels between 290–330 W for different sample thicknesses of 8–16 mm of honey. As the sample thickness was increased from 8 mm, a temperature gradient for the sample was developed along the sample axis. The color of honey was changed negligibly. Glucose and fructose barely increased, while sucrose and maltose barely decreased. The original honey flavor was preserved during the MVD drying of honey. Thus, MVD could have the potential in drying viscous liquids and syrups.

Leusink et al. [23] studied the dehydration of cranberries using MVD (30 mm Hg, 1.8 kW power for 30–35 min followed by 1.2 kW for 20–25 min), air-drying (55°C/60–65 h), and freeze-drying. It was found that MVD processed and freeze-dried cranberries preserved higher antioxidant activity and anthocyanin content than air-dried samples. Higher drying temperature and presence of oxygen led to deteriorative changes during air-drying.

Osmotic dehydration has been coupled with MVD by some researchers. In a study, Therdthai et al. [31] dried mandarin's cv. (*Sai-Namphaung*) by osmotic pretreatment prior to MVD. Osmotic solution was obtained by using sucrose and glycerol in ratios of 9:1, 7:3 and 5:5 (w/w). MVD parameters: power levels-960–1280 W at 2450 MHz, under vacuum pressure of 13.33 kPa. They also measured the dielectric properties of the solutions and investigated their effect on drying. The dielectric properties changed due to osmotic dehydration. Dielectric constant, ε' decreased, whereas the dielectric loss factor, ε'' suffered an increase. Loss tangent (ratio of ε'' and ε') increased. An increase in glycerol concentration led to increased ε'' and the drying rates. An increase in glycerol content led to a decrease in lightness and hue angle. Whereas, higher sucrose leads to more retention of β-carotene. Higher microwave power yielded lower β-carotene contents. Overall, osmotic pretreatment prior to MVD promoted faster drying, and better color, β-carotene content and texture in the dried products.

Lech et al. [21] used MVD drying as a finishing step to osmotic drying of pumpkins. Contrasting to the study by Therdthai et al. [31], they found that osmotic dehydration led to prolonged MVD drying times due to juice impregnation into the matrix. Also, L*, a* and b* values were decreased significantly by drying. A significant increase in the bioactivity in terms of polyphenolic content and antioxidants were also observed.

Tian et al. [32] studied the effects of process variables like microwave power, vacuum levels and on-off ratio on quality characteristics of lotus seeds (*Nelumbo nucifera*). They investigated the contributory effects of power (2–4 kW), vacuum (–70.0 to –90.0 kPa) and power on-off ratio (68/52 to 99/21 s) by a CCD and response surface model. They showed that high power, vacuum, and on-off ratio lead to lower drying times and lower shrinkage. Medium vacuum level at high microwave power and on-off ratio led to high rehydration ratio. Drying process was optimized at 3.2 kW power, –83.0 kPa vacuums, and an on-off ratio of 96/24, leading to a drying time of 9.5 min, 38.52% shrinkage ratio, 155% rehydration ratio, and whiteness index of 67.67. Microstructural studies revealed that MVD treated lotus seeds had more porous structure than hot air-dried seeds and were more attractive.

Legume-based products like *Dhokla* (a traditional Indian legume-based food product made of chickpea) was dried by Patel et al. [27], where they optimized its fermentation and MVD treatment conditions. Using response surfaces, they optimized the fermentation conditions of *dhokla* batter to 12.5 h time, 26.6°C, rice to gram ratio of 1.2 and moisture content of 65% wet basis. The chosen MVD conditions were batter thickness (10–17 mm), microwave power density (3.5–10 W/g), pulsating time, i.e., ratio of drying time to on time (1.3–2.0). Vacuum was set to –80 kPa. Optimized conditions were 17 mm batter, 10 W/g and 1.3 ratio, which yielded a product of bulk density 1014.22 kg/m³, rehydration ratio of 4.55, total color change of 9.57 and sensory acceptability score of 6.88. Moisture diffusivity values of 6.89×10^{-8} to 1.10×10^{-7} m²/s were obtained.

Zielinska and Michalska [33] used MVD process for drying blueberries and observed that drying reduced the phenolics and antioxidant capacity to 69 and 77%, respectively. Anthocyanins were also shown to decrease from

70–95%. A combined treatment of hot air at 90°C and MVD (1.3 W/g) led to the highest content of anthocyanins and antioxidant capacity in berries.

Marine products have also been investigated for drying by MVD. Recently in a study, Landymore et al. [20] dried krill (*Euphausia pacifica*) using different dehydration methods like air-drying (65°C/48 h in cabinet dryer until 14% wet basis moisture), MVD (60 min treatment at 33 mm Hg until a 7.88% wet basis moisture content) and freeze-drying. Significantly higher protein digestibility (*in vitro* and *in vivo*) was observed in MVD treated (comparable to freeze-dried krill) as compared to air-dry samples. In addition, higher available lysine was obtained for freeze-dried and MVD treated krill (8.5 mg/100 mg protein) as compared to air-dried samples (5.6 mg/100 mg protein).

Overall, MVD seems a promising dehydration technology which has been used for drying several biomaterials advantageously. In our research lab, we dehydrated strawberries (unpublished data) using air-drying (70°C), MVD (375 W at 25 mm Hg)) and freeze-drying. As seen in Figure 3.4, while the air-dried product was least acceptable, MVD treated strawberries looked to retain natural characteristics after the freeze-dried samples.

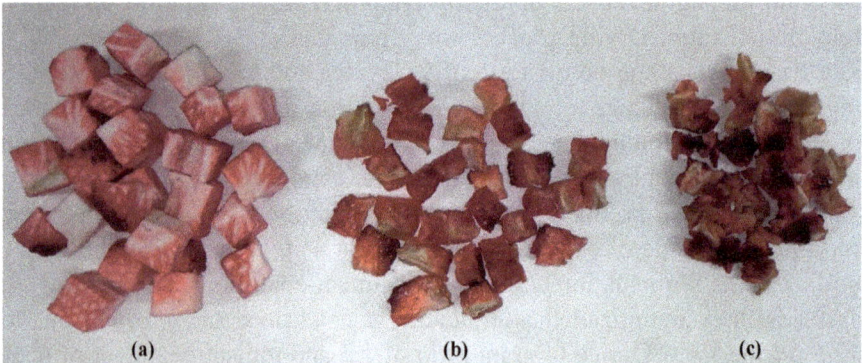

FIGURE 3.4   Strawberry cubes dried with different methods (a) freeze-dried; (b) MVD; (c) hot air-dried.

3.5   SUMMARY

MVD involves the application of microwave energy (up to 1.5 kW) for drying of food products under vacuum. The MVD drying kinetics has been modeled and studied by researchers by fitting drying behavior based on several empirical or well-known thin layer-drying models. However, the MVD drying rate can be enhanced by microwave power and the vacuum level, with the former having a larger effect on drying rates. Nevertheless, drying times are significantly reduced by MVD. In addition, the product quality has been shown better than convective air-dried samples in several studies, with the exception of some product burning at higher power levels. MVD is a promising food dehydration technology for the food industry. More research is warranted for its widespread commercialization and usage.

ACKNOWLEDGMENTS

Authors thank the Natural Sciences and Engineering Research Council of Canada (NSERC) for financial support through the Discovery Grants Program (RGPIN-2018-04735).

KEYWORDS

- food industry
- food preservation
- infrared
- microwave-assisted drying
- microwave radiation
- microwave vacuum dehydration

REFERENCES

1. Abano, E. E., Ma, H., & Qu, W., (2012). Influence of combined microwave-vacuum drying on drying kinetics and quality of dried tomato slices. *Journal of Food Quality, 35*(3), 159–168.
2. Cao, X., Zhang, M., Fang, Z., Mujumdar, A. S., Jiang, H., Qian, H., & Ai, H., (2017). Drying kinetics and product quality of green soybean under different microwave drying methods. *Drying Technology, 35*(2), 240–248.
3. Calín-Sánchez, Á., Kharaghani, A., Lech, K., & Figiel, A., (2015). Drying kinetics and microstructural and sensory properties of black chokeberry (*Aronia melanocarpa*) affected by drying method. *Food and Bioprocess Technology, 8*(1), 63–74.

4. Clary, C. D., Mejia, M. E., Wang, S., & Petrucci, V. E., (2007). Improving grape quality using microwave vacuum drying associated with temperature control. *Journal of Food Science, 72*(1), E023–E028.

5. Clary, C. D., Wang, S., & Petrucci, V. E., (2005). Fixed and incremental levels of microwave power application on drying grapes under vacuum. *Journal of Food Science, 70*(5), E344–E349.

6. Cui, Z. W., Sun, L. J., Chen, W., & Sun, D. W., (2008). Preparation of dry honey by microwave-vacuum drying. *Journal of Food Engineering, 84*(4), 582–590.

7. Cui, Z. W., Xu, S. Y., & Sun, D. W., (2004). Microwave-vacuum drying kinetics of carrot slices. *Journal of Food Engineering, 65*(2), 157–164.

8. Cui, Z. W., Xu, S. Y., Sun, D. W., & Chen, W., (2005). Temperature changes during microwave-vacuum drying of sliced carrots. *Drying Technology, 23*(5), 1057–1074.

9. Drouzas, A. E., & Schubert, H., (1996). Microwave application in vacuum drying of fruits. *Journal of Food Engineering, 28*(2), 203–209.

10. Drouzas, A. E., Tsami, E., & Saravacos, G. D., (1999). Microwave/vacuum drying of model fruit gels. *Journal of Food Engineering, 39*(2), 117–122.

11. Erle, U., & Schubert, H., (2001). Combined osmotic and microwave-vacuum dehydration of apples and strawberries. *Journal of Food Engineering, 49*(2/3), 193–199.

12. Figiel, A., (2010). Drying kinetics and quality of beetroots dehydrated by combination of convective and vacuum-microwave methods. *Journal of Food Engineering, 98*(4), 461–470.

13. Figiel, A., (2009). Drying kinetics and quality of vacuum-microwave dehydrated garlic cloves and slices. *Journal of Food Engineering, 94*(1), 98–104.

14. Giri, S. K., & Prasad, S., (2007). Drying kinetics and rehydration characteristics of microwave-vacuum and convective hot air-dried mushrooms. *Journal of Food Engineering, 78*(2), 512–521.

15. Horuz, E., & Maskan, M., (2015). Hot air and microwave drying of pomegranate (*Punica granatum* L.) arils. *Journal of Food Science and Technology, 52*(1), 285–293.

16. Jaya, S., & Durance, T. D., (2007). Effect of microwave energy on vacuum drying kinetics of alginate-starch gel. *Drying Technology, 25*(12), 2005–2009.

17. Kardum, J. P., Sander, A., & Skansi, D., (2001). Comparison of convective, vacuum, and microwave drying chlorpropamide. *Drying Technology, 19*(1), 167–183.

18. Kiranoudis, C. T., Tsami, E., & Maroulis, Z. B., (1997). Microwave vacuum drying kinetics of some fruits. *Drying Technology, 15*(10), 2421–2440.

19. Kraus, S., Sólyom, K., Schuchmann, H. P., & Gaukel, V., (2013). Drying kinetics and expansion of non-predried extruded starch-based pellets during microwave vacuum processing. *Journal of Food Process Engineering, 36*(6), 763–773.

20. Landymore, C., Durance, T. D., & Singh, A. P., (2019). Comparing different dehydration methods on protein quality of krill (*Euphausia pacifica*). *Food Research International, 119*, 276–282.

21. Lech, K., Figiel, A., Michalska, A., Wojdyło, A., & Nowicka, P., (2018). The effect of selected fruit juice concentrates used as osmotic agents on the drying kinetics and chemical properties of vacuum-microwave drying of pumpkin. *Journal of Food Quality, 9.* doi: https://doi.org/10.1155/2018/7293932.

22. Lech, K., Siudek, M., Michalska, A., & Figiel, A., (2016). Drying kinetics and textural properties of pears dehydrated by combined method with application of vacuum-microwaves. *Advances of Agricultural Sciences Problem Issues, 11*, 23–29.

23. Leusink, G. J., Kitts, D. D., Yaghmaee, P., & Durance, T., (2010). Retention of antioxidant capacity of vacuum microwave dried cranberry. *Journal of Food Science, 75*(3), C311–C316.

24. Moses, J. A., Norton, T., Alagusundaram, K., & Tiwari, B. K., (2014). Novel drying techniques for the food industry. *Food Engineering Reviews, 6*(3), 43–55.

25. Mujumdar, A. S., & Xiao, H. W., (2019). *Advanced Drying Technologies for Foods* (p. 246). Boca Raton, FL, USA: CRC Press.

26. Nahimana, H., & Zhang, M., (2011). Shrinkage and color change during microwave vacuum drying of carrot. *Drying Technology, 29*(7), 836–847.

27. Patel, D. N., Sutar, P. P., & Sutar, N., (2013). Development of instant fermented cereal-legume mix using pulsed microwave vacuum drying. *Drying Technology, 31*(3), 314–328.

28. Reis, F. R., (2014). *Vacuum Drying for Extending Food Shelf-Life* (p. 72). New York: Springer Briefs in Applied Sciences and Technology, Springer.

29. Shaheen, M. S., El-Massry, K. F., El-Ghorab, A. H., & Anjum, F. M., (2012). Microwave applications in thermal food processing. *The Development and Application of Microwave Heating*, 3–16.

30. Song, X. J., Zhang, M., Mujumdar, A. S., & Fan, L., (2009). Drying characteristics and kinetics of vacuum microwave-dried potato slices. *Drying Technology, 27*(9), 969–974.

31. Therdthai, N., Zhou, W., & Pattanapa, K., (2011). Microwave vacuum drying of osmotically dehydrated mandarin cv. (*Sai Namphaung*). *International Journal of Food Science and Technology, 46*(11), 2401–2407.

32. Tian, Y., Zhang, Y., Zeng, S., & Zheng, Y., (2012). Optimization of microwave vacuum drying of lotus (*Nelumbo nucifera* Gaertn.) seeds by response surface methodology. *Food Science and Technology International, 18*(5), 477–488.

33. Zielinska, M., & Michalska, A., (2016). Microwave-assisted drying of blueberry (*Vaccinium corymbosum* L.) fruits: Drying kinetics, polyphenols, anthocyanins, antioxidant capacity, color, and texture. *Food Chemistry, 212*, 671–680.

CHAPTER 4

THERMOELECTRIC REFRIGERATION TECHNOLOGY FOR PRESERVATION OF FRUITS AND VEGETABLES

PRASAD CHAVAN, GAGANDEEP KAUR SIDHU, and
MOHAMMED SHAFIQ ALAM

ABSTRACT

The food market is experiencing about 30% losses of fruits and vegetables due to a lack of scientific cold storage infrastructure. Thermoelectric devices have the edge over vapor compression refrigeration system (VCRS) in terms of its low cost, the higher level of compactness, noiselessness, and environment friendliness. This chapter illustrates the principle of thermoelectric refrigeration along with structure, material of module, and working of various prototypes as environmental friendly refrigeration technology. Thermoelectric refrigeration technology has a wider scope for application in cold preservation of fruits and vegetables during transit. The developed system will contribute to the mitigation of environmental hazards caused by emissions from the VCRS and to facilitate cold chain management of perishable commodities.

4.1 INTRODUCTION

Global performance on food production has impressively increased between 1960 and 2015 after the adoption of mechanization and green revolution in agriculture. Total food production has increased by threefold that shifted diet rate as well as a significant reduction in global malnutrition from 23.2% in 1990–1992 to 14.9% in 2014–2015 [39]. Reduction in malnutrition has not only contributed to reducing the malnutrition but also anticipation towards United Nation's Millennium and Sustainable Development Goals one and two, i.e., zero hunger and no poverty by increasing the employment in agriculture and related sector. Conversely to this, due to reduced malnutrition, health awareness and change of working style, life expectancy at birth has increased significantly, which caused to drastically global population rise from 2.53 billion in 1950 to 7.32 billion in 2015 [30]. In addition, with the current pace of population growth, it is estimated that the global population will reach 9.1 billion by 2050, which requires an increase in agriculture production by 70% (4,400 billion tons) to meet the food demand [63].

Among the various agricultural challenges such as reduced availability of irrigation water, depleting soil quality, unfavorable climate, and uncertainty of rainfall, global food wastage is recognized as a major challenge for sustainable development. According to FAO, almost one-third of all food produced was lost, and the amount is estimated to 1.3 billion tons per year [25]. These wastages and losses occur at each stage of the food supply chain (FSC) from production to end consumers and at each stage of transition from freshly harvested to the storage, postharvest processing, retail, and consumption [11], as shown in Figure 4.1.

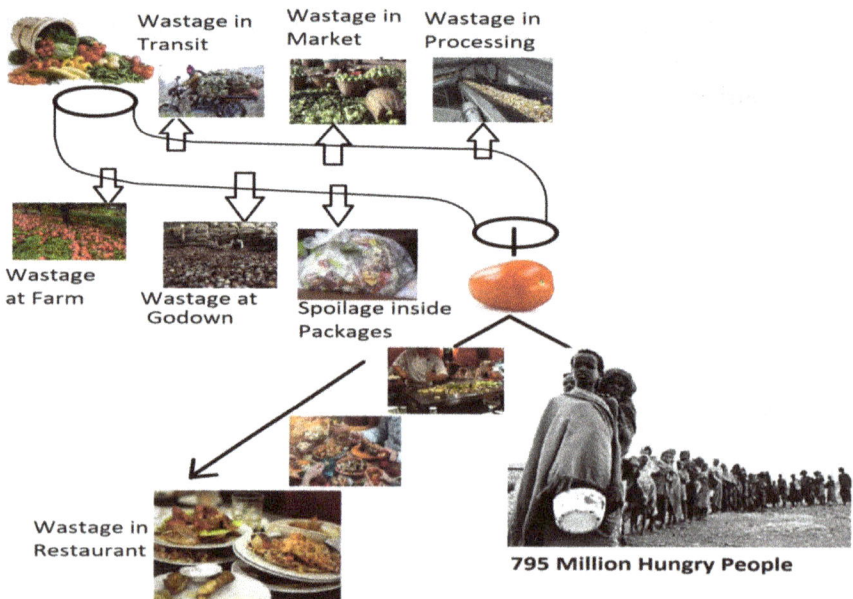

FIGURE 4.1   Stages of food wastage.

As one-third of world food is perishable in nature, therefore it requires the food to be stored at optimum and controlled temperature and humidity, not only to reduce the physiological loss in weight but also to supply the consumers with safe and wholesome products [34]. The refrigeration (cooling, chilling, and freezing) technique stops or remarkably reduces detrimental changes that occur in perishable foods. These changes can be classified as microbiological (microorganism and fungus growth), physiological (ripening, respiration, etc.), biochemical (lipid oxidation, browning reactions, and pigment degradation), and physical (loss in weight and texture) [36]. It is long proven that the storage of food at low temperatures extends the shelf-life either without or with the least occurrence of these detrimental changes.

Lorentzen [45] stated that the rate of respiration and ripening could be greatly reduced by lowering the temperature of the product. An exhaustive study conducted by various researchers concluded that refrigerated storage of food slows the chemical changes, inhibit microbial growth, extend shelf-life and maintain the quality of food [2, 30, 36]. Sirolli et al. [61] suggested refrigerated storage in combination with active modified atmospheric packaging (MAP: 7% O_2 and 0% CO_2) extended the shelf-life of apple up to 35 days without attaining spoilage yeast with the best retention of firmness and color.

This chapter focuses on the principle of thermoelectric refrigeration along with its structure and working of various prototypes as eco-friendly refrigeration technology.

4.2 GLOBAL SCENARIO OF FOOD LOSSES AND REFRIGERATION FACILITIES

It is estimated that there are 1.5 billion domestic refrigerators and freezers and 4 million refrigerated vehicles, including vans, trucks, semi-trailers, and trailers are in service worldwide [34]. However, the International Institute of Refrigeration in 2010 suggested that out of 6,300 million tons of perishable food products, only 400 million tons of perishables were preserved using refrigeration techniques [34]. It was also observed that there is wider inequality in food losses account between developed (9%) and developing (23%) countries due to lack of refrigeration facilities for safe food storage. A striking example of this is India, where only 4% of countries' perishable produce is transported by refrigerated condition compared UK's 90% [5].

The difference in the population, availability of refrigerated storage facilities, and food losses in developed and developing countries are presented in Table 4.1. Therefore, in order to maintain the cold chain of perishable produce, it is vital to improve the refrigeration facilities during food in transit. Additionally, it is an inconvenience for small shops, hotels, restaurants, bars, and vegetable vendors to sustain in future without adequate refrigeration facilities.

TABLE 4.1   Refrigeration Capacity and Food Loss Difference Among Developed and Developing Countries

Parameter	World Population	Developed Countries	Developing Countries
Population in 2009, billion	6.83	1.23	5.60
Refrigerated storage capacity, $m^3/1000$ inhabitants	52	200	19
Number of domestic refrigerators, per 1000 inhabitants	172	627	70
Overall food losses (%)	25	10	28
Losses of fruit and vegetables	35	15	40
Losses of perishable food due to lack of refrigeration facility (%)	20	9	23

4.3   REFRIGERATION AND CLIMATE CHANGE

One of the minor components of the atmosphere, i.e., ozone layer has crucial importance on the homeostasis of the temperature of the earth and maintains the ecological balance. Ozone layer is sandwiched between the stratosphere (85–90%) and troposphere [58]. This layer, with its concentration of 10 ppm, acts as an umbrella to absorb the ultraviolet (UV) radiations from the sun, which are lethal to life and prevent them from reaching to earth. However, after the emergence of industrialization threefold increase in population, there is great concern about the depletion of the ozone layer and rise of earth's surface temperature, i.e., global warming. Out of various factors that responsible for this ecological loss, refrigeration industry shares a greater contribution.

Refrigerants as one of the main constituent of refrigeration, air conditioning, and heat pump system are the working fluids that absorb heat from higher temperature region and release to the lower temperature region [67]. There is prolonged history since the first mechanical production of cooling in 1834 of development of refrigerants which will be more efficient for heat transfer and has less environmentally harmful. Ethyl Ester was the very first kind used refrigerant after which natural refrigerants such as ammonia, CO_2

and various chlorofluorocarbons (CFCs), hydrofluorocarbons (HFCs) and hydrochlorofluorocarbons (HCFCs) were employed [16].

The CFCs, HCFCs, and HFCs refrigerant is very stable, friendly to the human body and act as a good refrigerant. However, as they belong to the chemical group of halogenated hydrocarbons, they contain fluorine, which is responsible for ozone depletion and has very high global warming potential (GWP). It was estimated that almost 20% of the global warming impact was attributed due to leakage of refrigerants into the atmosphere. The leakage varies from one system to another, i.e., domestic refrigerators (2%) and mobile air conditioning (37%) [17].

HCFC-22 is commonly used refrigerant and leaves HFC-23 as a by-product during its production with 14,800 GWP as that of CO_2. These hydrocarbons leave to the upper atmosphere immediately or after some years and break down the ozone molecules thus deplete the ozone layer.

Over the last eight decades, CFC gases have been extensively used in the refrigeration system, which has not only high ozone depletion potential (ODP) but also persist in the atmosphere for a longer time more than a century. Moreover, as global warming is caused by heterogeneous phenomena, it is highly complicated to assess the contribution of refrigerants to global warming. Though, there is lack of data about the share of CFCs in the atmosphere, rising concern about the impact of greenhouse gas emission from the refrigerants has put forth various curative measures time to time which are discussed below:

- **1987: Montreal Protocol:** Phase-out of CFCs by end of 2010.
- **1992: UNFCCC:** Introduction of HFCs.
- **1997: Kyoto Protocol:** HFCs under control reduce the GHG emission in three-phase.
- **2007:** Accelerated phase-out of HCFCs (Montreal protocol), adopt climate-friendly technologies.
- **2015: Paris Climate Agreement:** Maintain the global temperature rise within 2°C compared to pre-industrial period. Effort should be made to maintain below 1.5°C.
- **2016: Kigli Amendment:** HFCs phase-out in stages—adoption of HFOs.

4.3.1 KIGLI AMENDMENT

Kigli Amendment is the outcome of the 28th Meeting of United Nation's Parties to the Montreal Protocol, at which all the member countries committed to significantly reduce the production and consumption of HFCs and avoid the emission of over 70 billion tons of CO_2 equivalent by 2050. It also mandates to all the HCFC-22 manufacturing facilities to collect and destroy the HFC-23 by-product from 2020. The agreement scheduled is in accordance with the Paris Climate Agreement, which commits to pursue the efforts to limit the average global temperature rise to 1.5°C. Under its common but differentiated responsibilities (CBDR), the agreement is scheduled to come into force from 2019 for developed countries and 2024 for developing countries. Moreover, unlike previous measures, Kigli Amendment is legally binding multilateral agreement which has the provision of penalty on failure countries which makes it a crucial contributor to the world's commitment to the Paris Climate Agreement to avoid dangerous climate change [68].

4.4   CHALLENGES FOR FOOD SECURITY AND CLIMATE CHANGE

According to Global Nutrition Report 2016 (From Promise to Impact-Ending Malnutrition by 2030), one out of three persons is suffering from malnutrition, and 45% of the population under 5 years child death are caused due to malnutrition [27]. It was estimated that out of world population of 7 billion, 2 billion, and 800 million people are suffering from malnutrition and calorie deficiency, respectively. Additionally, the Global Hunger Index (The Inequalities of Hunger) of 2017 reported that 13% of the world population is undernourished with countries from South and Southeast Asia, and South of African Sahara is critically suffering from hunger [26].

Data of regional variation suggests that indigenous peoples, rural dwellers, ethnic minorities, and poor which are isolated from the global industrialized world are more vulnerable to hunger and malnutrition, which led to the emergence of new diseases and increase the burden on the economy. In order to tackle this menace, increasing the productivity of agriculture with sustainable agricultural practices followed by safe storage and transportation maintaining the nutritional quality of this produce is vital in the humanitarian context. The agricultural market in the underdeveloped and rural community is controlled by retail fruit and vegetable shops and street vendors but has a shortage or non-availability of compatible refrigerated transport infrastructure. Development of small capacity multi-crop refrigerated carts for farmers and vegetable vendors would not only maintain the cold chain and quality of the produce but also help to fetch impressive price in the market and help to reduce the burden of food loss.

In addition to this is global warming caused by the emission of greenhouse gases to which refrigerants have remarkable share had made the environment vulnerable. The outcomes of climate change are frequent such as uncertainty of rainfall, heat waves, and frequent hurricanes in America and cyclones in South East Asia, which again led the aggrieved people in more danger. The contradictory nature of food transportation, storage facilities and food wastage and their effect on malnutrition and hunger is presented in Figure 4.2.

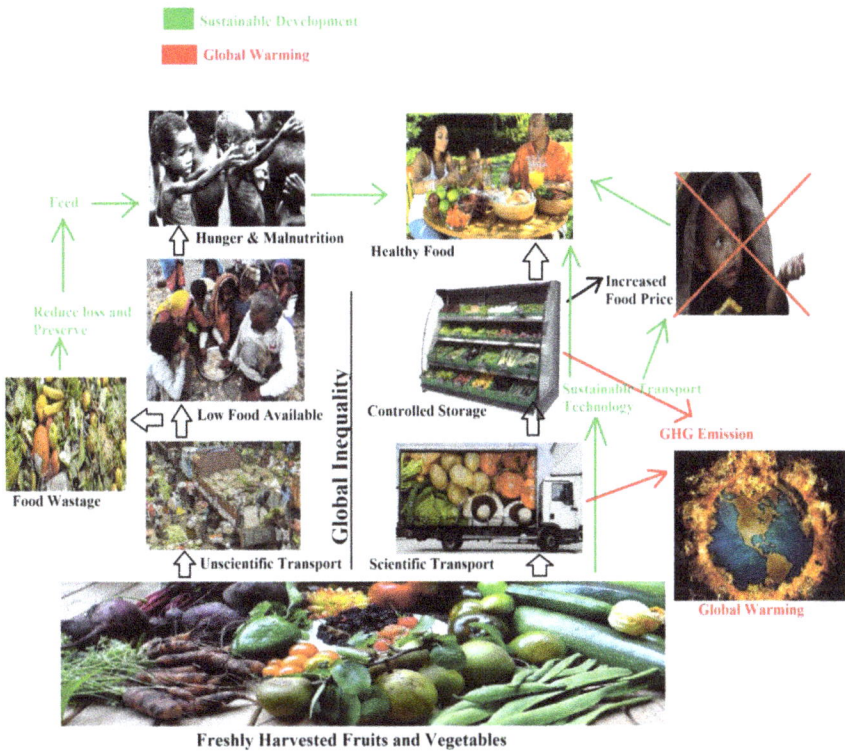

FIGURE 4.2 Diagrammatic presentation of the challenge of food security and global warming.

In accordance with the global hunger and climate change, the world is exposed to major challenges of increasing food production and reducing food losses, thus widening the infrastructure of cold storage for the rising population and at the same time reduction of disastrous hydrocarbon refrigerants. So, it is great challenge before the researchers to find the alternative refrigeration technology which will be eco-friendly, energy-efficient, compatible during the transition, solar operated and adaptable for local market so that the commitment to the Sustainable Development Goals, i.e., eradication of hunger, malnutrition, poverty will be fulfilled and health of rural population will be improved. Efforts made in this direction are discussed as follows,

4.4.1   SOLAR POWERED VENDING CART

Indian Agricultural Research Institute (IARI): New Delhi has developed a solar-powered vending cart which is capable of safe storage of fruits and vegetables for four to five days by controlling the surrounding environment [57]. The system is based on evaporative cooling principle with solar power obtained from the photovoltaic conversion is used for operating two fans for forced air circulation and LED lamp. The developed system is capable of reducing the temperature of the storage chamber by 8.1–11.2°C and increasing the humidity by 15–25%, thus it helps to maintain the freshness in terms of color, texture, and appearance of the produce.

4.4.2   PUSHCART WITH LOW ENERGY FOOD STORAGE SYSTEM

Evaporative low energy storage system was developed by Venu [69] with a storage capacity of 30–45 kg of vegetables. The basic structure of the system is of *almirah* type geometry in which water baths are placed at top and bottom. In order to make the system heat-insulated, coir pith and charcoal was used, whereas, gunny cotton was covered on four sides, which absorb water from the upper water bath by capillary action and give an evaporative cooling effect. The system is capable of reducing the storage temperature by 8–10°C and increasing humidity to 40–45% when the ambient atmospheric condition was 38°C and 35–45% relative humidity (RH). The practical applicability of the cooling system showed that the system is useful for tomatoes, carrots, beans, okra, amaranths during summer and winter seasons and increased shelf-life of vegetables up to 8–12 days compared 2–4 days under ambient condition.

The designed cooling systems effectively preserve the quality of fresh produce; these systems are environmentally controlled and are not suitable for crops requiring storage temperature below 15°C. In order to prevent the moisture loss from fruits and vegetables, these are required to be stored at RH above 90% which is attainable in the evaporative cooling system. Therefore, there is further scope for development of temperature and RH controlled self-regulatory mechanical refrigeration system which will act as an alternative to VCRS and environmentally sustainable.

Adekomaya et al. [3] and James et al. [36] have briefly explained the alternative refrigeration techniques (Thermoelectric, Eutectic, Thermo-tunnel, Magnetic, Ejector, etc.), with their development stages, expected efficiency, coefficient of performance (COP) and potential areas of application (Table 4.2).

TABLE 4.2 Summary of Alternative Refrigeration Techniques

Technology	Developmental State	COP (%)	Refrigeration Capacity Developed	Potential Area of Application
Thermoelectric	Initial stage (Low-cost low-efficiency system available)	10–15	Up to 20 KW	Refrigerators for truck, Restaurant, and mini bar, Small scale food transport
Thermoacoustic	R&D stage	20	Few watts	Domestic and commercial refrigerators, freezers, and cabinets
Themotunnel	Experimental	No data	–	–
Magnetic	Prototype	20%	Up to 540 W	Low capacity stationary and mobile refrigeration systems
Ejector	Bespoke steam ejector systems available	Up to 0.3	Few kW-60 MW	Food processing; refrigerated transport

4.5 THERMOELECTRIC SYSTEM

Thermoelectric devices are semiconductors that convert electric power into heat energy on the basis of the Peltier effect and heat energy into electric power based on the Seebeck effect [37, 64]. Thermoelectric devices possess remarkable features and advantages, including size, shape, lightweight, simplicity, no external moving parts, ability to heat and cool simultaneously with a single module, precise temperature control, all of due to which it became solid and highly compactness made it applicable at limited space availability. Moreover, no involvement of working fluid (refrigerants as in case of vapor compression refrigerator, wider temperature range and longer life span has increased its applicability as environment-friendly technology with high reliability [33, 48, 62, 71].

4.5.1 DESIGN OF THERMOELECTRIC MODULE

The thermoelectric module is composed of the number of P-type and N-type semiconductors joined together using highly conducting strips usually made of coppers and connected electrically in series and thermally in parallel [55]. The design construction of thermoelectric module is shown in Figure 4.3(a). The whole assembly is sandwiched between two insulate substrates like ceramic to prevent contact from each other and provide rigidity. The whole assembly is connected to the electric DC power supply. Heat sinks are attached to the hot side of the module for the dissipation of heat. While, the multistage type thermoelectric modules as shown in Figure 4.3(b) is the pyramid shape stacking of single-stage modules so the stages closer to the heat sink should transfer the heat from the heat source to the heat sink more efficiently and increase the performance of the module [38].

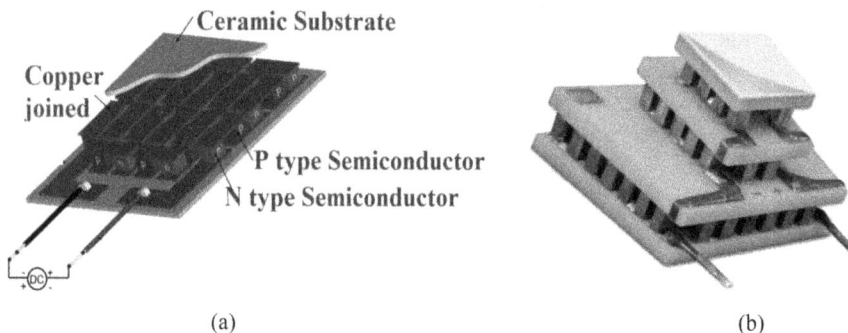

FIGURE 4.3   (a) Basic design of thermoelectric module; (b) multistage thermoelectric module.

4.5.2 PRINCIPLE OF THERMOELECTRIC

The thermoelectric module is based on the principle of *the Peltier Effect* after the name of French scientist Jean Peltier who stated that *"when an external electric current is made to flow the circuit of the thermoelectric module, temperature difference establishes on two sides of the junction and heat is absorbed at the cold side called active side and rejected to the hotter side"* [76]. Opposite to this is called Seebeck effect.

As shown in Figure 4.4, when DC electric current is applied across the terminals of the modules, the electron starts to move from low energy P-type element to the high-energy N-type element through high conductive joint material and absorbs the heat from the surrounding making side colder.

These electron further moves upward through a conduit of N-type element carrying the absorbed heat and jumps from N-type element to P-type element releasing the heat into surrounding making the other side hotter. The temperature at the colder side is noted as T_c, while the temperature at the hotter side is noted as T_h [23].

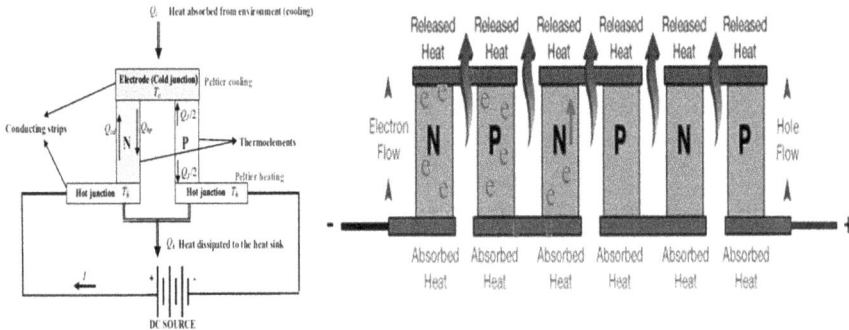

FIGURE 4.4   Working principle of thermoelectric module.

The temperature difference created across the junction is directly proportional to the amount of electric energy supplied to the module and with changing intensity and direction of the current, cooling and heating can be easily achieved [44]. The other parameters responsible for cooling are electrical and thermal conductivity of thermoelectric materials, the temperature of the environment surrounding hotter and colder side, electrical resistance at the contact between hot and cold side and heat dissipation capacity of heat sink at the hotter side [13].

The performance of the thermoelectric cooler can be determined by COP, figure of merit and cooling capacity (Q_c) of the cooler which can be calculate by following formula [23].

The single-stage devices are applicable still temperature difference between the hot and cold side is small, say for example, 70°K, as the temperature increases beyond the maximum temperature, the effectiveness of the single-stage refrigeration system reduces whereas multistage devices come to play [14]. Yu and Wang [74] have optimized the COP of cascades multistage module for maximum Q_c. Their computation optimized that the three-stage thermoelectric module increased approximately 35.8% of Q_c compared to single stage thermoelectric device. Furthermore, a mathematical model was developed by Yu and Wang [74] for the determination of cooling rate q_i per unit area of module for each ith stage of module, depending on COP of the

i^{th} stage and cooling rate q_i per area for the whole module (I stages) as given in Eqn. (1),

$$q_i = q_I[1+COP_I^{-1}][1+COP_{I-1}^{-1}] \ldots [1+COP_{I-i}^{-1}] \tag{1}$$

Goldsmid [29] also reiterated that multistage thermoelectric module is advantageous over single-stage module in which each stage removes heat extracted from previous stage and joules heat developed by that stage itself.

4.5.3   SOME IMPORTANT TERMINOLOGIES

1. **Thermoelectric Figure of Merit (ZT):** It depends on three parameters viz.; electrical conductivity (σ), thermal conductivity (k) and Seebeck coefficient (α) and indicate whether the material is good thermoelectric cooler. The temperature is considered as the absolute parameter and hence dimensionless parameter is product ZT and calculated using Eqn. (2).

$$ZTm = \frac{\alpha^2 \sigma T}{k} \tag{2}$$

2. **Cooling Capacity (Q_c):** of the refrigerator is the energy balance at the cold side and can be calculated from Eqn. (3),

$$Q_c = \alpha IT_c - KT - \frac{1}{2}R_e I^2 \tag{3}$$

where; $Q_g = \alpha IT_c$ – Heat pumped at cold junction (I = input current); $Q_d = K\Delta T$ – Heat flow conducted from hot junction to cold junction ($\Delta T = T_h - T_c$); $Q_j = R_e I^2$ – Joule heat depending on the electric resistance R_e.

3. **Coefficient of Performance (COP):** Efficiency of any refrigeration system is measured in terms of its COP, which is the ratio between Q_c and electric power consumption (P).

$$COP = \frac{Qc}{P}$$

For thermoelectric refrigeration $P = R_e I^2 + \alpha I \Delta T$.

In terms of figure of merit, COP of thermoelectric refrigeration system can be calculated using Eqn. (4) [77].

$$[COP]_{max} = \frac{T_c}{T_h - T_c} \frac{\sqrt{1 + ZT_m} - \dfrac{T_h}{T_c}}{\sqrt{1 + ZT_m} + 1} \tag{4}$$

where; T_c = cold side temperature; T_h = hot side temperature.

The key aspect for optimization of the material for thermoelectric application is maximization of its figure of merit which can be achieved by maximizing power factor ($\alpha^2\sigma$) and minimization of thermal conductivity (K). From the above equation, it is clear that COP of thermoelectric cooling is largely depend upon the temperature at cold and hot side and figure of merit of the module. Various scientists have made efforts for classification of semiconductor materials on the basis figure of merit as follows:

- **ZT = 1:** Bismuth telluride (Bi_2Te_3)-Most common thermoelectric material.
- **ZT = 2–4:** Thermoelectric refrigerator compatible with vapor-compression refrigerator.
- **ZT = 6:** Thermoelectric refrigerator is capable of cooling to cryogenic temperature ($-196°C$) from room temperature.

The detailed analysis of parameters affecting the performance of thermoelectric refrigeration such as COP, Q_c and cooling rate are discussed by Esfahani et al. [24].

4.5.4 MATERIALS FOR THERMOELECTRIC MODULE

The semiconductor materials having a high figure of merit, good electric properties, low thermal conductivity, and high Seebeck effect are most desirable for development of thermoelectric devices [51, 77]. The first generation devices earlier 200 before had ZT value 1.0 with the conversion efficiency of 4 to 5%. The second-generation devices developed after the 1990s were introduced with nanostructures to increase the ZT to 1.7 and power conversion efficiency to 11%–15%. Third generation devices under development are bulk thermoelectric with ZT increased to 1.8 and power conversion efficiency to 15%–20% [76].

Alloys of Bi_2Te_3, Bi_2Se_3 and Sb_2Te_3 with ZT = 1 were the conventionally used material for thermoelectric refrigeration. Riffat and Ma [54] specified that $(Bi,Sb)_2(Se,Te)_3$ which is the alloy of bismuth telluride works in the temperature range of −120°C to 230°C. Vast numbers of materials are tested to form the alloy to increase the figure of merit of the thermoelectric module having the figure of merit value of 1.2–2.2 at a temperature in the range of 600–800°K. It was observed that thermoelectric coolers having $ZT = 1$ works at 10% of Carnot cycle whereas 30% of Carnot efficiency can be achieved by increasing the ZT up to 4.

4.5.5   COOLING APPLICATIONS OF THERMOELECTRIC DEVICE

The effectiveness of any cooling device is measured in terms of its COP, which is the ratio of amount of cooling to the energy demand. The principle of thermodynamics indicates that maximum efficiency (Carnot efficiency) cannot be exceeded. Hence the real efficiency of any cooling device is expressed in terms of Carnot efficiency. Till 1999, the thermoelectric devices were able to operate at about 10% of Carnot efficiency as against 30% and 90% for domestic kitchen refrigerators and room cooling air conditioners, respectively [20]. Owing to its low figure of merit and low COP, thermoelectric cooling has not widened its area of application and restricted to space missions, scientific instruments, and medical equipment, cooling of electric instruments where space is the limitation and higher COP is not a preliminary criteria. However, with exhaustive research and development (R&D) of advanced opportunities, new applications for the thermoelectric cooling are emerging which can be classified as:

1. **Civil Market:** Cooling small enclosures, portable, and domestic refrigerators, icebox and picnic baskets [8, 18, 19, 49, 56, 70].
2. **Medical Applications:** Cooling of laboratory and scientific instruments [53].
3. **Industrial Applications:** Dissipation of head and temperature control in electric circuits [13, 75, 78].
4. **Automobile Applications:** Automobile mini-refrigerators, heating, and cooling of seat and air conditioning [15].

4.5.5.1   THERMOELECTRIC REFRIGERATOR

The colder side of the thermoelectric module exposed into the cooling chamber transfer the heat from the heat source to the heat sink at the hotter side and further to the environment. It consisted of a cooling cabinet, thermoelectric module sandwiched between two heat exchangers for efficient transfer of heat, d. c. power supply along with the temperature controller. As electric energy is supplied to the assembly, heat is pulled from the cooling cabinet (similar to the evaporator in vapor compression refrigeration system), conveyed to the other side through free electrons in semiconductor module (electrons plays the role of refrigerant) and dissipated through heat sink (heat exchanger) at the hot side [49]. Thermoelectric refrigeration system is capable of accelerated cooling as temperature reduction from 27°C to 5°C in about 44 min was stated by Twaha et al. [65].

The simplest design of thermoelectric cooler employs single stage thermoelectric module with fan-cooled fin-type heat sink on the hot side. When the system comes under operation, there exist three effects *viz.* Joules heating due to electric current, Thomson heat due to temperature gradient between the hot and cold side of the semiconductor and heat migration from high temperature heat sink to low temperature cooler space. This are the major factors responsible for the least COP of the thermoelectric refrigerator [14].

Moreover, the properties of the material of semiconductor, heat transfer area of heat exchanger, thermal resistivity of the cabinet material, etc., also affects the performance of the refrigerator. Therefore, to prevent this reverse migration of heat, it is vital to faster dissipation of heat from the hotter side to the heat exchanger and maintains the temperature of the hotter side of the module at optimum range. Two decades ago, it was found that at efficient dissipation of heat from hotter side of Peltier, with one degree reduction in temperature difference between hotter side and ambient temperature increases COP of the thermoelectric refrigerator by 2.3% [21]. In order to achieve this objective and increase the energy efficiency of the refrigerator, continuous efforts were made in the design of refrigerator. Experimental investigation proved that COP of single-stage refrigerator decreases with increase in the temperature gradient between hot and cold side of semiconductor.

Chen et al. [14] observed that, at $\theta = 1.1$ the COP of single-stage refrigerator was 1.24 and with an increase of θ to 1.3, the COP reduced to 0.102. It was further noted that COP reduced to zero at $\theta=1.5$. In such circumstances, multistage semiconductors showed better performance with COP up to 0.201. Luo et al. [46] stated that optimum cooling load and optimum COP increases with an increase of total heat transfer surface area. Min and Rowe [49] designed a refrigeration system having cabinet capacity of 40 liters with a solid heat exchanger at the hot side for the fast rejection of heat from the hot side. The COP of around 0.3 to 0.5 was attained at an operating temperature of 5°C when the ambient temperature of about 25°C. The water vapor and ice formation on the cold side surface inside the cabinet reduce the heat transfer coefficient, which causes to reduction in effectiveness of the evaporator and consequently the COP of the refrigerator.

The device was developed by Astrain et al. [9] based on thermosyphon with fluid (R-141b) phase change material (TSF) to improve the heat dissipation from the hotter side of Peltier. The thermosyphon device was a hermetically sealed chamber-containing refrigerant at its bottom. The hot side of the Peltier attached at bottom of backside of TSF, whereas fins and fan was placed on the other higher side. Absorbing heat from the hotter side of Peltier, refrigerant get vaporized and rise higher inside chamber where it condensed by contacting with fins and fan and again drops in the lower portion. Thus getting maximum area for heat dissipation, the hotter side of Peltier remains at optimum temperature. Incorporation of TSF in refrigeration system increased the COP of the system by 36% compared to fin based heat sink device at 303°K. Though the results obtained were quite impressive, owing to its ODP, R-141b was listed under the phase-out list of Montreal Protocol with "Worst First" approach [66].

4.5.5.2   RECENT DEVELOPMENT FOR EFFICIENCY IMPROVEMENT IN THERMOELECTRIC REFRIGERATION SYSTEM

With growing awareness of environment degradation, changing regulation at United Nations (UN) and evolution of nanostructure engineering, thrust for development of thermoelectric materials having maximum figure of merit (ZT) has increased at the same time various advanced design for application of thermoelectric principle in refrigeration were explored.

4.5.5.2.1   Material for Thermoelectric Devices

Ohtaki [52] claimed that though Bi_2Te_3/Sb_2Te_3 superlattice works extraordinary (ZT = 2.4) at 300°K, these materials were toxic in nature and not stable in higher temperature range. In the context of this, metal oxide having good stability and environmentally friendliness has gain attention. In recent years, p-type $(Ca_3CO_4O_9)$ and n-type $(ZnO, SrTiO_3$ and $CaMnO_3)$ were studied with slit increment in value of figure of merit (ZT<1.0) [73]. Dura et al.

[22] prepared bulk nanostructure samples of $La_{0.875}Sr_{0.125}COO_3$ with grain size ranging between 19 nm and 0.7 μm by subsequent mechanical milling. The nano-structuring had a strong influence over electron and thermal conductivity resulted into 30% increase in figure of merit (ZT) vis-a-vis microstructure materials.

Simultaneously, efforts made by some researchers [10] revealed that amorphous silicon (a-Si) thin-film enhanced ZT by 7 order reaching 0.64 ± 0.13, by arsenic implantation and low temperature activation. As the highest material available has reached a ZT value of 3, achieving ZT 4 is a great challenge [32]. Thus uses of metal oxide as well as other material of thermoelectric properties have opened new area for the development of more efficient thermoelectric devices.

4.5.5.2.2 Design of Heat Sink

The second line of action for increasing of thermal efficiency of the thermoelectric refrigeration system is the improvement of heat dissipation from the hot side of Peltier using modified heat sink or evaporators. Nagy and Buist [50] have modeled the heat distribution pattern of thermoelectric refrigerator and concluded that higher thermal resistance of heat sink and thermal resistance between module and heat sink are responsible factor for lower performance of thermoelectric refrigerator which need to be improved.

Astrain et al. [7] undertaken a comparison of conventional fan-cooled fin type heat sink, evaporator type coil permitting water circulation with fan cooling and liquid phase change material based heat pipe. For evaluation of COP of refrigerator, the electric power consumption for operating auxiliary equipments such as fan and pump were also taken into consideration which significantly influence energy consumption and applicability of the system. Results revealed that difference of COP at the Peltier module alone and total COP including auxiliary devices was of value 13%. Due to no requirement of heat pump, 0.656 COP was obtained by heat pipe type device. A prototype of a mini channel water-cooled thermoelectric refrigerator has also been made, which was found to be effective in generating low temperatures below zero degree [28]. The design consists of a mini-channel heat sink at the hot side of the Peltier while heat dissipater was faced at cold side and a water flow rate of 1.5 L/min, temperature of –0.1°C was attained at the end of 2 hours with COP of 0.26.

4.5.5.2.3   *Photovoltaic: Thermoelectric Cooler*

The twenty-first century is the era of modernization with using renewable and green sources of energy. Among some renewable sources, photovoltaic use has gained remarkable importance due to its durability, movability, compactness with no adverse effect from environmental conditions such as heatwave or rainfall. In order to make the thermoelectric refrigeration system economically viable, research is in progress to develop a thermoelectric refrigerator driven by photovoltaic-based free solar energy.

Liu et al. [41] developed a patented design for operating a thermoelectric module powered by a solar cell. In order to cool the battery of photovoltaic assembly which gets excessively heated, thermoelectric module powered by photovoltaic electricity were employed. The prototype of solar-powered thermoelectric refrigerator was developed by Dai et al. [19] to achieve temperature inside cooling chamber in the range of 5–10°C with 15°C temperature drop and has COP of 0.3. Detailed review of photovoltaic powered thermoelectric applications is provided by Xi et al. [72].

Sabah and Abdul-Wahab [56] claimed that photovoltaic powered thermo-electric cooler developed for desert people reduces temperature from 27°C to 5°C within 44 min and operated at COP of 0.16. Principle of solar-powered thermoelectric cooler was used by Shetty et al. [60] for cooling water. The temperature of water was reduced by 10°C within 60 min when the Peltier was powered by the battery of 12 V, 7 Ah charged using photovoltaic solar cell. Calculated COP for the assembly was 0.095, however, the battery was capable of working continuously for 3 hrs once fully charged.

4.5.6   *OTHER APPLICATIONS OF THERMOELECTRIC DEVICE*

4.5.6.1   *AIR CONDITIONING*

It is noted that globally, almost 59% of commercial and household energy consumption is attributed to air conditioning and water heating appliances [6]. In conventional air conditioning devices, heat from the air condensation is exhausted to the outside, which has made irreversible damage to the environment, disturbing its heat balance and pollution. Additionally, a large amount of extra energy is consumed for meeting demand of hot water [42]. Nowadays, more attention is paid on development of green building technology. So, effective utilization of both side of Peltier that is hotter side for water heating and cold side for space heats is proposed as an alternative.

Some of the application of thermoelectric modules in this direction, such as thermoelectric ventilator, domestic air conditioners, and thermoelectric cooled ceiling were already commercialized [31, 40, 43].

- **Space Cooling and Water Heating Mode:** This system is applicable in summer wherein air is circulated through heat sink at cold face using fan and moved down to enter into the room. At the same time, to dissipate the heat from the hotter side, water is circulated through the evaporator, which in term gets heated and can be used for other appliances.
- **Space Heating:** The system is operation in winter by simply changing the direction of electricity supplied to the thermoelectric module and the above results can be vice versa obtained.

Prototype of design and algorithm for its operation in simultaneous model was developed and experimentally tested by Liu et al. [44]. Due to dual mode of operation, remarkable COP was achieved, which was as high as 4.51 in water heating and space cooling mode and 3.01 in space heating mode. Manikandan et al. [47] operated a thermoelectric based space cooling system at pulse rate and observed that temperature achieved at the cold side of Peltier in this design was lower than that of a steady-state condition. The reason for this difference was that when pulses were applied, the sudden drop in temperature at the cold side of Peltier and peak is achieved later on, followed by steady value due to Joules heating, and Fourier heat transfer. The optimized pulse current ratio and the pulse width was 2 and 5 s at which cooling power and COP were increased by 23.3% and 2.12%, respectively.

Irshad et al. [35] claimed an optimum temperature difference of 6.8°C for a room of volume 9.45 m³ when thermoelectric combined air duct system powered by the photovoltaic system was used in which COP of achieved was 1.15 with Q_c of 517.24 W. Shen et al. [59] had investigated the internal temperature distribution pattern by establishing energy transfer model.

4.5.6.2 WATER DISTILLATION

As only 1% of the earth's water is available in drinkable form and demand for this life-saving commodity increasing day by day, the distillation of seawater has gain imperative solution, and wider research has been undertaken in this direction [1, 12]. In general, heat rejected at the hot side of the Peltier is wasted. The application of the Peltier principle facilitates best utilization of energy in two ways. The heat generated at the hotter side was used for heating the water contained in the distillation chamber. The water vapor generated were risen up and allowed to condensate in the condenser to which cold side was faced [4].

In order to achieve more efficient desalination of water, aluminum heat sink fixed at cold side while heat exchanger was used for cooling the hot side of the Peltier. Additionally, unevaporated water was circulated from the top of the unit over a copper heat exchanger to the bottom basin using a small pump. Developed prototype system produced 28.5 ml (equivalent 678 ml/ m^2) per hour with total electricity consumption of 0.0324 kW h or 0.00114 kW h/ml. Esfahani et al. [24] used thermoelectric devices to increase the temperature difference between evaporating and condensing units. Average daily recovery using this principle was 1.2 lit/m^2 with 13% efficiency in winter days, however, the cost per liter for this distillation plant is lower than sun-tracking solar still. The scope for further improvement proposed was that to use thermoelectric distillation assisted with solar radiated evaporation and optimization of design.

4.6   SCOPE OF THERMOELECTRIC IN COLD CHAIN MANAGEMENT

As the thermoelectric devices have the least efficiency in terms of COP, there exist limitation in the use of this technology for application in large appliances. The efficiency of commercial thermoelectric module remained low against that of the kitchen refrigerator, which works on VCRS. A wider scope remains for efficient dissipation of heat from the hot side which another alternative. Above stated applications of thermoelectric module witness that more attention needs to pay for the design of thermoelectric based small size refrigerators. Features of thermoelectric module such as small size, compactness, and lightweight make a great opportunity for the design of small size mobile thermoelectric refrigerator to overcome the complicated design of VCRS. A similar effort has been made by Astrain et al. [7] for the development of cold storage for fruits and vegetables, which shown impressive results and scope for future development. More research is needed to optimize the capacity of refrigeration compatible with thermoelectric module, shelf-life, and quality of fruits and vegetables stored under thermoelectric refrigeration condition.

4.7 SUMMARY

In order to balance the ever-increasing population, reduction of food wastage in supply chain management is the priority, world leader under the title of UN Sustainable Development Program in 2015 have adopted several goals to be accomplished by 2030. Some targets in this are directly related to zero hunger, good health and wellbeing, and climate action. Though more attention is paid to the establishment of community based cold storage facilities, cold supply chain management is vital and need to be highlighted. Attention is needed to be paid at the horticulture crop market in underdeveloped countries, which remained unstructured with no scientific facilities for preservation of fruits and vegetables during transit. Simultaneously, more concern also needs to be paid at changing climate scenario and ozone depletion due to release of greenhouse gases from the refrigeration system. Being an environmentally friendly technology, design of small size thermoelectric mobile cold storage for farmers and rural vegetables vendors would have significant advantages such as maintenance of quantity and quality of produce, reduces food losses, better price for farmers and vendors and contribution to mitigation of global warming.

KEYWORDS

- chlorofluorocarbons
- coefficient of performance
- food cold chain
- food waste
- global warming
- thermoelectric refrigeration

REFERENCES

1. Abad, H. K., Ghiasi, M., Mamouri, S. J., & Shafii, M. B., (2013). A novel integrated solar desalination system with a pulsating heat pipe. *Desalination, 15*(311), 206–210.
2. Abbas, K. A., Saleh, A. M., Mohamed, A., & Lasekan, O., (2009). The relationship between water activity and fish spoilage during cold storage: A review. *Journal of Food, Agriculture and Environment, 7*(3/4), 86–90.

3. Adekomaya, O., Jamiru, T., Sadiku, R., & Huan, Z., (2016). Sustaining the shelf-life of fresh food in cold chain-burden on the environment. *Alexandria Engineering Journal, 55*, 1359–1365.

4. Al-Madhhachi, H., & Min, G., (2017). Effective use of thermal energy at both hot and cold side of thermoelectric module for developing efficient thermoelectric water distillation system. *Energy Conversion and Management, 133*, 14–19.

5. Renie, S., (2011). *Anonymous Country: India's Cold Chain Industry, US Commercial Service, Department of Commerce.* http://www.academia.edu/5952490/IndiasCold-Chain_Industry (accessed on 18 January 2021).

6. Anonymous, (2012). *Building Energy Data Book.* US Department of Energy. DOI: https://openei.org/doe-opendata/dataset/buildings-energy-data-book/resource/3edf59d2-32be-458b-bd4c-796b3e14bc65 (accessed on 11 March 2021).

7. Astrain, D., Aranguren, P., Martínez, A., Rodríguez, A., & Pérez, M. G., (2016). Comparative study of different heat exchange systems in a thermoelectric refrigerator and their influence on the efficiency. *Applied Thermal Engineering, 103*, 1289–1298.

8. Astrain, D., Vian, J. G., & Albizua, J., (2005). Computational model for refrigerators based on Peltier effect application. *Applied Thermal Engineering, 25*, 3149–3162.

9. Astrain, D., Vian, J. G., & Dominguez, M., (2003). Increase of COP in the thermoelectric refrigeration by the optimization of heat dissipation. *Applied Thermal Engineering, 23*, 2183–2200.

10. Banerjee, D., Vallin, Ö., & Samani, K. M., (2018). Elevated thermoelectric figure of merit of N-type amorphous silicon by efficient electrical doping process. *Nano Energy, 44*, 89–94.

11. Beausang, C., Hall, C., & Toma, L., (2017). Food waste and losses in primary production: Qualitative insights from horticulture. *Resources, Conservation Recycling, 126*, 177–185.

12. Byrne, P., Fournaison, L., & Delahaye, A., (2015). Review on the coupling of cooling, desalination, and solar photovoltaic systems. *Renewable and Sustainable Energy Review, 31*(47), 703–717.

13. Chein, R., & Huang, G., (2004). Thermoelectric cooler application in electronic cooling. *Applied Thermal Engineering, 24*(14/15), 2207–2017.

14. Chen, J., Zhou, Y., Wang, H., & Wang, J. T., (2002). Comparison of the optimal performance of single- and two-stage thermoelectric refrigeration Systems. *Applied Energy, 73*, 285–298.

15. Choi, H. S., Yun, S., & Whang, K., (2007). Development of a temperature-controlled car-seat system utilizing thermoelectric device. *Applied Thermal Engineering, 27*(17/18), 2841–2849.

16. Ciconkov, R., (2018). Refrigerants: There is still no vision for sustainable solutions. *International Journal of Refrigeration, 86*, 441–448.

17. Coulomb, D., (2008). Refrigeration and the cold chain serving the global food industry and creating a better future: Two key IIR challenges for improving health and environment. *Trends in Food Science and Technology, 19*, 413–417.

18. Dai, Y. J., Wang, R. Z., & Ni, L., (2003). Experimental investigation and analysis on a thermoelectric refrigerator driven by solar cells. *Solar Energy Materials Solar Cells, 77*, 377–391.

19. Dai, Y. J., Wang, R. Z., & Ni, L., (2003). Experimental investigation on a thermoelectric refrigerator driven by solar cells. *Renewable Energy, 28*, 949–959.

20. DiSalvo, F. J., (1999). Thermodynamic cooling and power generation. *Science, 285*, 703–706.

21. Domìnguez, M., Garcia, D., Esarte, J., Astrain, D., & Vian, J. G., (1999). Possibilities of efficiency improvement in the thermoelectric systems. *Journal Thermoelectricity, 2,* 31–40.

22. Dura, O. J., Andujar, R., Falmbigl, M., & Rogl, P., (2017). The effect of nanostructure on the thermoelectric figure-of-merit of $La_{0.875}Sr_{0.125}COO_3$. *Journal Alloys Compounds,* doi: 10.1016/j.jallcom. 2017.03.335.

23. Enescu, D., & Virjoghe, E. O., (2014). A review on thermoelectric cooling parameters and performance. *Renewable and Sustainable Energy Review, 38,* 903–916.

24. Esfahani, J. A., Rahbar, N., & Lavvaf, M., (2011). Utilization of thermoelectric cooling in a portable active solar still: Experimental study on winter days. *Desalination, 269,* 198–205.

25. Food and Agricultural Organization (FAO), (2011). *Global Food Losses and Food Waste. Extent, Causes and Prevention.* FAO, Rome. http://www.fao.org/docrep/014/mb060e/mb060e.pdf (accessed on 18 January 2021).

26. Global Hunger Index (GHI), (2017). *The Inequalities of Hunger.* Washington, DC/Dublin/Bonn.

27. Global Nutrition Report (GNR), (2016). *From Promise to Impact: Ending Malnutrition by 2030* (p. 182). International Food Policy Research Institute, Washington DC, USA.

28. Gökçek, M., & Şahin, F., (2017). Experimental performance investigation of mini channel water cooled-thermoelectric refrigerator. *Case Studies Thermal Engineering, 10,* 54–62.

29. Goldsmid, J. H., (2010). Introduction to thermoelectricity. *Book Series in Material Science* (p. 278). Berlin Heidelberg: Springer.

30. Govindan, K., (2017). Sustainable consumption and production in the food supply chain: A conceptual framework. *International Journal of Production Economics*, http://dx.doi.org/10.1016/j.ijpe (accessed on 18 January 2021).

31. Han, T., Gong, G., Liu, Z., & Ling, Z., (2014). Optimum design and experimental study of a thermoelectric ventilator. *Applied Thermal Engineering, 67*(1/2), 529–539.

32. Harman, T. C., & Walsh, M. P., (2005). Nanostructured thermoelectric materials. *Journal of Electronic Materials, 34*(5), 19–22.

33. Huen, P., & Daoud, W. A., (2017). Advances in hybrid solar photovoltaic and thermoelectric generators. *Renewable and Sustainable Energy Review, 72,* 1295–302.

34. International Institute of Refrigeration (IIR), (2015). *The Role of Refrigeration in the Global Economy* (p. 14). 29th Informatory Note on Refrigeration Technologies, Paris, France. https://iifiir.org/en/fridoc/138763 (accessed on 18 January 2021).

35. Irshad, K., Habib, K., Basrawi, F., & Saha, B. B., (2017). Study of a thermoelectric air duct system assisted by photovoltaic wall for space cooling in tropical climate. *Energy, 19,* 504–522.

36. James, S. J., & James, C., (2010). The food cold-chain and climate change. *Food Research International, 43,* 1944–1956.

37. Kane, A., Verma, V., & Singh, B., (2017). Optimization of thermoelectric cooling technology for an active cooling of photovoltaic panel. *Renewable and Sustainable Energy Review, 75,* 1295–1305.

38. Karimi, G., Culham, J. R., & Kazerouni, V., (2011). Performance analysis of multi-stage thermoelectric coolers. *International Journal of Refrigeration, 34,* 2129–2135.

39. Keating, B. A., Herrero, M., & Carberry, P. S., (2014). Food wedges: Framing the global food demand and supply challenge towards 2050. *Global Food Security, 3,* 125–132.

40. Kim, Y. W., Ramousse, J., & Faisse, G., (2014). Optimal sizing of a thermoelectric heat pump (THP) for heating energy-efficient buildings. *Energy Buildings, 70,* 106–116.

41. Levinson, L. M., (1993). *Solar Powered Thermoelectric Cooling Apparatus* (Vol. 5, pp. 197–291). US Patent.

42. Liu, Z. B., Zhang, L., & Gong, G. C., (2015). Review of solar thermoelectric cooling technologies for use in zero energy buildings. *Energy Buildings, 102*, 207–216.

43. Liu, Z. B., Zhang, L., & Gong, G. C., (2014). Experimental evaluation of a solar thermoelectric cooled ceiling combined with displacement ventilation system. *Energy Conversion Manage, 87*, 559–565.

44. Liu, Z. B., Zhang, L., Gong, G. C., Luo, Y. Q., & Meng, F. F., (2015). Experimental study and performance analysis of a solar thermoelectric air conditioner with hot water supply. *Energy Buildings, 86*, 619–625.

45. Lorentzen, G., (1978). Food preservation by refrigeration: A general introduction. *International Journal of Refrigeration, 1*(1), 13–26.

46. Luo, J., Chen, L., Sun, F., & Wu, C., (2003). Optimum allocation of heat transfer surface area for cooling load and cop optimization of a thermoelectric refrigerator. *Energy Conversion Manage, 44*, 3197–3206.

47. Manikandan, S., Kaushik, S. C., & Yang, R., (2017). Modified pulse operation of thermoelectric coolers for building cooling applications. *Energy Conversion Manage, 140*, 145–156.

48. Meng, F. K., Chen, L. G., & Sun, F. R., (2010). Multivariable optimization of two-stage thermoelectric refrigerator driven by two-stage thermoelectric generator with external heat transfer. *Indian Journal of Pure and Applied Physics, 48*(10), 731–742.

49. Min, G., & Rowe, D. M., (2006). Experimental evaluation of prototype thermoelectric domestic-refrigerators. *Applied Energy, 83*, 133–152.

50. Nagy, M. J., & Buist, R. J., (1994). Effect of heat sink design on thermoelectric cooling performance. In: *AIP Conference Proceedings* (Vol. 316, No. 1, pp. 147–149).

51. Nolas, G. S., Poon, J., & Kanatzidis, M., (2006). Recent developments in bulk thermoelectric materials. *MRS Bull., 31*(3), 199–205.

52. Ohtaki, M., (2011). Recent aspects of oxide thermoelectric materials for power generation from mid-to-high temperature heat source. *Journal of Ceramic Society of Japan, 119*, 770–775.

53. Putra, N., (2010). The characterization of a cascade thermoelectric cooler in a cryosurgery device. *Cryogenics, 50*, 759–764.

54. Riffat, S. B., & Ma, X., (2003). Thermoelectrics: A review of present and potential applications. *Applied Thermal Engineering, 23*, 913–935.

55. Rowe, D. M., (2006). *Thermoelectrics Handbook-Macro to Nano* (p. 1022). Boca Raton (FL): CRC Press Taylor & Francis.

56. Sabah, A., & Wahab, A., (2009). Design and experimental investigation of portable solar thermoelectric refrigerator. *Renewable Energy, 34*, 30–34.

57. Samuel, D. V. K., Sharma, P. K., & Sinha, J. P., (2016). Solar-powered evaporatively cooled vegetable vending cart. *Current Science, 111*(12), 2020–2025.

58. Sarbu, (2014). A review on substitution strategy of non-ecological refrigerants from vapor compression-based refrigeration, air-conditioning, and heat pump systems. *International Journal of Refrigeration, 46*, 123–141.

59. Shen, L., Tu, Z., Hu, Q., Tao, C., & Chen, H., (2017). The optimization design and parametric study of thermoelectric radiant cooling and heating panel. *Applied Thermal Engineering, 112*, 688–697.

60. Shetty, N., Soni, L., Manjunath, S., & Rathi, G., (2016). Experimental analysis of solar powered thermoelectric refrigerator. *International Journal of Mechanical and Production Engineering, 4*(8), 99–102.

61. Siroli, L., Patrignani, F., & Serrazanetti, D., (2014). Efficacy of natural antimicrobials to prolong the shelf-life of minimally processed apples packaged in modified atmosphere. *Food Control, 46*, 403–411.

62. Su, Y., Lu, J., & Huang, B., (2018). Freestanding planar thin-film thermoelectric micro-refrigerators and the effects of thermal and electrical contact resistances. *International Journal of Heat and Mass Transfer, 117*, 436–446.

63. Sun, F., Yun, D. A. I., & Yu, X., (2017). Air pollution, food production, and food security: A review from the perspective of food system. *Journal of Integrative Agriculture, 16*(12), 2945–2962.

64. Tritt, T. M., (2002). Thermoelectric materials: Principles, structure, properties and applications. In: *Encyclopedia of Materials: Science and Technology* (pp. 1–11). Kidlington, Oxford, UK: Elsevier Science Ltd.

65. Twaha, S., Zhu, J., Yan, Y., & Bo, L., (2016). A comprehensive review of thermoelectric technology: Materials, applications, modeling, and performance improvement. *Renewable and Sustainable Energy Review, 65*, 698–726.

66. UEPA, (2018). *United States Environmental Protection Agency.* https://www.epa.gov/ods-phaseout/phaseout-class-ii-ozone-depleting-substances#common-cfcs-and-their-uses (accessed on 18 January 2021).

67. UNEP, (1994). *Report of the Refrigeration, Air-Conditioning, and Heat Pumps Technical Options Committee.* United Nations Environment Programme, Nairobi, Kenya.

68. UNFCCC, (2016). *EIA Briefing to the 22nd Conference of the Parties (CoP22) to the United Nations Framework Convention on Climate Change (UNFCCC).* Marrakech, Morocco.

69. Venu, S. A., (2012). *Development of a Pushcart with Low Energy Storage System for Vegetable Vending.* University of agricultural sciences, Bangalore, India.

70. Vian, J. G., & Astrain, D., (2009). Development of a thermoelectric refrigerator with two-phase thermosyphons and capillary lift. *Applied Thermal Engineering, 29*, 1935–1940.

71. Wang, J., Wang, Y., Su, S., & Chen, J., (2017). Simulation design and performance evaluation of a thermoelectric refrigerator with in homogeneously doped nanomaterials. *Energy, 121*, 427–432.

72. Xi, H., Luo, L., & Fraisse, G., (2007). Development and applications of solar-based thermoelectric technologies. *Renewable and Sustainable Energy Review, 11*, 923–936.

73. Yinong, Y., Tudu, B., & Tiwari, A., (2017). Recent advances in oxide thermoelectric materials and modules. *Vacuum, 146*, 356–374.

74. Yu, J., & Wang, B., (2009). Enhancing the maximum coefficient of performance of thermoelectric cooling modules using internally cascaded thermoelectric couples. *International Journal of Refrigeration, 35*, 32–39.

75. Zhang, H. Y., Mui, Y. C., & Tarin, M., (2010). Analysis of thermoelectric cooler performance for high power electronic packages. *Applied Thermal Engineering, 30*, 561–568.

76. Zhang, X. L., & Zhao, D., (2015). Thermoelectric materials: Energy conversion between heat and electricity. *Journal of Materiomics, 1*, 92–105.

77. Zhao, D., & Tan, G., (2014). Review of thermoelectric cooling: materials, modeling and applications. *Applied Thermal Engineering, 66*(1/2), 15–24.

78. Zhou, Y., & Yu, J., (2012). Design optimization of thermoelectric cooling systems for applications in electronic devices. *International Journal of Refrigeration, 35*, 1139–1144.

CHAPTER 5

ELECTROSPINNABILITY OF FOOD-GRADE BIOPOLYMERS

B. G. SEETHU, R. DEVARAJU, B. RAJUNAIK,
F. MAGDALINE ELJEEVA EMERALD, HEARTWIN A. PUSHPADASS,
B. SURENDRA NATH, and LAXMANA N. NAIK

ABSTRACT

Nanoencapsulation of bioactive compounds is an emerging technology for better assimilation and sustainable release of bioactive constituents. Among various techniques for nanoencapsulation of bioactives, electrospinning is a one-step technique that produces clean fibers of nano-scale range, functionality, etc., which could be used for various food applications. Polysaccharides- and protein-based biopolymers (such as pullulan, inulin, carrageenan, fructo-oligosaccharides (FOS), whey protein isolate (WPI), soy protein isolate (SPI), zein, and their combinations) derived from plants, animals, and microorganisms, could be used as wall materials for encapsulation of various natural bioactive compounds.

5.1 INTRODUCTION OF ENCAPSULATION TECHNIQUES

Nanoencapsulation of bioactive compounds improves their stability in gastric conditions and bioavailability. The type of nanoencapsulation technique decides the particle size, solubility, encapsulation efficiency, and releasing mechanism of the encapsulated bioactive. Various techniques of nanoencapsulation are coacervation, emulsification, nano-precipitation, inclusion complexation, supercritical fluid technique, emulsification-solvent evaporation, and electrospinning. Amongst them, electrospinning could be considered as one of the best alternative techniques to encapsulate bioactives and to enhance their functional properties, controlled-release, and bioavailability. As the process is carried out under ambient conditions, heat-sensitive bioactive compounds could be encapsulated. Electrospun fibers have many structural and functional merits such as high efficiency of encapsulation, sustained release of encapsulated material, and greater thermal, light, and storage stability.

Electrospinning uses a high potential electric field for the production of nanofibers from polymers. The electrospinning setup consists of a high voltage power supply, blunt-ended stainless steel capillary spinneret, and a grounded collector. Factors like polymer concentration, feed rate, voltage applied, distance between spinneret and collector, viscosity of feed solution, and solvent, etc., affect the formation of nanofibers and their morphology [15]. Therefore, screening and selection of biopolymers should be based on their viscoelastic characteristics and electrospinnability. For food applications, the selected biopolymer should be non-toxic, biocompatible, biodegradable, and sustainable. The commonly used solvents for nanoencapsulation of bioactive compounds in food applications are water, ethanol, and in some cases, acetic acid.

Electrospinnability of biopolymers depends on solution parameters, process parameters, and ambient conditions. Properties of biopolymers such as distribution of charged group, purity, molecular weight, degree of deacetylation, etc., may change depending on its source [38]. Similarly, the secondary and tertiary structure of proteins limits their electrospinnability. For this reason, globular proteins like soy protein isolate (SPI) and whey protein isolate (WPI) are used along with other natural or synthetic fiber-forming polymers for electrospinning due to the inability of their molecular chains to entangle during electrospinning. However, such combinations cannot be used in foods.

The use of biopolymers for the production of nanofibers is challenging as most of them are poorly soluble in water or even in organic solvents due to their crystalline nature. The biopolymers also may possess extreme electrical conductivity, surface tension, and viscosity that are not suitable for electrospinning. The number of carbohydrates and proteins that could be electrospun into nanofibers is very less. The selection of solvent is also important because the feed solution parameters are governed by the solvent used to dissolve the polymer or the bioactive compound. The bioactive compound also should be soluble in the selected solvent. Teo and Ramakrishna [44] opined that the morphology and diameter of electrospun nanofibers were affected by the nature of feed solution, comprising of the selected biopolymer(s), solvent, and bioactive compound.

In this chapter, the process parameters (such as applied potential difference, feed rate, and distance between spinneret and collector, and properties of biopolymer solutions governing electrospinnability of carbohydrates and proteins into nanofibers) are discussed in detail. The electrohydrodynamics of formation of nanofibers is explained in terms of dimensionless numbers. The fiber-forming properties of various biopolymers are discussed and are correlated with electrohydrodynamics and physical properties.

5.2 PROPERTIES OF BIOPOLYMER SOLUTIONS AFFECTING ELECTROSPINNING

5.2.1 CONCENTRATION OF BIOPOLYMER

For electrospinning and formation of uniform and bead-less fibers, an optimal concentration of feed solution is necessary [7]. Studies showed that at low concentration of polymer solutions, the number and concentration of beads increased [9]. Li et al. [27] reported that at 10% concentration (or lower) of zein, only droplets formed on the spinneret tip. As the zein concentration increased to 20%, uniform, and continuous fibers were formed because of enhanced molecular chain entanglement and interaction between the polymer and solvent. Yao et al. [53] also observed that the mean diameter and morphology of zein nanofibers were related to the polymer concentration used. However, as the polymer solution concentration increased beyond a specific level, thick, and ribbon-like fibers were formed. For example, clean gelatin nanofibers were formed at concentrations in the range of 5 to 12.5% (w/v) using 2,2,2-trifluoroethanol as solvent [19].

An increase in concentration of gelatin from 5 to 12.5% also resulted in an increase in the diameter of nanofibers. However, an increase in gelatin concentration beyond 12.5% (w/v) caused difficulty in electrospinning due to an increase in viscosity and accumulation of fluid at the tip of the spinneret despite applying higher voltage. Gelatin formed smooth bead free nanofibers at 7.5% (w/v) with a mean fiber diameter of 140 nm.

5.2.2 VISCOSITY AND VISCOELASTIC PROPERTIES

Molecular weight, concentration, structure, and conformation of polymer, and the solvent used influence the viscosity of feed solution [11]. The optimum range of viscosity for electrospinning is about 150–600 m.Pa.s. Viscosity also increases due to molecular entanglement caused by an increase in the molecular weight of polymer. To form a stable liquid jet and clean nanofibers, sufficient entanglement of polymer chains is necessary before evaporation of solvent. For example, casein, and chitosan could not be electrospun owing to high elasticity and insufficient chain entanglement, respectively even though they possessed sufficient viscosity. On the other hand, polymer solutions of optimal viscosity eject electrospun fibers to the collector in a continuous form with minimum diameter. Very high viscosity of feed solution makes it difficult to pump (the solution), besides leading to drying of the droplet at spinneret tip.

As, formation of continuous and clean fibers is difficult in low viscosity solutions, polyethylene oxide (PEO) was used along with low viscous biopolymers such as collagen, silk, and soluble eggshell membrane proteins to increase their viscosity to a sufficient level for electrospinning. Sullivan et al. [42] blended whey protein with PEO to facilitate the production of nanofibers as whey proteins alone could not form fibers in their native or denatured form due to low viscosity. At whey protein to PEO ratio of 3:1, the diameter of the formed nanofibers ranged between 312 and 690 nm, depending on the electrospinning conditions. Stijnman et al. [41] reported that electrospinnable polysaccharides showed no shear thinning behavior at low shear rates but had slight shear thinning at high shear rates. Polymer solutions with a higher degree of pseudoplasticity form only droplets during electrospinning.

5.2.3 ELECTRICAL CONDUCTIVITY

High electrical conductivity of feed solution is favorable as it improves the repulsion charges that can overcome surface tension forces on the pendant droplet at the spinneret tip, thereby lowering the critical voltage required for electrospinning. Higher electrical conductivity also imparts more surface charge to the spinning polymer jet, resulting in increased electrostatic repulsion forces responsible for inducing bending, instability, and stretching, which are mandatory for the formation of submicron fibers [11]. The electrical conductivity of feed solution could be improved by the addition of salt, polyelectrolytes, or ionic surfactants, or using select proteins, which dissociate into ions or ionic groups when dissolved in the solvent. For example, sodium alginate nanofibers were formed by reducing the repulsive forces amongst polyanionic sodium alginate molecules by the addition of PEO [30]. Similarly, bead-free fibers of pullulan were formed at low concentrations by improving its electrical conductivity using sodium salts [26].

5.2.4   SURFACE TENSION

The composition of solvent used to dissolve the feed polymer has an important role on electrospinnability of the polymer and the formation of continuous bead-free fibers. This is because the surface tension of the polymer solution is dependent on the solvent used [18]. The optimum range of surface tension for electrospinning is about 30 to 70 mN/m. For formation of smooth fibers, lower surface tension is desirable as it reduces the electric field strength required [36].

In order to reduce the surface tension and to improve the electrospinnabilty of the polymer solution, as well as, to reduce the diameter of electrospun nanofibers, surfactants are used. For example, electrospinnability of polyvinyl pyrrolidone at 48% (w/w) was enhanced by the addition of Triton RX-100 as non-ionic surfactant [50]. Similarly, tween 80 at a concentration of 1% (w/w) was used in gelatin solutions to reduce the surface tension before electrospinning [10]. Such an addition caused only a slight change in viscosity of feed solution, and did not affect the interaction of gelatin molecules, and resulted in the clean formation of bead-free fibers without any change in their diameter. Also, it was reported that the addition of acetic acid (10 to 90%) decreased the surface tension of chitosan solutions of different molecular weights due to an increase in net charge density of the solution, providing more ions for repulsion [55].

5.3   INFLUENCE OF PROCESS PARAMETERS ON SPINNABILITY OF BIOPOLYMERS

5.3.1   APPLIED VOLTAGE

The major factor in electrospinning is the need to apply high voltage to the feed solution. Change in voltage affects the electrical field strength between the spinneret and collector, and ultimately the strength of the drawing force. Once the right voltage is applied to the feed solution, the electrospinning process is initiated when the electrostatic force in the solution overcomes the surface tension forces [32]. Lin et al. [28] reported that application of less than optimum voltage resulted in electrospun fibers of comparatively larger diameter and beads were formed. Zhao et al. [54] observed that the crystallinity of electrospun fibers increased with increasing voltage until a critical point, beyond which crystallinity decreased with further increase in applied voltage. The critical voltage required for initiation of fiber formation is given by Eqn. (1).

$$V_c^2 = 4\frac{H^2}{L^2}\left(ln\frac{2L}{R} - 1.5\right)(0.117\pi\gamma R) \tag{1}$$

where; V_c is the critical voltage, H is the distance between electrodes, L is the length of spinneret, γ is the surface tension of solution, and R is the jet radius.

5.3.2 FEED RATE

The feed rate decides the volume of solution available at the spinneret tip during electrospinning. For a particular applied voltage above the minimum level, there is a corresponding feed rate for formation of a stable Taylor cone [37]. Demir et al. [9] observed a Power-law relationship between feed rate and applied voltage. That is, flow rate was proportional to the third power of the applied voltage. Ballengee and Pintauro [4] and Megelski et al. [33] reported that fibers with relatively large diameters were commonly electrospun at high feed rates, whereas, lower feed rate frequently yielded uniform electrospun fibers. Zong et al. [56] postulated the reasons behind this in terms of differences in velocity of the polymer jet, and drying time of the jet.

5.3.3 DISTANCE BETWEEN SPINNERET AND COLLECTOR

The elongation of polymer solution into a liquid jet takes place in the space between the spinneret tip and collector, which subsequently gets segregated into nanofibers and are deposited on the collector. Inadequate drying of fiber could be attributed to lack of sufficient distance between the spinneret tip and collector [5]. Increasing the tip to collector distance (TCD) result in longer flight times, and solvent evaporation time, which lead to reduction in bead formation and mean fiber diameter [3, 28].

5.4 ELECTRODYNAMICS AFFECTING SPINNABILITY AND FORMATION OF NANOFIBERS

The electrospinnability of a polymer solution is usually evaluated by its dimensionless numbers. Surface tension, density, and viscosity, of the polymer solution are used to calculate the dimensionless numbers. Those governing parameters of electrospun nanofibers are Ohnesorge (Oh), Deborah (De), Berry (Be), and Weber (We) numbers. For screening and selection of polymers, these numbers should be computed, and the conditionality of electrospinning has to be ascertained before electrospinning. In addition, in order to understand the solution jet breakup and its distance, it is necessary to look for these set of dimensionless numbers that do not depend on velocity scale. The conditions to be satisfied for the formation of smooth bead-less fibers during electrospinning are De$\geq$1, Oh$\geq$1, and De$\geq$Oh$\geq$1 [39].

5.4.1 DEBORAH NUMBER (DE)

The Deborah number (De) is defined as the ratio of relaxation time of polymer solution to Rayleigh instability growth time for inertia-capillary break-up of an inviscid jet (Eqn. (2)). It indicates the extent of elasticity of the polymer solution [39]. It is used to describe free surface viscoelastic flows. Rayleigh time scale shows the speed of electrospinning process.

$$De = \frac{\lambda}{\sqrt{d^3 \rho / \gamma}} \qquad (2)$$

where; d is the diameter of the spinneret, λ is the relaxation time, ρ is the density, and γ is the surface tension of polymer solution.

Polymer solutions with large 'De' could form bead-free uniform fibers. In contrast, solutions with 'De'<1 would experience droplet breakup, which leads to electrospraying of the feed rather than electrospinning it into fibers.

5.4.2 OHNESORGE NUMBER (OH)

Ohnesorge number (Oh) is the ratio of viscous to inertial and surface tension forces (Eqn. (3)), which is used to describe the breakup of viscoelastic jets. It is influenced by the orientational forces, and thus, reflects the viscous nature of polymer solution. It has a minimum value at which the nanofibers get stabilized, and a maximum at which viscous effects resist the spinnability of the polymer.

$$Oh = \frac{\eta}{\sqrt{d\rho\gamma}} \qquad (3)$$

where; d is the diameter of the spinneret, λ is the relaxation time, ρ is the density, and γ is the viscosity of polymer solution.

5.4.3 WEBER NUMBER

Weber number (We) is the ratio of inertial to surface stresses, representing the kinetic energy of the jet relative to tension dissipation. It indicates the point of jet breakup. It is used for the optimization of radius of curvature for strong stretching force and fast extension speed. When 'We' is higher, the radius of curvature of a viscous fluid jet becomes more loosely coiled due to decrease in surface tension, and vice versa. The radius of jet formed is thin at larger 'We.'

5.4.4 BERRY NUMBER (BE)

Berry number (Be) is the product of intrinsic viscosity and polymer solution concentration. The electrospinnability of a polymer and diameter of nanofibers formed depends on 'Be.' It represents viscous force as compared to its resilience, caused by the formation of entangled network between molecular chains. It is thus related to the number of chain entanglements, and its value should be more than 1. Polymer solutions of higher concentrations with low molecular weights resulted in 'Be' more than 1, exhibiting high viscosity. However, in such conditions, the molecular chains of the polymer may not be able to form entanglements due to their shorter length. In addition, at lower concentration and higher molecular weight, entanglements could not be formed between molecular chains, and the diameter of nanofibers increased with increase in 'Be' [6].

5.4.5 TROUTON RATIO (TR)

Trouton ratio (Tr) is the relationship between extensional and shear viscosities. Newtonian liquids show Tr ratio of about 3. Mun et al. [34] studied the effect of molecular weight and concentration on Tr ratio of PEO solution. At higher concentrations, the Tr ratio increased with molecular weight, whereas at lower concentrations, it was independent of molecular weight. Splitting of the jet reduced at higher Tr ratio, while higher molecular weight of PEO resulted in higher diameter.

5.5 BIOPOLYMERS SUITABLE FOR ELECTROSPINNING

5.5.1 POLYSACCHARIDE-BASED POLYMERS

5.5.1.1 PULLULAN

Pullulan is a straight-chain polysaccharide consisting of repeating units of α-(1→6) linked maltotriose, with the three glucopyranose units linked by α-(1→4) glycosidic bonds [46] (Figure 5.1). It is produced extracellularly by the fungus *Aureobasidium pullulans*. Its myriad applications include use as, texturizer for tofu, dietary gum, sausage, ham, and as a wall material for protecting flavors [24].

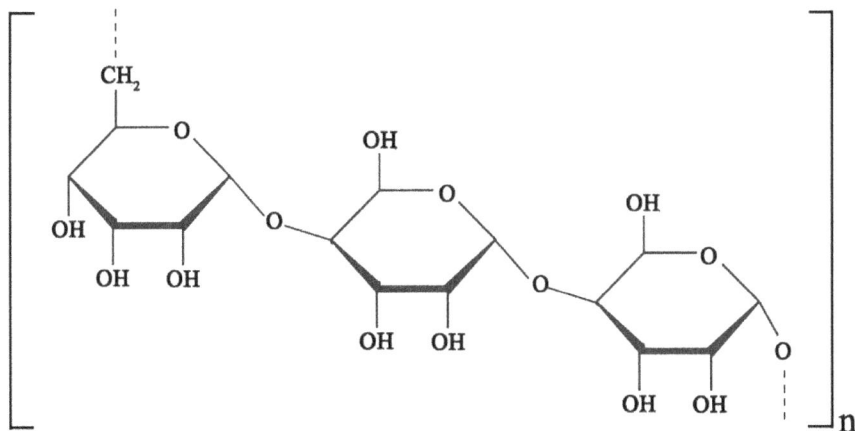

FIGURE 5.1 Structure of pullulan.

Pullulan in powder form is amorphous, odorless, tasteless, and white in color. It is soluble in water and forms a viscous solution, but insoluble in organic solvents [43]. Pullulan is non-hygroscopic, and decomposes at 250–280°C. It has good film-forming characteristics, and bears the ability to form nanoparticles, nanofibers, and flexible coatings. It is a non-toxic, non-carcinogenic, and biodegradable polymer [46]. The molecular weight lies in the range of 50 to >2000 kDa [24, 31]. It is slowly digestible in the gastrointestinal (GI) tract, and can be administered to the extent of 150 g per day [51].

5.5.1.1.1 Electrospinnability of Pullulan

In our study, before electrospinning, the viscosity of pullulan solutions of 10, 12, 14, and 16% (w/w) concentrations were determined, and it was found to be 90.4, 192.0, 413.1, and 745.0 m.Pa.s, respectively (Table 5.1). The effects of electrical conductivity, viscosity, and surface tension of the polymer solution, and process factors such as voltage applied and feed rate in the selected range of 18–24 kV and 0.25–1.0 mL/h on formation of electrospun nanofibers were studied. The most significant parameter for formation of electrospun nanofibers was found to be the viscosity of feed solution. At 10% concentration, the viscosity of pullulan solution was less than 100 m.Pa.s. Low viscosity was observed at this concentration of pullulan due to less entanglement of molecular chains. Concentrations in the range of 12–16% produced uniform nanofibers without beads. Above 16% concentration, electrospinning was difficult owing to drying of the jet at the tip of the spinneret.

Thus, it could be stated that the optimal viscosity of pullulan solution for electrospinning would be in the range of 100–745 m.Pa.s. The critical voltage required to electrospun pullulan solutions of different concentrations was calculated using Eqn. (1) as 16.39, 15.59, and 14.70 kV for 10, 12, and 14% concentrations, respectively. Meanwhile, formation of Taylor cone at the spinneret tip and elongation of jet were observed at the calculated critical voltage for all concentrations.

The electrospun nanofibers of 10 to 16% (w/w) pullulan solutions were observed through scanning electron microscopy (SEM). Pullulan solution of 10% concentration yielded nanofibers with beads (Figure 5.2(A)). In contrast, at concentrations of 12% (w/w) and above, clean fibers without beads were obtained (Figure 5.2(B–D)). The mean fiber diameter at 12, 14, and 16% concentration was 72.96, 102.62, and 115.0 nm, respectively (Table 5.2).

Beads formed at lower pullulan concentration due to low viscosity, wherein surface tension forces were dominant effect. The absence of beads in nanofibers obtained from pullulan at concentrations above 10% was attributed to the optimal level of viscosity and surface tension. The surface tension decreased from 71.67 N/m at 10% concentration to 70.5, 57.6, and 50.4 N/m for 12, 14, and 16% concentrations, respectively.

TABLE 5.1 Properties of Polymer Solutions

Polymer	Concentration (%) (w/w)	Ratio	Viscosity (m.Pa.s)	Electrical Conductivity (µS/cm)	Surface Tension (mN/m)
Inulin	40	–	28.65	1230	68.47
	50	–	59.15	880	63.28
	60	–	93.74	370	55.71
	70	–	186.53	220	43.39
FOS	50	–	98.27	120	58.71
	60	–	193.61	60	52.36
	70	–	327.53	20	40.29
Zein	15	–	50.49	300	48.96
	18	–	100.26	280	36.71
	21	–	163.50	250	28.57
	24	–	378.20	210	19.86
Pullulan	10	–	90.40	110	71.67
	12	–	192.00	90	70.50
	14	–	413.10	70	57.60
	16	–	745.00	60	50.40
WPI	20	–	17.43	2470	46.35
SPI	16	–	28.44	1306	18.23
WPI to pullulan	20 to 16	50:50	260.93	1200	48.17
	20 to 16	60:40	136.57	1470	45.49
	20 to 16	75:25	56.65	1770	42.91
SPI to pullulan	16 each	50:50	125.51	6630	39.10
	16 each	60:40	80.00	7530	29.64
	16 each	75:25	40.00	9960	23.22

NOTE: FOS: Fructo-oligosaccharides; WPI: Whey protein isolate; SPI: Soy protein isolate.

The optimal surface tension of pullulan solution for electrospinning was recommended to be 50 to 70 mN/m. The viscosity, electrical conductivity, and surface tension of pullulan at different concentrations are summarized in Table 5.1.

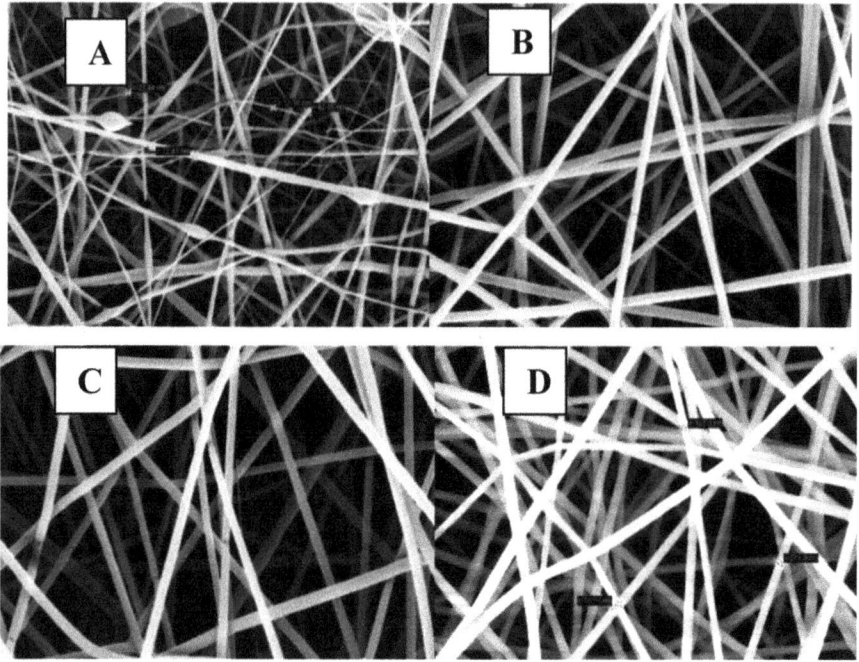

FIGURE 5.2   SEM micrographs of pullulan nanofibers obtained at (A) 10%; (B) 12% pullulan; (C) 14%; and (D) 16% at 25 KX magnifications.

In our study, it was observed that the diameter of nanofibers was increased as the concentration of pullulan increased from 12 to 16% (Figure 5.2). The electrical conductivity of pullulan solution decreased from 110 to 60 μS/cm as the concentration increased from 10 to 16%. The optimal electrical conductivity range for electrospinning of pullulan solution (12–16%) was found to be 90 to 60 μS/cm. Lower concentrations of pullulan could be electrospun into nanofibers by decreasing the electrical conductivity. Alternatively, another biopolymer could be added to pullulan to reduce its electrical conductivity and increase the viscosity so that the combination could be spun into nanofibers.

Liu et al. [29] observed a reduction in electrical conductivity when pullulan was added to pectin solution, which improved the spinnability. Similarly, Wang et al. [49] made electrospun nanofibers from 4% (w/v) pullulan solution by reducing its electrical conductivity with aqueous starch sodium-palmitate. Qin et al. [35] developed fast dissolving oral films from electrospun pullulan-chitosan composite biopolymers. In this study, the ratio of pullulan and chitosan affected the conductivity and viscosity of solution,

TABLE 5.2 Evaluation of Various Biopolymers for Electrospinning

Polymer	Solvent	Concentration (%) (w/w)	Ratio	Feed Rate (mL/h)	Voltage (kV)	Fiber Forming Ability	Mean Fiber Diameter (nm)
Inulin	Water	40	—	0.25, 0.5, 1.0	16, 20, 24	Spraying	—
		50	—	0.25, 0.5, 1.0	16, 20, 24	Spraying	—
		60	—	0.25, 0.5, 1.0	16, 20, 24	Spraying	—
		70	—	0.25, 0.5, 1.0	16, 20, 24	Spraying	—
FOS	Water	50	—	0.25, 0.5, 1.0	16, 20, 24	Fiber	—
		60	—	0.25, 0.5, 1.0	16, 20, 24	Fiber	—
		70	—	0.25, 0.5, 1.0	16, 20, 24	Fiber	—
Zein to FOS	Ethanol and water (70:30 ratio)	20	50:50	0.25, 0.5, 1.0	16, 20, 24	Spraying	—
		25	50:50	0.25, 0.5, 1.0	16, 20, 24	Spraying	—
		20	60:40	0.25, 0.5, 1.0	16, 20, 24	Fiber + Spraying	—
		25	60:40	0.25, 0.5, 1.0	16, 20, 24	Fiber + Spraying	—
		20	70:30	0.5 and 1.0	16, 20, 24	Fiber	—
		25	70:30	0.5 and 1.0	16, 20, 24	Fiber	—
		20	80:20	0.5 and 1.0	16, 20, 24	Fiber	—
		25	80:20	0.5 and 1.0	16, 20, 24	Fiber	—
Zein	Ethanol and acetic acid (70:30 ratio)	15	—	0.5 and 1.0	16, 20, 24	Fiber	124.00
		18	—	0.5 and 1.0	16, 20, 24	Fiber	86.00
		21	—	0.5 and 1.0	16, 20, 24	Fiber	115.00

TABLE 5.2 *(Continued)*

Polymer	Solvent	Concentration (%) (w/w)	Ratio	Feed Rate (mL/h)	Voltage (kV)	Fiber Forming Ability	Mean Fiber Diameter (nm)
		24	—	0.5 and 1.0	16, 20, 24	Fiber	186.00
		26	—	0.25, 0.5, 1.0	16, 20, 24	Not spinnable	—
Pullulan	Water	10	—	0.5, 0.75, 1.0	18, 21, 24	Beaded fibers	34.73
		12	—	0.5, 0.75, 1.0	18, 21, 24	Fiber	72.96
		14	—	0.5, 0.75, 1.0	18, 21, 24	Fiber	102.62
		16	—	0.5, 0.75, 1.0	18, 21, 24	Fiber	115.00
WPI	Water	20	—	0.5, 0.75, 1.0	18, 21, 24	Spraying	—
SPI	1% NaOH	16	—	0.5, 0.75, 1.0	15, 18, 21	Spraying	—
WPI to pullulan	Water	20 to 16	50:50	0.60, 1.0	13, 18, 23	Fiber	76.00
		20 to 16	60:40	0.60, 1.0	13, 18, 23	Beaded fibers	—
		20 to 16	75:25	0.60, 1.0	13, 18, 23	Beaded fibers	—
SPI to pullulan	1% NaOH and salt	16 each	50:50	0.50, 1.0	13, 18, 23	Beaded fibers	—
		16 each	60:40	0.50, 1.0	13, 18, 23	Beaded fibers	—
		16 each	75:25	0.50, 1.0	13, 18, 23	Beaded fibers	—

NOTE: FOS: Fructo-oligosaccharides, WPI: Whey protein isolate, SPI: Soy protein isolate.

which consequently influenced the morphology of nanofibers. However, the influence of electrical conductivity on the morphology of fibers was less marked as compared to surface tension and viscosity.

As discussed before, the spinnability of a polymer is influenced by the dimensionless numbers. The dimensionless numbers and critical voltage for electrospinning of pullulan at different concentrations are presented in Table 5.3. The 'Oh' was less than 1 for 10% pullulan solution, whereas it was greater than 1 at concentrations above 12%. However, 'De' was more than the 'Oh' for all the concentrations studied. The 'Be' gives the required concentration for continuous chain entanglement during electrospinning, and it should be more than 1 as well. In this study, the 'Be' was calculated to be more than 1 for all concentrations of pullulan tested. 'We' decreased from 32.20 to 24.71 as the concentration of pullulan increased from 10 to 16%. Higher 'We' is indicative of higher velocity of liquid jet, which could lead to instability and beaded nanofibers. Thus, it could be stated that the dimensionless numbers have a considerable influence on the electrospinnability of polymers.

TABLE 5.3 Dimensionless Numbers and Critical Voltage for Electrospinning of Pullulan and Zein

Electro-Hydrodynamic Parameter	Concentration of Pullulan Solution (%)				Concentration of Zein Solution (%)			
	10	**12**	**14**	**16**	**15**	**18**	**21**	**24**
Berry number (Be $\geq$ 1)	24.33	51.07	64.94	75.42	15.36	22.86	33.97	45.01
Deborah number (De $\geq$ Oh $\geq$ 1)	25.00	26.43	27.42	29.16	47.53	48.57	50.28	52.27
Ohnesorge number (Oh $\geq$ 1)	0.45	1.09	1.68	2.16	0.74	1.38	2.24	3.27
Weber number (We)	32.20	29.52	27.26	24.71	36.21	28.93	23.74	18.97
Critical voltage (V_c) (kV)	16.39	15.95	14.70	13.23	11.21	12.38	13.86	15.42

Many authors reported the electrospinnability of pullulan alone and in combination with other synthetic or natural biopolymers in water and in organic solvents for various applications. Sun et al. [43] reported the electrospinnability of pullulan in aqueous solution with a mean fiber diameter of 100 to 700 nm. Kong and Ziegler [25] produced pullulan nanofibers using dimethyl sulfoxide (DMSO)/water as a solvent, with diameter ranging from nanometer to micrometers. Electrospun nanofibers were also produced from pullulan using 95% formic acid as solvent [2].

In this study, increased pullulan content led to a decrease in electrical conductivity and an increase in apparent viscosity. The mean diameters of fibers ranged from 227 to 352 nm. Similarly, Li et al. [26] achieved clean bead-free nanofibers from 8% (w/v) pullulan solution with mean diameter of 124 ± 34 nm and 154 ± 36 nm in the presence of 0.20 M sodium chloride and 0.05 M sodium citrate, respectively for food-grade applications.

Pullulan can form hydrogen bonds with proteins [17]. Therefore, it could be combined with proteins, and electrospun into nanofibers. Aceituno-Medina et al. [1] developed hybrid pullulan-based fibers with amaranth protein isolate having diameter of 200 to 310 nm with improved thermal stability. Drosou et al. [12] produced electrospun nanofibers from pullulan-WPI, and studied the effects of feed rate, applied voltage, and distance between the spinneret and collector on their morphology and thermal stability. The addition of pullulan to WPI led to increased viscosity and decreased electrical conductivity, which were conducive for the formation of uniform nanofibers with mean diameter of 231 nm. The produced nanofibers could be used as potential encapsulation matrices for bioactives.

Karim et al. [22] produced composite pullulan fibers ranging from 50 to 500 nm diameter using 20% pullulan solutions containing montmorillonite to the extent of 1–10%. The clay was added to enhance the tensile properties and heat stability of pullulan. Similarly, pullulan and β-cyclodextrin emulsions were electrospun into nanofibers encapsulating R-(+)-limonene [16]. Controlled release of limonene was achieved by changing the RH of the fibers.

Tomasula et al. [45] made caseinate nanofibers for food applications using pullulan as carrier material. Pullulan and calcium caseinate concentrations were in the range of 5–15% and 3–15%, respectively. At pullulan to calcium caseinate ratio of 1:2, the electrospun nanofibers formed had the least fiber diameter of 172 ± 43 nm. Similarly, when pullulan-sodium caseinate solution was electrospun, co-existence of both fibers and beads was observed. Thus, pullulan seems to be a biopolymer which as promising potential as wall material for encapsulating bioactive ingredients in the form of electrospun nanofibers. In addition, it could be concluded that pullulan in the concentration range of 12–16% could easily be electrospun into nanofibers with diameters of about 100 nm. Such encapsulated bioactive substances in the form of nanofibers of about 100 nm diameter could be explored towards food applications so that their unique characteristics were utilized to its maximum potential.

5.5.1.2 INULIN

Inulin is a natural carbohydrate present in the roots of chicory and other plants. In 2018, the United States Food and Drug Administration (USFDA) approved inulin as a functional ingredient in food products. It has a high molecular weight of 6179 g/mol, and is classified as a neutral polysaccharide because of its oligomeric and polymeric sequence of fructose with glucose being connected to the last fructose unit (Figure 5.3).

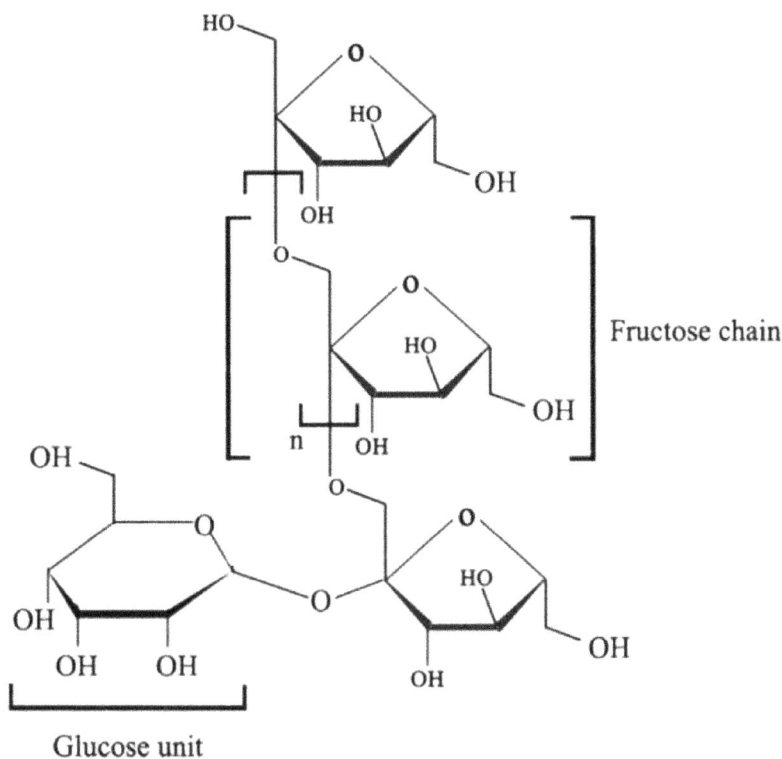

FIGURE 5.3   Structure of inulin.

In our study, inulin solutions of different concentrations were prepared using water as solvent, and their viscosity, surface tension, and electrical conductivity were determined. Electrospinning of inulin solutions was done at feed rate of 0.25, 0.5, and 1.0 mL/h, an applied voltage of 16, 20, and 24 kV. However, at all conditions, electrospraying was observed without any signs of fiber formation. Lack of fiber formation could be ascribed to the higher surface tension (68.47–43.39 mN/m) and electrical conductivity (1230–220 µS/cm) of inulin solutions.

Efforts were thus made to reduce the surface tension of inulin solutions by adding tween 80 as surfactant. When these solutions were again electrospun, only very thin nanofibers were obtained along with spraying, which could be attributed to the lower viscosity (28.65–186.53 m.Pa.s) and higher electrical conductivity (1300–300 µS/cm), causing the droplet to stretch too much and break. The nanofibrous mat was oily due to the presence of tween 80, and the nanofibers disappeared from the mat due to hygroscopicity of inulin as time elapsed. Thus, it could be stated that inulin is difficult to be electrospun into fibers, but it could be electrosprayed easily.

5.5.1.3 CARRAGEENAN

Carrageenan is a water-soluble polymer with a straight chain of partially sulfonated galactans that are obtained from red seaweed. In our study, carrageenan solutions of 0.5 and 1% concentrations were evaluated for their electrospinnability. Surface tension, electrical conductivity, and viscosity were measured, and the various dimensionless numbers were computed to validate the inequality conditions of electrospinning. The viscosity of carrageenan solutions ranged from 14.37 to 47.5 m.Pa.s, which was not in the optimal range for electrospinning. The electrical conductivity and surface tension of 0.5 and 1% solutions were 2.9 and 4.3 mS/cm and 58.8 and 57.0 mN/m, respectively. The inequality conditions for fiber formation were also not satisfied.

At applied voltage range of 15–24 kV, feed rate 0.2–1.0 mL/h and at 10 cm distance from the spinneret to collector, only electrospraying of the carrageenan solution occurred. The liquid jet from the feed solutions broke up consistently due to low viscosity. The concentration of carrageenan was so low that continuous chain entanglement of the molecules did not occur, which affected the formation of fibers. In addition, the higher electrical conductivity of carrageenan caused the solution to stretch, thereby causing break-up of the droplet at the tip of the spinneret. Stijnman et al. [41] reported that 1% κ-and λ-carrageenan having viscosity of 24.7 and 106.6 m.Pa.s, respectively, and with molecular weights of 845 and 3036 kDa, respectively were not electrospinnable into nanofibers. In contrast, at concentrations above 1%, carrageenan formed a thick gel, which could not be drawn from the spinneret, and thus, it was not electrospinnable. Hence, carrageenan was found to be difficult to be electrospun with the range of concentrations and process conditions studied.

Fan et al. [14] fabricated carboxy methyl κ-carrageenan/alginate blended nanofibers by electrospinning with aqueous $CaCl_2$ and ethanol as solvents. The authors determined the mechanical and water retention properties, and reported that the nanofibers treated with aqueous silver nitrate exhibited good antibacterial activity.

Similarly, Jauri and Razak [21] prepared electrospun polylactic-acid (PLA) nanofiber coated with carrageenan and polyethylene glycol. It was reported that carrageenan improved the hydrophobicity of nanofiber mat. However, these nanofibers could not be used for food applications. From these observations, it could be stated carrageenan needed an electrospinnable co-polymer to increase the chain entanglements in the polymer, without contributing to an increase in viscosity but lowering the surface tension. It is left to the scope of the readers to identify such biopolymers that could be electrospun into nanofibers along with carrageenan for food applications.

5.5.1.4 FRUCTO-OLIGOSACCHARIDES (FOS)

Fructo-oligosaccharides (FOS) belong to the group of oligosaccharides with a degree of polymerization between 2 and 10 (normally 5) (Figure 5.4). It is obtained through partial hydrolysis of inulin with inulinase. The molecular weight of FOS ranges from 505 to 828 g/mol.

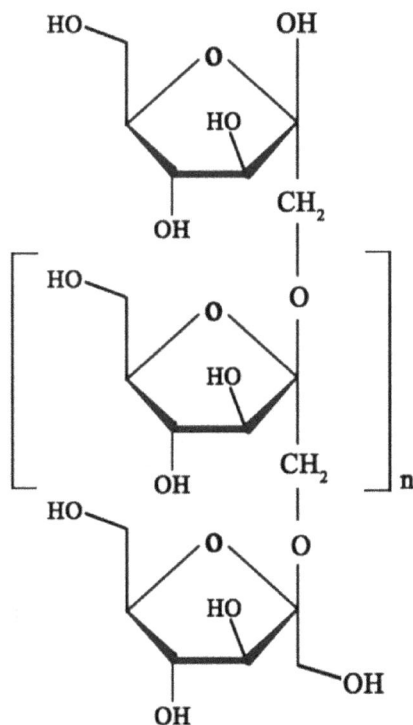

FIGURE 5.4   Structure of fructo-oligosaccharides.

FOS solutions of 50, 60, and 70% were prepared using water as solvent, and the viscosity, electrical conductivity, and surface tension were determined. The FOS solutions were subjected to electrospinning at the same process conditions. Surprisingly, FOS yielded very good nanofibers because of its optimal viscosity (98.27–327.53 m.Pa.s), and electrical conductivity (120–20 μS/cm) as compared to inulin. However, as in the case of inulin, the FOS nanofibers were very hygroscopic, and they disappeared from the aluminum foil. SEM micrographs (Figure 5.5) confirmed that the nanofibers disappeared within two days (time gap between electrospinning and SEM imaging). Thus, FOS alone is not suitable for electrospinning into nanofibers.

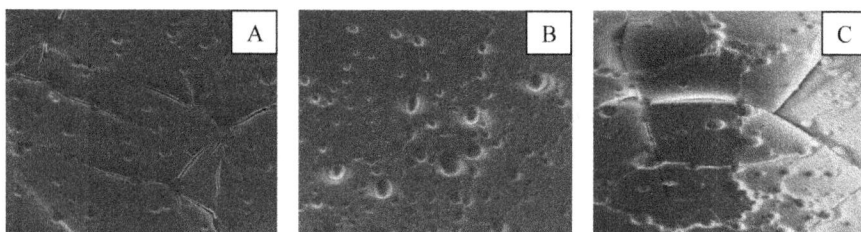

FIGURE 5.5   SEM micrographs of FOS nanofibers at (A) 50%; (B) 60%; and (C) 70% concentrations at 25KX magnifications.

5.5.2   PROTEIN-BASED POLYMERS

5.5.2.1   WHEY PROTEIN ISOLATE (WPI)

In our study, WPI and pullulan of 20 and 16% concentrations, and blended in the ratio of 50:50, 60:40, and 75:25, were used to produce nanofibers. The influence of surface tension, electrical conductivity, and viscosity on fiber formation was evaluated at different ratios of WPI and pullulan. It was found that an increase in addition of WPI to pullulan decreased the viscosity of feed solution to less than 60 m.Pa.s (75:25 ratios). As the proportion of WPI increased to 60:40 and 75:25, dripping of the feed solution was observed at the spinneret tip because the viscosities were less than the minimum of 150 m.Pa.s.

Moreover, the inequality conditions of dimensionless numbers (De≥1, Oh≥1≥De, Be≥1) for formation of fibers were not satisfied at 60:40 and 75:25 combinations. In contrast, at WPI to pullulan ratio of 50:50, clear nanofibers without beads were obtained. At this concentration, the viscosity, surface tension, and electrical conductivity of the feed solution were found to be 260 m.Pa.s, 48.2 N/m and 1.2 mS/cm, respectively. The mean diameter of nanofibers was 76 nm (Table 5.2).

A few researchers were successful in electrospinning WPI into nanofibers with the use of other synthetic polymers as well. A mixture of WPI and PEO solutions was used to fabricate nanofibers by electrospinning [8]. The ratio of PEO and WPI was altered so as to obtain nanofibers from different concentrations of polymers. Solution properties such as viscosity, electrical conductivity, and surface tension were studied as a function of PEO-WPI ratio. Solutions with viscosity below 415 m.Pa.s, surface tension below 55 mN/m, and electrical conductivity above 527 µS/cm yielded only beaded fibers. On the other hand, at viscosity and surface tension above these limits, clean, smooth, and bead-free fibers with diameters ranging between 227 ± 36 and 264 ± 66 nm were obtained.

Sullivan et al. [42] also electrospun WPI into nanofibers in combination with PEO. At WPI to PEO ratio of 3:1, fibers of mean diameter ranging from 312 to 690 nm were produced, depending on polymer composition. Vega-Lugo et al. [48] studied the effect of pH on electrospinning of WPI with PEO. The authors reported that the alterations in the secondary structure of proteins and polymer solution properties induced by pH changes significantly influenced the electrospinning behavior of WPfI-PEO solutions.

5.5.2.2 SOY PROTEIN ISOLATE (SPI)

Soy proteins are one of the cheapest vegetable proteins with a wide range of use. In our study, SPI was subjected to electrospinning at different concentrations (12–16% w/w) and at applied voltage range of 15–24 kV and feed rate of 0.2–1.0 mL/h. Soy proteins alone could not be electrospun at all the tested conditions owing to their very low viscosity (19.0–31.0 m.Pa.s). In order to facilitate fiber formation by electrospinning, it is necessary to unwind the coiled structure of SPI. Therefore, SPI was denatured using a combined thermal and alkaline treatment using aqueous NaOH at 80°C.

Alternatively, in our study, pullulan was used along with SPI to increase the viscosity of the latter, and to enable formation of the fiber network. The concentration of SPI and pullulan was fixed at 16% each, and different ratios of SPI and pullulan namely 50:50, 60:40 and 75:25 were investigated for their spinnability. Amongst them, 50:50 ratio of SPI and pullulan formed clean fibers. Similar to WPI, as the proportion of SPI increased in the feed solution, its viscosity decreased, and formation of clean fibers was not possible.

The electrospinnability of SPI with different solvents and along with other polymers has been attempted. Vega-Lugo et al. [47] evaluated various formulations of SPI and PEO for their electrospinning behavior and fiber morphology. When 0.8% PEO was added to it, SPI could easily be electrospun into nanofibers. This is because PEO functioned as a co-spinning polymer and improved the spinnability of SPI. Similarly, Xu et al. [52] prepared nanofibrous membrane by electrospinning using SPI and PEO dissolved in 1,1,1,3,3,3-hexafluoro-2-propanol. Khabbaz et al. [23] also prepared electrospun nanofibrous mats and casting films using SPI and polyvinyl alcohol (PVA). The prepared mats and films were evaluated for their physical, chemical, mechanical, and biological properties. However, these electrospun nanofibers cannot be used for food applications.

5.5.2.3 ZEIN

Zein is a prolamine-rich protein that contains a higher portion of hydrophobic amino acids such as proline and glutamine (Figure 5.6). This protein is extracted from the endosperm of maize (corn). Due to its hydrophobic nature, zein is widely used for drug-delivery and food applications [40]. The USFDA classifies zein as generally recognized as safe (GRAS) polymer for food applications. It has a high molecular weight ranging from 22 to 27 kDa [13].

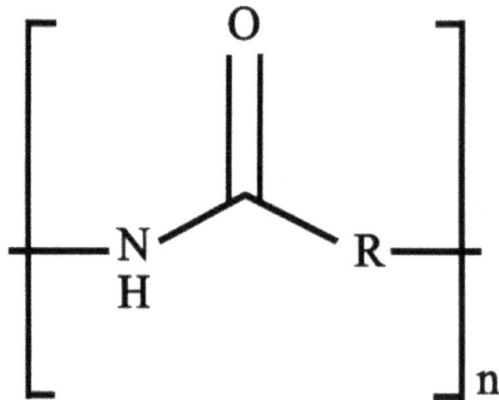

FIGURE 5.6 Structure of zein.

Zein at 15, 18, 21, 24, and 26% (w/w) concentrations was prepared using 70% ethanol and 30% acetic acid as solvent, and they were evaluated for its electrospinning properties. The applied voltage varied from 16 to 24 kV, and the feed rate was kept at 0.5 and 1.0 mL/h. The distance between spinneret and collector was kept as 15 cm. Zein at 15% concentration yielded only beaded fibers (Figure 5.7(a)), while zein at 18 to 24% concentrations produced clean nanofibers without beads (Figure 5.7(b)–(d)). Zein could not form clean bead free nanofibers at 15% concentration because of its lower viscosity (50.49 m.Pa.s) and dominant surface tension (48.96 mN/m) forces. The degree of network entanglement of polymer chains was also insufficient for fiber formation, and the voltage applied produced beaded fibers primarily due to Rayleigh instability (capillary wave break-up).

At concentrations of 18% and above, the electrospun nanofibers produced were clean and bead free owing to the optimal viscosity (100.26–378.20 m.Pa.s), electrical conductivity (280–210 µS/cm) and surface tension (36.71–19.86 mN/m). As the concentration of zein increased, the viscoelastic forces became more dominant, overcoming the surface tension forces. The lowest mean fiber diameter of 86 nm was obtained at 18% concentration of zein, and it increased to 186 nm as the concentration of zein increased to 24% (Table 5.2). At 26% concentrations and above, zein was not electrospinnable due to its very high viscosity (926.83 m.Pa.s).

The dimensionless numbers and critical voltage for electrospinning of zein at different concentrations are presented in Table 5.3. The 'Be' calculated for all concentrations were more than unity. When the concentration of zein was 15%, the 'Oh' was less than unity. This might be due to the higher surface tension of zein, which dominated the viscous force. It also indicated that the nanofibers produced at 15% concentration might have beaded structure. However, 'Oh' calculated at 18, 21, and 24% concentrations was more than unity, suggesting that these concentrations satisfied the electrospinning conditions.

Regardless of the concentration, 'De' was greater than unity, and it was greater than 'Oh.' 'We' decreased from 36.21 to 18.97 as concentration of zein increased from 15 to 24%. Higher 'We' is indicative of higher velocity of liquid jet, which could lead to instability and beaded nanofibers. Critical voltage indicates the requirement of minimum voltage to initiate fiber formation. When the concentration of zein solution increased from 15 to 24%, the critical voltage required for initiation of fiber formation increased from 11.21 to 15.42 kV. This was expected because denser molecular chains at higher concentrations of zein required more electric charge to be drawn to the collector.

Combinations of zein and FOS were tried at ratios of 50:50, 60:40, 70:30, and 80:20, and at concentrations 20 and 25%. Ethanol and water at 70:30 ratios were used as solvent for the preparation of zein+FOS solutions. Only spraying, and thin fibers with spraying were observed at ratios of 50:50 and 60:40, respectively. However, at zein to FOS ratios of 70:30 and 80:20, apparently clean nanofibers were formed. However, SEM micrographs revealed that the nanofibers were heavily beaded in nature (Figure 5.8) regardless of the concentration tried.

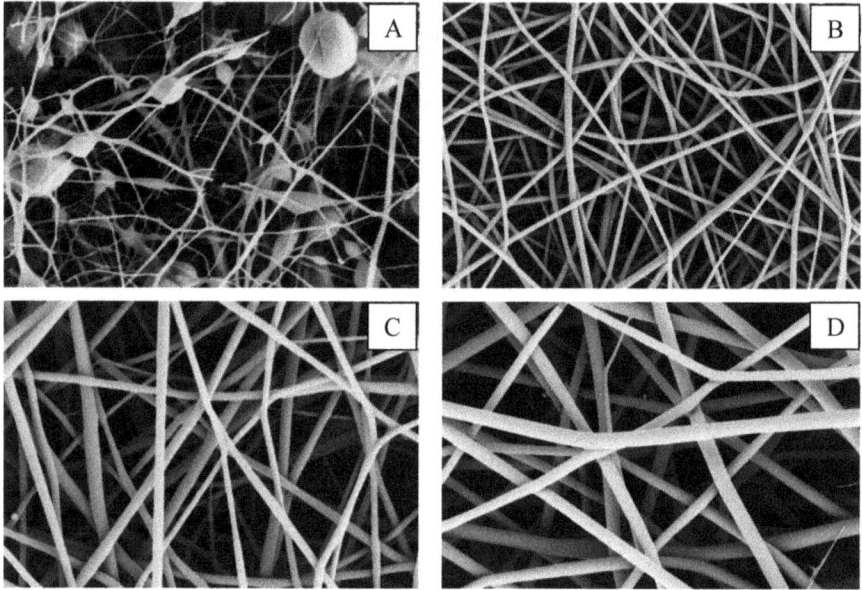

FIGURE 5.7   SEM micrographs of zein nanofibers at (A) 15%; (B) 18%; (C) 21%; and (D) 24% concentrations at 25KX magnifications.

5.6 SUMMARY

The influence of polymer solution properties and process factors on electrospinning of biopolymers was studied. The optimal range of viscosity, surface tension, and electrical conductivity for electrospinning of various biopolymers are discussed. The electrospinnability of polysaccharide-based and protein-based polymers were evaluated in terms of their solution properties and dimensionless numbers affecting their electrohydrodynamics of spinning. Amongst them, pullulan and zein were found to be easily electrospinnable into nanofibers, and hence, could be recommended as wall materials for nanoencapsulation of bioactives. FOS and inulin were electrospinnable, but due to their hygroscopicity, the nanofibers disappeared from the mat as time lapsed. WPI and SPI could not form clean and bead-free nanofibers due to their unfavorable viscosities, surface tension, and electrical conductivity, and require a copolymer, such as, zein or pullulan to form fibers. Biopolymers that possess optimal range of viscosity and fulfill the inequality conditions of electrospinning could form uniform, clean, and bead-free nanofibers.

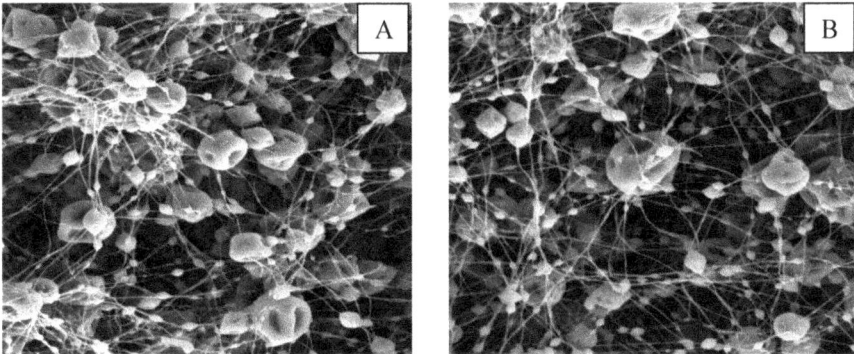

FIGURE 5.8 SEM micrographs of zein and FOS nanofibers at (A) 70:30; and (B) 80:20 ratios at 25KX magnifications.

ACKNOWLEDGMENTS

The authors sincerely thank DPRP Division, Department of Science and Technology, New Delhi and National Agriculture Science Fund of Indian Council of Agricultural Research, New Delhi, India for providing the instrumentation facilities for conducting this study.

KEYWORDS

- bioactives
- electrospinning
- fructo-oligosaccharides
- nanoencapsulation
- nanofibers
- polymer

REFERENCES

1. Aceituno-Medina, M., Mendoza, S., Lagaron, J. M., & López-Rubio, A., (2013). Development and characterization of food-grade electrospun fibers from amaranth protein and pullulan blends. *Food Research International, 54*(1), 667–674.
2. Aceituno-Medina, M., Mendoza, S., Lagaron, J. M., & López-Rubio, A., (2015). Photoprotection of folic acid upon encapsulation in food-grade amaranth (*Amaranthus hypochondriacus L.*) protein isolate-pullulan electrospun fibers. *LWT-Food Science and Technology, 62*(2), 970–975.
3. Ahn, Y. C., Park, S. K., Kim, G. T., Hwang, Y. J., Lee, C. G., Shin, H. S., & Lee, J. K., (2006). Development of high efficiency nanofilters made of nanofibers. *Current Applied Physics, 6*(6), 1030–1035.
4. Ballengee, J. B., & Pintauro, P. N., (2011). Morphological control of electrospun Nafion nanofiber mats. *Journal of the Electrochemical Society, 158*(5), B568–B572.
5. Barhate, R. S., Loong, C. K., & Ramakrishna, S., (2006). Preparation and characterization of nanofibrous filtering media. *Journal of Membrane Science, 283*(1/2), 209–218.
6. Basu, S., Gogoi, N., Sharma, S., Jassal, M., & Agrawal, A. K., (2013). Role of elasticity in control of diameter of electrospun PAN nanofibers. *Fibers and Polymers, 14*(6), 950–956.
7. Bhardwaj, N., & Kundu, S. C., (2010). Electrospinning: A fascinating fiber fabrication technique. *Biotechnology Advances, 28*(3), 325–347.
8. Colin-Orozco, J., Zapata-Torres, M., Rodríguez-Gattorno, G., & Pedroza-Islas, R., (2015). Properties of poly(ethylene oxide)/whey protein isolate nanofibers prepared by electrospinning. *Food Biophysics, 10*(2), 134–144.
9. Demir, M. M., Yilgor, I., Yilgor, E. E. A., & Erman, B., (2002). Electrospinning of polyurethane fibers. *Polymer, 43*(11), 3303–3309.
10. Deng, L., Kang, X., Liu, Y., Feng, F., & Zhang, H., (2017). Effects of surfactants on the formation of gelatin nanofibers for controlled release of curcumin. *Food Chemistry, 231*, 70–77.
11. Drosou, C. G., Krokida, M. K., & Biliaderis, C. G., (2017). Encapsulation of bioactive compounds through electrospinning/electrospraying and spray drying: A comparative assessment of food-related applications. *Drying Technology, 35*(2), 139–162.
12. Drosou, C., Krokida, M., & Biliaderis, C. G., (2018). Composite pullulan-whey protein nanofibers made by electrospinning: Impact of process parameters on fiber morphology and physical properties. *Food Hydrocolloids, 77*, 726–735.
13. Elzoghby, A. O., Elgohary, M. M., & Kamel, N. M., (2015). Implications of protein-and peptide-based nanoparticles as potential vehicles for anticancer drugs: Chapter 6. In: Donev, R., (ed.), *Protein and Peptide Nanoparticles for Drug Delivery* (pp. 169–221). Waltham, USA: Academic Press.
14. Fan, L., Peng, K., Li, M., Wang, L., & Wang, T., (2013). Preparation and properties of carboxymethyl κ-carrageenan/alginate blend fibers. *Journal of Biomaterials Science, 24*(9), 1099–1111.
15. Fong, H., Chun, I., & Reneker, D. H., (1999). Beaded nanofibers formed during electrospinning. *Polymer, 40*(16), 4585–4592.
16. Fuenmayora, C. A., Mascheronia, E., Cosioa, M. S., Piergiovannia, L., Benedettia, S., Ortenzic, M., & Manninoa, S., (2013). Encapsulation of R-(+)-limonene in edible electrospun nanofibers. *Chemical Engineering, 32*, 1771–1776.

17. Gounga, M. E., XU, S. Y., & Wang, Z., (2010). Film-forming mechanism and mechanical and thermal properties of whey protein isolate-based edible films as affected by protein concentration, glycerol ratio, and pullulan Content. *Journal of Food Biochemistry, 34*(3), 501–519.
18. Haghi, A. K., & Akbari, M., (2007). Trends in electrospinning of natural nanofibers. *Physica Status Solidi, 204*, 1830–1834.
19. Huang, Z. M., Zhang, Y. Z., Ramakrishna, S., & Lim, C. T., (2004). Electrospinning and mechanical characterization of gelatin nanofibers. *Polymer, 45*(15), 5361–5368.
20. Hurtado-Lopez, P., & Murdan, S., (2006). Zein microspheres as drug/antigen carriers: A study of their degradation and erosion in the presence and absence of enzymes. *Journal of Microencapsulation, 23*(3), 303–314.
21. Jauri, A., & Razak, S. I. A., (2018). Effects of PEG-polysaccharide coating on electrospun PLA nanofiber. *Materials Science and Engineering, 440*(1), 1–4.
22. Karim, M. R., Lee, H. W., Kim, R., Ji, B. C., Cho, J. W., Son, T. W., & Yeum, J. H., (2009). Preparation and characterization of electrospun pullulan/montmorillonite nanofiber mats in aqueous solution. *Carbohydrate Polymers, 78*(2), 336–342.
23. Khabbaz, B., Solouk, A., & Mirzadeh, H., (2019). Polyvinyl alcohol/soy protein isolate nanofibrous patch for wound-healing applications. *Progress in Biomaterials, 8*(3), 185–196.
24. Kimoto, T., Shibuya, T., & Shiobara, S., (1997). Safety studies of a novel starch, pullulan: Chronic toxicity in rats and bacterial mutagenicity. *Food and Chemical Toxicology, 35*(3–4), 323–329.
25. Kong, L., & Ziegler, G. R., (2014). Rheological aspects in fabricating pullulan fibers by electro-wet-spinning. *Food Hydrocolloids, 38*, 220–226.
26. Li, R., Tomasula, P., De Sousa, A., Liu, S. C., Tunick, M., Liu, K., & Liu, L., (2017). Electrospinning pullulan fibers from salt solutions. *Polymers, 9*(1), 32.
27. Li, Y., Lim, L. T., & Kakuda, Y., (2009). Electrospun zein fibers as carriers to stabilize (−) epigallocatechin gallate. *Journal of Food Science, 74*(3), 233–240.
28. Lin, Y., Yao, Y., Yang, X., Wei, N., & Li, X., (2008). Preparation of poly(ether sulfone) nanofibers by gas-jet/electrospinning. *Journal of Applied Polymer Science, 107*(2), 909–917.
29. Liu, S. C., Li, R., Tomasula, P. M., Sousa, A. M., & Liu, L., (2016). Electrospun food-grade ultrafine fibers from pectin and pullulan blends. *Food and Nutrition Sciences, 7*, 636–646.
30. Lu, J. W., Zhu, Y. L., Guo, Z. X., Hu, P., & Yu, J., (2006). Electrospinning of sodium alginate with poly(ethylene oxide). *Polymer, 47*(23), 8026–8031.
31. Madi, N. S., Harvey, L. M., Mehlert, A., & McNeil, B., (1997). Synthesis of two distinct exopolysaccharide fractions by cultures of the polymorphic fungus *Aureobasidium pullulans*. *Carbohydrate Polymers, 32*(3/4), 307–314.
32. Mazoochi, T., & Jabbari, V., (2011). Chitosan nanofibrous scaffold fabricated *via* electrospinning: The effect of processing parameters on the nanofiber morphology. *International Journal of Polymer Analysis and Characterization, 16*(5), 277–289.
33. Megelski, S., Stephens, J. S., Chase, D. B., & Rabolt, J. F., (2002). Micro-and nano-structured surface morphology on electrospun polymer fibers. *Macromolecules, 35*(22), 8456–8466.
34. Mun, R. P., Byars, J. A., & Boger, D. V., (1998). The effects of polymer concentration and molecular weight on the breakup of laminar capillary jets. *Journal of Non-Newtonian Fluid Mechanics, 74*(1–3), 285–297.

35. Qin, Z. Y., Jia, X. W., Liu, Q., Kong, B. H., & Wang, H., (2019). Fast dissolving oral films for drug delivery prepared from chitosan/pullulan electrospinning nanofibers. *International Journal of Biological Macromolecules, 137*, 224–231.

36. Ramakrishna, S., Teik-Cheng, L., & Kazutoshi, F., (2005). *An Introduction to Electrospinning and Nanofibers* (p. 396). Singapore: World Scientific Publishing.

37. Rutledge, G. C., Li, Y., Fridrikh, S., Warner, S. B., Kalayci, V. E., & Patra, P., (2000). *Electrostatic Spinning and Properties of Ultrafine Fibers* (pp. 98–101). National Textile Center, Technical Report.

38. Schiffman, J. D., & Schauer, C. L., (2008). A review: Electrospinning of biopolymer nanofibers and their applications. *Polymer Reviews, 48*(2), 317–352.

39. Seethu, B. G., Pushpadass, H. A., Emerald, F. M. E., Nath, B. S., Naik, N. L., & Subramanian, K. S., (2020). Electrohydrodynamic encapsulation of resveratrol using food-grade nanofibers: Process optimization, characterization, and fortification. *Food and Bioprocess Technology, 13*, 341–354.

40. Shukla, R., & Cheryan, M., (2001). Zein: The industrial protein from corn. *Industrial Crops and Products, 13*(3), 171–192.

41. Stijnman, A. C., Bodnar, I., & Tromp, R. H., (2011). Electrospinning of food-grade polysaccharides. *Food Hydrocolloids, 25*(5), 1393–1398.

42. Sullivan, S. T., Tang, C., Kennedy, A., Talwar, S., & Khan, S. A., (2014). Electrospinning and heat treatment of whey protein nanofibers. *Food Hydrocolloids, 35*, 36–50.

43. Sun, X. B., Jia, D., Kang, W. M., & Cheng, B. W., (2013). Research on electrospinning process of pullulan nanofibers. *Applied Mechanics and Materials, 268*, 198–201.

44. Teo, W. E., & Ramakrishna, S., (2006). Review on electrospinning design and nanofiber assemblies. *Nanotechnology, 17*(14), 89–106.

45. Tomasula, P. M., Sousa, A. M., Liou, S. C., Li, R., Bonnaillie, L. M., & Liu, L. S., (2016). Electrospinning of casein/pullulan blends for food-grade applications. *Journal of Dairy Science, 99*(3), 1837–1845.

46. Trinetta, V., & Cutter, C. N., (2016). Pullulan: A suitable biopolymer for antimicrobial food packaging applications: Chapter 30. In: Barros-Velázques, J., (ed.), *Antimicrobial Food Packaging* (Vol. 1, pp. 385–397). Cambridge, USA: Academic Press Inc.

47. Vega-Lugo, A. C., & Lim, L. T., (2008). Electrospinning of soy protein isolate nanofibers. *Journal of Biobased Materials and Bioenergy, 2*(3), 223–230.

48. Vega-Lugo, A. C., & Lim, L. T., (2012). Effects of poly(ethylene oxide) and pH on the electrospinning of whey protein isolate. *Journal of Polymer Science Part B: Polymer Physics, 50*(16), 1188–1197.

49. Wang, H., & Ziegler, G. R., (2019). Electrospun nanofiber mats from aqueous starch-pullulan dispersions: Optimizing dispersion properties for electrospinning. *International Journal of Biological Macromolecules, 133*, 1168–1174.

50. Wang, S. Q., He, J. H., & Xu, L., (2008). Non-ionic surfactants for enhancing electro-spinability and for the preparation of electrospun nanofibers. *Polymer International, 57*(9), 1079–1082.

51. Wolf, B. W., (2005). U.S. Patent No. 6,916,796, Washington, DC-USA; Patent and Trademark Office.

52. Xu, X., Jiang, L., Zhou, Z., Wu, X., & Wang, Y., (2012). Preparation and properties of electrospun soy protein isolate/polyethylene oxide nanofiber membranes. *ACS Applied Materials and Interfaces, 4*(8), 4331–4337.

53. Yao, C., Li, X., & Song, T., (2006). Electrospinning and crosslinking of zein nanofiber mats. *Journal of Applied Polymer Science, 103*(1), 380–385.

54. Zhao, S., Wu, X., Wang, L., & Huang, Y., (2004). Electrospinning of ethyl-cyanoethyl cellulose/tetrahydrofuran solutions. *Journal of Applied Polymer Science, 91*(1), 242–246.

55. Ziani, K., Henrist, C., Jérôme, C., Aqil, A., Maté, J. I., & Cloots, R., (2011). Effect of nonionic surfactant and acidity on chitosan nanofibers with different molecular weights. *Carbohydrate Polymers, 83*(2), 470–476.

56. Zong, X., Kim, K., Fang, D., Ran, S., Hsiao, B. S., & Chu, B., (2002). Structure and process relationship of electrospun bioabsorbable nanofiber membranes. *Polymer, 43*(16), 4403–4412.

PART II
Advances in Food Processing and Preservation Techniques

CHAPTER 6

TECHNOLOGIES FOR SHELF-LIFE ENHANCEMENT OF HERBS AND LEAFY VEGETABLES

R. S. GAUDHAM, ROHIT KUMAR, RAJASREE RANJIT,
ARUN SHARMA, PRAMOD K. PRABHAKAR, and NEELA EMANUEL

ABSTRACT

The major challenge of the utilization of herbs and leafy vegetables is to conserve their quality and to retard the process of spoilage. Herbs and leafy vegetables share common factors, which affect their shelf-life and play a key role in quality deterioration and contamination caused by the activity of mycotoxins, respiration rate of the produce, water activity of the produce, rate of ethylene production or the maturation process, development or stage of physical damages on the produce and water loss of the produce. This chapter includes discussions on thermal techniques, nonthermal techniques, and storage based techniques.

6.1 INTRODUCTION

Shelf-life is the time duration for which food product is fit for consumption after harvesting and processing and exhibits no significant deterioration of quality. Shelf-life is the maximum duration of time recommended for a product to be stored and must no longer be kept on a store shelf after the expiration of its shelf-life. The shelf-life of the produce is directly related to its quality. Losses in general can be either controllable or uncontrollable. Quality loss is controllable, when suitable methods are adopted. In this competitive world, shelf-life is defined by the quality of the produce and it also influences the selling price of the food produce. In order to satisfy the human demand and needs, the availability of the food produces must be on optimum level, which can be achieved by prolonging or extending the shelf-life of the food materials.

The shelf-life of a commodity varies depending on internal factors involving changes in the biological mechanism of the produce and external factors involving change in the environment, processing, and handling of the produce. The level of processing of the produce has a direct relationship on the preservation or shelf-life of the produce. The preservation of food has its origins dating back to the period of discovery of fire. These methods have undergone changes and improvements. The developments in science provided the technical background and knowledge for better preservation methods. The expiration date is a must, which acts as a guideline on normal and expected quality of the food material/produce and its exposure to environmental factors including moisture and temperature.

The food processing technology is defined as "an efficient/skillful way of making or achieving something related to food material." Over the days with the development of science, the technologies have improved and innovated. Methods involving the minimum occurrence of loss or degradation in quality of the produce with minimum capital input for processing are the most desirable factor in the field of development of techniques. Practically, the loss of quality cannot be stopped due to the constant biological changes but when the best suitable method is followed or employed, this can be controlled to a reasonable food extent.

This chapter provides the basic knowledge on food preservation and deals with the techniques for the extension of shelf-life of herbs and leafy vegetables.

6.1.1   HERBS

A plant or part of a plant valued for its medicinal, aromatic qualities or savory is called herb [48]. The major difference between herb and spice is its culinary use. Herbs are commonly referred to leafy green part of plants (it may be fresh or can be dried form), on the other hand spice is a kind of product belonging to a plant part (mostly dried), which includes seeds, bark, and roots [39]. According to IJAM (International Journal of Ayurvedic Medicine), India is the second-largest exporter of medicinal plants with an overall production of herbs around 1,25,000 tons/year. According to IJAM statistics of highly traded medicinal plants, the top 10 herbs based upon their medicinal and economic value are:

- Amlaki (*Emblica officinalis*);
- Aswagantha (*Withania somnifera*);
- Bilva (*Aegle marmeloscorr*);
- Brahmi (*Bacopa monneri*);
- Catharanthus (*Vinca rosea*);
- Haritaki (*Terminalia cbebula*);
- Pippali (*Piper longumlinn*);
- Satavari (*Asparagus racemosus*);
- Svarnapatri (*Cassia angustifolia*);
- Vasa (*Adhatoda vasica*).

6.1.2 LEAFY VEGETABLES

Leafy vegetables are leaves of plant, which are eaten as vegetable or may be accompanied by tender petioles and shoots. Among all vegetables, the leafy vegetables have a specific place in terms of nutritional requirements as they are a good source of vitamins like ascorbic acid, riboflavin, beta-carotene, etc., and of minerals like calcium, iron, and zinc. They include non-nutritive health beneficial compounds like phytochemicals and antioxidants. Antioxidants are vital in the food of human diet and act as an anticancer agent [43]. Vitamin A deficiency is a serious issue in various countries across the world. The study indicates that about 250 to 500 thousand persons in Southeast Asian and African countries are becoming partially or blind yearly and worldwide about two billion humans are facing insufficiency in iron intake [25]. The top 10 healthiest leafy vegetables are:

- Broccoli (*Brassica oleracea var. italica*);
- Cabbage (*Brassica oleracea var. capitata*);
- Chard (*Beta vulgaris subsp. vulgaris*);
- Collards (*Brassica oleracea*);
- Dandelion greens (*Traxacum officinalea*);
- Kale (*Brassica oleracea var. sabellica*);
- Lettuce (*Lactuca sativa*);
- Parsley (*Petroselinum crispum*);
- Spinach (*Spinacia oleracea*);
- Turnip Greens (*Brassica rapa subsp. rapa*).

Both herbs and leafy vegetables are essential parts of achieving a balanced diet. Saturated fat and cholesterol are relatively less in these foods.

These are also good source of protein, magnesium, phosphorus, and calcium, vitamin A, C, and K, thiamin, dietary fiber, vitamin B6, riboflavin, iron, folate, potassium, and manganese. They also contain Chlorophyll, which is intimately associated with human blood. It acts as a detoxifier and cleanser to the blood. Chlorophyll helps in the fight against infection and helps to overcome skin problems and it is also anti-inflammatory. Leafy green vegetables like kale, broccoli, and spinach are packed with magnesium and calcium, which are compulsory for the proper functioning of teeth and bones.

Phytonutrients are natural chemicals in foods that are not only vital for bodily function but act as a preventive step in decreasing cancer risk and increasing overall health benefits. Phytonutrients or phytochemicals found in leafy green vegetables are beta-carotene, lutein, and zeaxanthin. Consuming sufficient fiber in daily diet is essential for proper digestion and bowel health and cardiovascular health. Consumption of basic required limit of leafy vegetables lowers the risk of diabetes, which is achieved by regulating the blood sugar levels in the body. They also reduce the risk of heart disease, obesity, and related illnesses.

6.2 MAJOR CHALLENGES

The lack of knowledge about preservation and processing techniques of herbs and green leafy vegetables makes it difficult for their complete utilization. The processing in some cases may not fall under the economic range, which also adds to the issue of quick deterioration of quality in terms of shelf-life. Chemical, physical, and biological hazards potentially contaminate the fresh produce. Chemical hazards involve pesticides used on the farm and chemicals used during packaging. Physical hazards may include metal piece, dust, wood, and sand during harvesting. Biological hazards comprise of microbiological contaminants like pathogens available in the soil, *Salmonella*, *Escherichia coli*, and other pathogens [13, 17, 35]. Hence proper cleaning and surface treatment is required for elimination of such hazards.

The main challenge of utilization of herbs and leafy vegetables is to extend their shelf-life. Rapid quality deterioration occurs due to water loss, spoilage pathogens, and metabolism being active as soon as the produce is harvested. One of the biggest challenges is to supply good quality herbs and leafy vegetables throughout the year. This challenge can be minimized by processing the herbs and leafy vegetables without losing their nutritional values. In reducing quality losses of leafy vegetables, the postharvest technologies play an important role [36, 38]. Postharvest technologies help to preserve the product quality and safety until the produce reaches the consumers. Various packing and storage techniques have already been developed for shelf-life improvement and quality of produce.

Developing countries follow traditional techniques and storage technologies of low cost for the preservation of produce. Methods involving the minimum occurrence of loss or degradation in the quality of the produce with minimum capital input for processing are the most desirable factor being looked forward in the field of development of techniques.

6.3 QUALITY OF HERBS AND LEAFY VEGETABLES

Quality of any product is a degree of excellence or a high standard or economic value. The appearance is the most common quality parameter of herbs and green leafy vegetables, such as color, flavor, shape, texture, maturity, tenderness, free from defects and pale coloring, etc. Quality includes both safety of the produce and originality of produce [1].

The quality parameters of herbs and leafy vegetables vary from product to product (commodity), consumer's acceptance index and its intentional use, e.g., fenugreek, and coriander are well recognized for their aroma and also utilized as flavoring agent while lettuce and broccoli due to their textural properties are used in salads and basil, amlaki are recognized for its medicine value [6, 9, 45].

6.3.1 QUALITY LOSSES

The quality of leafy vegetables rapidly deteriorates after harvesting due to biological growth processes, which take place inside plant-like cell division, synthesis of protein. The postharvest quality depends on factors starting from the field until it reaches the table [58]. Biotic factors involve changes due to genetically and physiological related factors. On the other hand, abiotic factors involve changes due to the handling, storage conditions. Both factors can be controlled or restricted to a limited extent utilizing various technologies, such as, refrigeration, packaging, etc.

Some biochemical reactions, which include the production of ethylene, Millard reaction, and vitamins degradation, occur along with quality losses after harvesting. The crop with harvest injuries results in producing ethylene causing the development of yellowing of leaves [18]. Chilling injury caused by large ice crystal formation in cells may also result in the production of ethylene. Oxidation of polyphenols is the main source of browning or color change in the plants. This enzyme converts the phenolic compounds in the presence of oxygen into quinines. Quinines further undergo polymerization to make insoluble polymers called melanin.

Another, important spoilage organisms for quality deterioration in leafy vegetables and herbs after the harvesting are bacteria and molds. Injured areas in the leaves are the best site for the growth of microbes [57]. The healthy leaves get contaminated by the action of microbes from unhealthy leaves, which further accelerates the spoilage of harvested produce. The farm operations and climatic conditions also affect the quality of the product.

6.3.2 FACTORS AFFETCTING QUALITY DETERIORATION

6.3.2.1 MYCOTOXIN CONTAMINATION

During harvesting and sun drying, the herbs are more susceptible to mold contamination. Molds growing under favorable conditions produce metabolites that are toxic not only to human beings but also to the animals. These toxic metabolites are called mycotoxins. There are two secondary metabolites (mycotoxins) for a single fungal species, i.e., there maybe 20,000 to 300,000 unique mycotoxins. Among those mycotoxins, the most known are Aflatoxins, Ochratoxin A, T-2 toxin deoxynivalenol (DON) according to the World Health Organization (WHO).

Consumption of contaminated herbs and leafy vegetables results in health hazard in consumers. The presence of toxic fungal (mold) species like *Aspergillus*, *Penicillium*, and *Alternaria* in powdered samples of *Emblica officinalis* (commonly called Aamla) and *Terminalia chebula* (commonly called Haritiki) revealed that these herbs are not suitable for direct consumption [13, 61].

6.3.2.2 RESPIRATION

The quality deterioration of herbs and leafy vegetables even after harvesting is attributed to metabolic (biological) activity and process of respiration [44]. During the respiration, the sugars along with other cell components present in the plant-cells are broken down into water and carbon dioxide in the presence of oxygen and resulting in the formation of energy. The respiration process consists of:

$$C_6H_{12}O_6 + 6O_2 \rightarrow 6CO_2 + 6H_2O + energy$$

Following losses of quality have been observed:

- Loss of flavor and texture;
- Loss of nutrition;
- Need of extensive ventilation due to release of carbon dioxide;
- Release of heat, which raises refrigeration cost.

At the time of photosynthesis, there is accumulation of the solar energy in form of molecule of sugar within the leaves by the chloroplast. During the glycolysis cycle, the energy is released and energy is supplied to perform the biochemical process of the plant. Rate of respiration varies on the type of commodity or produce, maturity state of commodity, chemical composition of the commodity and variety. In comparison to other agricultural produces, herbs, and leafy vegetables generally show a sudden change in respiration after harvesting [49].

6.3.2.3   WATER ACTIVITY

Water activity is used to determine the quality and stability of the herbs or leafy vegetables in reference to rate of deteriorative reactions, chemical properties or physical properties and microbial growth. The term water activity refers to partial vapor pressure of water in a substance divided by the partial vapor pressure of pure water at the same temperature. The scale of water activity varies from zero (which represents bone dry) to one (which represents pure water). Water activity level ranges from 0.2 for food produce, which is very dry, to 0.99 in the case of fresh food produce.

For a produce with good shelf-life either the acidity level (pH) or the water activity (a_w) level or combination of pH should be analyzed. The growth of hazardous pathogens can be contained by reducing or lowering the water activity of produce.

6.3.2.4   EFFECT OF ETHYLENE PRODUCTION

Ethylene (C_2H_4) is involved in the growth regulation, senescence, and ripening even at low concentrations (ppm: parts per million). It has been determined that continuous prolonged exposure to ethylene, even at a very low concentration, will cause considerable loss in fresh produce. There are other sources of ethylene that include decay, physical damages, increase in temperature and stress caused by water, produce chilling, which can result a shelf-life loss of 10–30% [11].

6.3.2.5   PHYSICAL DAMAGES

Physical damage generally includes bruising (damage caused due to abrasion or pressure), surface, and vibration damages including crushing of leaf and tearing of leaf results in leakage of phenolic compounds and become viable sites for browning and accelerates water loss and production of ethylene. It has been determined that moisture loss increases by 3–4 times due to mechanical damages. Other types of following injuries occur due to improper storage conditions:

- **Chilling Injury:** It involves damage to produce due to the surrounding storage temperature above the freezing point (32°F or 0°C). Chilling injury is most common to plants belonging to the tropical or the subtropical regions. Leaves may turn to purple or reddish in case of chilling injury and sometimes may wilt. In an experiment, Chinese cabbage was stored at different temperatures and it was found that patchy papery necrosis was more severe at 2°C and 0°C, while almost no injury was noted at 20°C [15]. Growing tip, browning of leaves, and loss of glossy appearance of leaves are symptoms of chilling injury in leafy vegetables.
- **Freeze Injury:** It occurs due to ice crystals formation inside the tissues of produce, disrupting membranes and dehydrating cells. Bulk freezing is observed to occur when air at less than freezing point temperature of product moves inside into the storage area and displaces the warm air, resulting the temperature of plant produce to drop low causing ice crystals to form inside the tissues. At plant level when the temperature is at or below freezing, the temperature of plant is colder than air temperature.
- **Low Oxygen Injury:** It is one in which toxic concentrations of ethanol and acetaldehyde accumulates in plant matter resulting in development of discolored tissues and off-odors [57].

6.3.2.6   WATER LOSS

Water loss mainly occurs through transpiration in case of herbs and leafy vegetables. Both qualitative and quantitative losses occur due to the loss of water. The water present in intercellular region of leaves plays a major role in providing turgidity and crispness of food produce, which decreases with the increase of the water loss. Loss of appearance of the leaves, wilted, and shriveled is an outcome of water loss. Loss of 5–10% in fresh leafy vegetables gives appearance of wilted leaves. Degradation of nutrients may also happen due to loss of water (e.g., loss of vitamins and minerals). Water stress increases ethylene production and rate of respiration. In a study, water loss occurring on pakchoi was measured at temperature of 35°C with water loss rate of 2.8% per hour [5].

6.3.2.7   *IMPACT OF GAS COMPOSITION ON RESPIRATION RATE*

The rate of respiration directly affects composition of environmental gases. The oxygen level present in the atmosphere is proportional to rate of respiration occurring in the food product. By reducing the oxygen level, respiration rate become slow, which results in reduction of metabolic activity of produce.

As the O_2 concentration decreases, the rate of respiration is lowered as shown in Figure 6.1. When concentration of oxygen is changed from 2 to 0.5%, respiration rate falls, but the head damage percentage increases due to low concentration of oxygen (Figure 6.2) [37]. When oxygen concentration becomes is <2%, then a decrease in rate of respiration is observed followed by rise in injuries because of anaerobic conditions in the environment (Figure 6.2) [37].

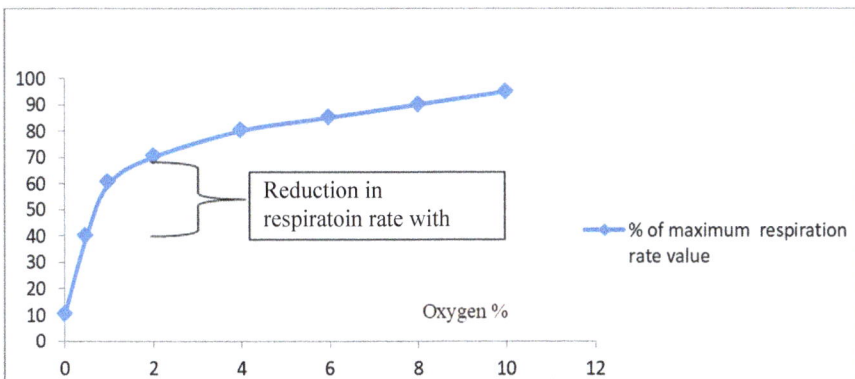

FIGURE 6.1   Comparison of respiration rate towards change in oxygen percentage [37].

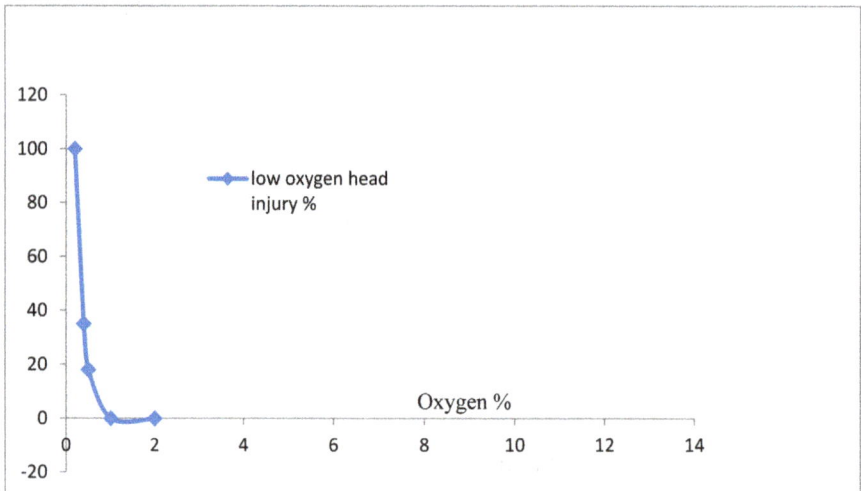

FIGURE 6.2   Comparison of oxygen injury to that of percentage of oxygen present [37].

6.4   TECHNIQUES TO EXTEND THE SHELF-LIFE

Any technique, which is capable of controlling or eliminating the microbial activity to a greater extent without altering the chemical constituent of the food (leafy vegetables or herbs) with a least economic cost of implementation, is known as sustainable technique. Methods involving minimum occurrence of loss or degradation in quality of the produce with minimum capital input for processing is the most desirable factor being looked forward in the field of development of techniques. Practically, the loss of quality cannot be completely stopped due to constant biological and environmental changes, but when the best suitable method is followed or employed this can be controlled to the best possible extent. Some of the most effective techniques which have been practically implemented and futuristic techniques are listed in Figure 6.3.

Thermal techniques provide information on methods, which involve extending shelf-life of the produces by effective application of suitable form of thermal energy. The methods under nonthermal techniques provide alternatives, which do not involve application of any form of thermal energy [4, 47]. Storage and packaging techniques provide details on possible ways of storage, which are effective and efficient in retarding the quality loss and retention of desired factors to the best possible extent for enhanced shelf-life of herbs and leafy vegetables.

Non-Thermal Techniques	Thermal Techniques	Storage and Packaging Techniques
•Irradiation •Gas Fumigation •Minimal Processing •Fermentation •Pickling Technology •Photosensitization •Ultravoilet Treatment	•Steam Treatment •Blanching Methods •Drying Methods	•Corrugated fibreboard packaging •Controlled Atmospheric Storage (CAS) •Modified Atmospheric Packaging (MAP) •Active Packaging •Cooling Technology •Nanocomposite Packaging

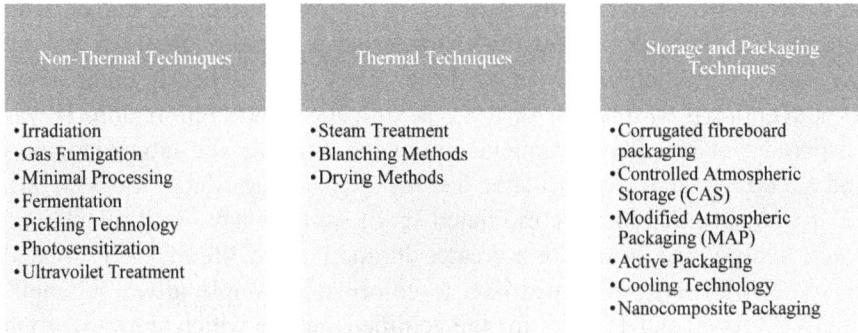

FIGURE 6.3 Classification of techniques for shelf-life extension of herbs and leafy vegetables.

6.4.1 NONTHERMAL METHODS

6.4.1.1 IRRADIATION

Irradiation refers to the process in which food products are introduced to ionizing radiation, which is responsible for releasing electrons in targeted food products from their atomic bonds without being in physical contact with it. Irradiation is a suitable method to preserve food produces, reducing foodborne illness, preventing the invasion of pests, and also eliminate/retards the sprouting and ripening of produce [32]. The first U.S. and British patent was issued in 1905 to use irradiation (ionizing radiation) to kill bacteria in foods. USFDA approved the food irradiation method of processing of pork, poultry, fruits, vegetables, spices, wheat, and red meat. A radiation dose of 5 KGy for fungal growths and a dosage of 10 KGy for bacterial is generally used to attain the commercial sterility [41, 42].

The Codex Alimentary Commission has approved the application of ionizing radiation (gamma rays) for decontamination of microorganisms in food products, including herbs and spices. Irradiation studies indicate that irradiation does not create any toxic changes within the food and hence it is safe. Gamma rays dose up to 18 kGy is allowed to be exposed to food material. The sensory attributes of most spices and herbs are maintained well in the range of 7.5 to 15 kGy [12].

The irradiation method is safe as it affects only the DNA part of the microorganism causing the destruction of the microbial population and there is no chemical constitutional change in the produce (herbs or leafy vegetables). Therefore, this method is also referred as cold sterilization. In an experiment, irradiation of diced celery on low dosage was studied with reference to shelf-life and quality characteristics. The dosage of 0.5 and 1.0 kGy and its irradiation effects on diced celery was studied in comparison to conventional treatments, such as: acidification and chlorination. It was observed that one kGy treatment eliminated *Listeria* like monocytogenes and *E. coli*, which were inoculated thereby showing the effectiveness [3, 50].

It was observed that the irradiated set of samples have well-maintained color, texture, and aroma for a greater duration. Shelf-life of 1 kGy treated celery was 29 days in comparison to chlorinated sample providing shelf-life of 22 days, and 15 days for the acidified sample which shows that the shelf-life of the produce got extended due to the irradiation treatment [27, 46]. Decontamination of medicinal plants and spices through the method of irradiation is safe and effective. The losses of bioactive substances are lower in irradiation technique than the fumigation techniques.

6.4.1.2 GAS FUMIGATION

Methyl bromide (MB) is one of the important fumigants for plant quarantine, but the Montreal Protocol forces to reduce the use of MB and its emissions into the atmosphere. The development of alternative treatments to MB is increasingly urgent. Ethylene oxide is commonly used to decontaminate herbs and leafy vegetables, with varying degree of composition depending upon the product and quantity. Usage of ethylene oxide in some countries like (Japan, United Kingdom) is prohibited, because it reacts with the organic components and leaves residues of ethylene chlorohydrins and ethylene bromohydrin on herbs.

Currently permitted level of residues is 50 ppm. Fumigating with the mixture of ethyl formate and carbon dioxide is also used commercially under the name of Vapormate™ (carbon dioxide 83.3% (w/w) and ethylene formate 16.7%) in various countries. This vapormate is considered a food additive, which is generally recognized as safe (GRAS). Methylcyclopropene (1-MCP) is a relatively new compound, marketed as EthylBloc ®. The gas is often used as a fumigant to prevent ethylene hormone activity, which is naturally secreted in the plants [53]. An experiment was performed on 1-MCP (1-methylcyclopropene) fumigation to see if it could assist in maintenance of quality of produce.

Commercial usage of 1-MCP with choy sum, Shanghai band, and spearmint has been justified. General visual quality and reduction in yellowing of leaves were observed in choy sum and Shanghai choy when they were fumigated with 1-MCP. At storage temperature of 12°C and 0°C, the various changes were studied. Generally, for all three of these crops, optimum storage temperature of 0°C is preferred and if 1-MCP is used commercially then it should be higher than the optimum handling temperature [60]. There is no specified fumigant, which can be used for all the variety of herbs and vegetables. The efficiency and level of function of various fumigants varies depending on the produce.

6.4.1.3 MINIMAL PROCESSING

Minimal processing refers to the fresh produce (leafy vegetables or herbs) that has been diced, shredded, sliced or peeled just before the packaging operation and varied from other methods of processing as it do not involve heating, therefore, plant tissue remains viable [24]. Research on market preferences of consumers showed that most people perceive or look out for fresh produce because they are more nutritious than the processed foods [33]. Lot of nutrient phytochemicals present in the food produce gets destroyed at the time of processing of produces [34].

It has been found that red lettuce variety can provide five times the amount of antioxidant content in comparison to leafy alike varieties, because of higher proportion of phenolic and anthocyanins [7]. Similarly, Spinach has been recognized for its high level of ascorbic acid [23]. Likewise, there are different herbs and leafy vegetables with its health benefits. Once the minimal processing is done, the cells beneath the surface of the food material are exposed to the external environment. It has been observed that some of the leafy vegetables show browning (color change), such as, chicory, and lettuce show browning during storage. On the other hand, there are certain leafy vegetables, like Red lettuce, show higher browning resistance process [14, 15].

Refrigerated storage is one of the methods to improve the shelf-life of the minimally processed produces. At a cool storage (<4C), the antioxidants and phenolic compounds are stable provided proper cooling temperature is maintained for the leafy vegetables [45]. It was also found that spinach leaves when stored at 4°C for 7 days, the flavonoids content did not change but an increase was observed on day 3 on those cut leaves [16]. The modified atmospheric packaging (MAP) is also effective in extending shelf-life of the produces.

6.4.1.4 FERMENTATION

The process of fermentation holds adequate for the preservation of leafy vegetables than herbs. In the presence of lactic acid bacteria (LAB), leafy vegetables and herbs produce acid and flavors. It has been found that nearly all leafy vegetables can be fermented with the help of LAB. There is an increase in level of vitamins and digestibility due to lactic acid fermentation [56]. According to FAO (Food and Agriculture Organization) the three common possible conditions of lactic acid fermentation are:

- **Salted Condition:** For each 100 kg of vegetable, 3 kg of dry salt is added. First washing is done and then these are placed in a fermentation container in the form of layer-arrangement at height of 2.5 cm each. Over each layer, salt is sprinkled. The same procedure is followed till container reaches three quarter level. Cloth is placed over which force is applied to help in the development of brine. The total time nearly takes 24 hours. Once the brine formation is over, the fermentation process is initiated and CO_2 is generated. Based on the ambient temperature, the fermentation takes about 7 to 28 days. Once the bubble formation stops, the product is mixed with oil and spices, if needed [56].
- **Non-Salted Condition:** Products are acted by LAB without using any salt or brine. To prevent the growth of yeast, the oxygen is restricted. There is a decrease in the pH level by the activity of LAB, which creates an adverse environment for the development of spoilage organisms.
- **Brine-Salted:** This method is preferred for leafy vegetables, which have low moisture content. The produce is immersed in the brine solution and is left out for fermentation. About 15% to 20% of solution of brine is made by preparing salt solution. If the salt concentration goes below 12%, there will not be proper fermentation. The salt concentration, brine solution temperature, types of microbe, and fermentable material availability, etc., are major factors, which can affect the fermentation process. The optimum temperature for proper fermentation ranges from 12.5 to 20°C [56].

6.4.1.5 PICKLING TECHNOLOGY

In pickled technology for leafy vegetables, usually edible acid with vinegar is added. In cabbage pickling, it is initially shredded to small pieces of about 1/32 inch and then citric acid is used to acidify the shredded leaves and vinegar, which is about 0.5–0.7% acid, and is sweetened to desirable level (5–10% sugar), then it is heated in boiling water at 74–77°C. Overall, the shredded cabbage with acid-sugar solution along with water in a bottle is followed by heat treatment and tap water cooling [30].

6.4.1.6 PHOTOSENSITIZATION

Photosensitization is a novel idea in food processing for removal of antimicrobials. Visible light and photoactive compounds (photosensitize) interact with the food content that leads to a reduction in microorganisms. It is sprinkled onto the surface of freshly produced food, where most of the surface microorganisms are bound with it. Different photocytotoxicity reactions arise through application of visible light to fruit and result in the death of microorganisms with no adverse effects on the properties of produce [8].

6.4.1.7 ULTRAVIOLET (UV) TREATMENT

Application of low frequency (100–400 nm) ultraviolet (UV) radiation is sufficient to inactivate the microorganisms and is differentiated into three categories: (i) near-UV radiation UV-A (315–400 nm), (ii) mid-range UV, i.e., UV-B (280–315 nm) and (iii) far-UV, i.e., UV-C (100–280 nm). UV-C (100–280 nm) is the most commonly used type of UV radiation for treatment of herbs and leafy vegetables because it indirectly or directly works as an antimicrobial agent. It leads to direct DNA destruction of bacteria and also increases the resistance mechanisms for spoilage microorganisms in various foods. Higher doses of UV-C are generally applied to protect foods that start degrading quickly after harvesting [50].

6.4.2 THERMAL TECHNIQUES

6.4.2.1 STEAM TREATMENT

In steam treatment of herbs and leafy vegetables, the air is evacuated followed by steam treatment of food produce with at 100 to 200°C, dried through hot air, and cooled quickly. Treatment time may vary from 5 to 15 minutes, depending on the type of produce and its quantity [55]. An experiment was conducted to compare the efficiency of component retention by various methods of treatment and it was found that the steam treatment has almost the same effect as the irradiation method [2].

6.4.2.2 BLANCHING

Blanching is mild heat treatment to deactivate enzymes present in herbs and leafy vegetables. Blanching process should be performed in an optimum level [52]. At the time of cold storage, there is a reaction between peroxidase (POD) and cold damage on produce. This kind of enzymatic activities can be eliminated by the process of blanching. In an experiment of destruction of virus present in herbs, steam blanching was done for duration of 2.5 min at 95° C to cause inactivation of infectious microbes [26].

6.4.2.3 DRYING

Drying reduces water content in agriculture produce, which helps in the improvement of shelf-life by lowering the respiration rate, enzyme activities, and rate of deterioration. The drying process also reduces packaging, transporting cost, storage, and handling cost of materials by converting them into dried product. Almost all the vegetables can be dried by employing any method, but not all will be of good taste and quality after drying. The end product quality depends on the temperature, velocity of airflow, time, RH, etc. The following are some of the common methods of drying:

- **Solar or Sun Drying:** It involves spreading the herbs or leafy vegetables on a clean floor. The effectiveness of solar drying depends on the environment and the geographical location. In an experiment, leaves of spinach were dried to a moisture content of 3.5 to 4% in direct sunlight (at 35°C), in shade (at 28°C) and using oven heat at 45°C followed by storing in polyethylene wrappers. The stored sample moisture, total chlorophyll content, moisture level, and ascorbic acid content were found out after subjecting to solar energy. There was a small moisture gain of spinach at storage period of 12 weeks with <1% in all 3 drying methods. The result with more nutrient retention was in sun drying compared to other two techniques [19].
- **Hot Air Drying:** It involves air velocity of 1.3 m/s, temperature of 40°C and relative humidity (RH) of 58–63% for general configuration. Hot

air drying is sometimes assisted with solar drying. In an experiment for performance evaluation of dryer, batches of about 200 kg each of four rosella variety flowers and batches made up of three lemon-grasses variety were used for experiment under controlled environment. The results show that the dried products were of high quality [19].

- **Low Humidity Air Drying:** It involves an air velocity similar or above than hot air drying but it is attached with a dehumidifier to reduce the level of humidity of inlet air. The airflow temperature on the produce was maintained at 40°C and RH 28–30%.

- **Microwave Drying:** Microwave power between 360 to 900 W is used generally to dry the food produce. The water molecules present inside the produce act as dipoles and their rotation performs the drying process. In microwave method of drying of parsley, it was observed that close to original fresh parsley, the experimented one maintained a good green color. When the setup was working at a power range of 900 W against 360 W, the time was reduced by 64% with superior produce quality [29].

6.4.3   STORAGE TECHNIQUES

The factors that influence the storage-life of the produce are temperature, moisture, ethylene, and other exhaust gases produced by respiration of food products and other environmental conditions. Ethylene production should be checked to prevent deterioration using preventive measures, such as, restriction of supply of fresh air to produce and usage of chemicals like potassium permanganate for storage of leafy vegetables [28]. The temperature of the storage environment should be reduced to lower the metabolism rate. A proper CO_2 to O_2 ratio should be maintained to control the respiration and ethylene production. Browning of the produce is prevented by subjecting to blanching or by water content reduction to less than 13% of the produce. Cold chain management and good handling practices should be incorporated in transportation of the produce to make sure the good quality for the market. Microbial action on the produce resulting in deterioration is influenced by humidity, atmospheric conditions, type of vegetable and temperature conditions.

Heat treatment (blanching and sterilization) helps in the enzyme inactivation and also destruction of microbes. However, there is a negative impact of blanching as herbs and leafy vegetables can lose acetic acid, flavor, and other vitamins. Conventionally, water is used as a medium for blanching but recent techniques use chlorinated water or microwave blanching to control the loss of nutrients and reduce microbial load in leafy vegetables. Pre-cooling is done to mitigate effects of heat during harvesting [19]. Pre-cooling can be done by using water or room cooling or vacuum cooling or forced air-cooling methods. Cooling rooms are provided with ventilation fans. Vacuum cooling is used during transportation of the produce [20, 24].

Chemicals can also be used to preserve the food produce but should not create any adverse effects on the human health. Organic acids (sorbic acid, phosphates) are some common chemical additives. Polythene films can be used to control the movement of moisture inside or out of the product from environment. It is necessary to take care of the product from developing off-flavor due to controlled atmosphere (CA).

The most important leafy vegetables in the market are spinach and lettuce, whose methods of storage practices and preservation can be adapted to almost all of leafy vegetables. Harvested produce should usually be shaded away from direct sunlight and then moved to the cooling houses immediately. Once the harvesting is done, the produce should be cleaned using chlorinated water. Free chlorine (@ 100 ppm) in freshwater reduces the microorganisms count by 98%. In case of herbs, the similar process is applicable but the procedure of processing and preservation varies from one another.

6.4.3.1 CONTAINER PACKAGING

Different types of containers are available for handling of herbs and leafy vegetables. The technique of packaging is selected based on economic and medicinal value and market demand of the produce. For export of high-valued herbs and leafy vegetables usually expensive containers (cartons, foam box, etc.), are used. Bamboo baskets are most commonly used for local markets and are available in different shapes and sizes [59]. Rigid containers (cartons and crates) are preferred by customers than the non-rigid containers.

- **Basket:** It is made up of woven materials like bamboo, rattan, or plastic strips.
- **Box:** It is constructed by corrugated fiberboard or Styrofoam. It may be either a carton that can be closed with top flaps or a two-piece box. The contents can be place-packed or bulk-filled.
- **Crate:** It is a wooden or plastic container. Wooden crates are generally wire-bounded and there are chances of collapse, as load exceeds the critical load capacity.

- **Plastic Containers:** These have the advantage of resistance to water and good strength for stacking telescoping construction and dividers. To reduce the transportation and mechanical injuries, container liners along the corner are designed. The packaging container is vented to maintain and lower temperature of storage of leafy vegetables. Cold air through the vents can be quickly circulated. Vents release the heat and gases produced in container by respiration of food produce.

6.4.3.2   CONTROLLED ATMOSPHERIC STORAGE (CAS)

Controlled atmospheric storage (CAS) involves maintaining of a fixed concentration of gasses around the product through careful monitoring and addition of inert gases. It can also be used to preserve the herbs and leafy vegetables, but it is costly than the modified atmospheric packaging [16].

6.4.3.3   MODIFIED ATMOSPHERIC PACKAGING (MAP)

MAP is a successful technique for retention of product freshness, extending shelf-life and also retaining the fresh green color and natural fresh taste, reducing water loss and reducing heat of respiration. Modified atmospheric packaging (MAP) is preferred because of its cost-effectiveness. MAP can be classified as passive and active MAP. Packing material should have high permeability and soft, such as, low-density polyethylene (LDPE) film is generally used as compared to high-density polymers (HDPE) [51]. Because LDPE has good CO_2 and O_2 permeability, ease of sealing, good appearance and tear resistance. Polyethylene bags are not rigid, and extra load can collapse the produce. In an experiment on Chinese kale, trimming, bruising, and weight loss were measured in three different quantities, i.e., 5, 10, and 15 kg per polyethylene bag and it was observed that the losses were increased with increasing product weight [10].

6.4.3.4   ACTIVE PACKAGING

It is a type of modified atmospheric packaging where the packaging material interacts with internal environment of gases and thus resulting in the improvement of shelf-life and also its quality [21, 54]. These technologies continuously modify the gaseous environment inside the packaging. This method is also called interactive packaging. Some of the common technologies of active packaging are:

- **Antimicrobial Preservatives:** These inhibit the microbial activity inside a packed food material and thereby are generally classified as antimicrobial agent contained in a film, which migrates to the surface of food and films effective without migration. The antimicrobial preservative films contain BHT (butylated hydroxytoluene), BHA (butylated hydroxyanisole), tocopherol, which help in the preservation and enhancement of shelf-life.
- **Carbon Dioxide (CO_2) Absorbers/Generators:** It can absorb the carbon dioxide developed inside the packing material. CO_2 absorbers (sachets) usually consist of calcium hydroxide and potassium hydroxide or sodium hydroxide, calcium oxide and silica gel. The basic reaction is [$Ca\ (OH)_2 + CO_2 = CaCO_3 + H_2O$]. Removal of carbon dioxide is necessary to avoid the busting of the packed food material. Some of the common carbon dioxide generators are ascorbic acid, ascorbate, and sodium hydrogen carbonate. These are mainly used to retard the development of bacteria's and molds.
- **Ethylene Absorbers:** These are hormones naturally produced by plants by ripening and respiration of the produce. It accelerates the process of respiration, resulting in senescence and maturity of the harvested herbs or leafy vegetables. The shelf-life of fresh produce can be increased by eliminating ethylene from the packaging. Potassium permanganate is commonly used to remove or absorb ethylene. Commonly used ethylene absorbers are: Activated carbon + Metallic (palladium) catalyst (sachet); Aluminum oxide and Potassium permanganate (sachet); and zeolites, clay (films).
- **Moisture Absorbers:** These can reduce the water activity in the product, thereby retarding the microorganism growth development in the food produce. Moisture absorbers also enhance the product appearance and freshness. Some of the most commonly used moisture absorbers are polyacrylates (sheets); silica gel (sachet); and propylene glycol (film).
- **Oxygen Scavengers:** These reduce or eliminate the oxygen present in the packaging and thereby they prevent oxidation of the produce. Oxygen scavengers are also helpful in preventing insect damage and development of molds and bacteria. Oxygen levels of 0.1% or lower is suitable to restrict mold development and mycotoxin production [22]. Some of the commonly used oxygen scavengers are iron powder, Ascorbic acid, etc.

6.4.3.5 COOLING TECHNOLOGY

Cooling can protect the quality of the produce. Cooling technology improves the shelf-life by decreasing the rate of respiration, transpiration rate and retardation in microorganism growth [40, 62]. Various methods of cooling of the fresh produce are: (1) avoiding direct solar energy; (2) making use of natural methods of cooling like cold water when naturally available, harvesting during early morning hours, using cool air during night; (3) mechanical refrigeration; (4) evaporative cooling method; and (5) post-harvest cooling. The quality of leafy vegetables and herbs is maintained by rapid cooling. Hydro cooling along with package icing can be used. Hydro cooling of kale at a temperature of 4°C in water for duration of 5–10 min and stored at 7°C showed reduction in yellowing, loss of water and extended shelf-life (Table 6.1) [16, 63].

TABLE 6.1 Classification and Application of Different Cooling Methods

Methods of Cooling	Commodities	Comment
Forced-air cooling (pressure cooling)	All types of herbs and leafy vegetables except susceptible to chilling injury	Rate of cooling very fast compared to room cooling method; cooling rates is uniform when used in proper condition. Venting of the container is critical to effectively cool. This method is efficient and also economical.
Hydro cooling	Most of herbs and leafy vegetables	Rate of cooling is fast and uniform. In the case of packed shipping containers, it varies at an extensive rate; daily proper cleaning is necessary;
Package icing	All types of herbs and leafy vegetables	Faster method of cooling; best for commodities which can tolerate ice and water contact; It's an efficient and economical method.
Room cooling	Most of herbs and leafy vegetables	Rate of cooling is slow in case of perishable commodities. Rate of cooling varies at a greater rate between the loads, containers, and pellets.
Top and channel-icing	All types of herbs and leafy vegetables	It's an irregular slow method; Containers which can tolerate water is needed.
Vacuum cooling	Most of herbs and leafy vegetables	For each 6°C cooling, it may cause about 1% weight loss. This equipment is costly and shipping containers should be water tolerant.

The RH can be controlled in the refrigerator by: (1) using humidifier, which helps in adding moisture to atmospheric air; (2) using air vents; (3) using moisture barrier and insulating the walls of storage room; and (4) using wet-floor. In developing countries, techniques of low cost, such as ice packaging and evaporate cooling are commonly used with temperature ranging from 35–40°C to 20–25°C [62].

6.4.3.6 NANOCOMPOSITE PACKAGING

Nanotechnology plays a vital role in designing the nanocomposite materials for the development of new packaging materials with improved mechanical and antimicrobial properties. With emerging trends in the food industry to avoid use of non-biodegradable packages, use of eco-friendly packaging materials is encouraged. Biopolymers and nanocomposite packaging materials are used to overcome this issue but low protection restricts their application in the food industry [31].

6.5 SUMMARY

This chapter focuses on techniques for enhancement of shelf-life of herbs and leafy vegetables. Thermal techniques such as steam treatment, drying, blanching, etc., can be implemented easily in comparison to techniques of nonthermal basis. In spite of undertaking any treatment of thermal or nonthermal means, there are chances of contamination of the produce after treatment and storage techniques are indispensable for enhancement of shelf-life. Suitable storage and packaging techniques need to be implemented to achieve a greater shelf-life. Techniques like MAP, CAS, active packaging are common and popular techniques which are used widely today along with recent advances like nano-composite packaging.

KEYWORDS

- controlled atmospheric storage
- generally recognized as safe
- herbs
- lactic acid bacteria
- leafy vegetables
- shelf-life

REFERENCES

1. Al, U., Wills, R. B. H., Bowyer, M. C., & Golding, J. B., (2019). Interaction of the hydrogen sulfide inhibitor, propargylglycine (PAG), with hydrogen sulfide on postharvest changes of the green leafy vegetable, Pak choy. *Postharvest Biology and Technology, 147,* 54–58.
2. Barba, F. J., Saraiva, J. M. A., Cravotto, G., & Lorenzo, J. M., (2019). *Innovative Thermal and Nonthermal Processing, Bioaccessibility and Bioavailability of Nutrients and Bioactive Compounds* (1st edn., p. 370). New York, USA: Woodhead Publishing.
3. Bardsley, C. A., Boyer, R. R., Rideout, S. L., & Strawn, L. K., (2019). Survival of *Listeria monocytogenes* on the surface of basil, cilantro, dill, and parsley plants. *Food Control, 95,* 90–94.
4. Bermúdez-Aguirre, D., & Barbosa-Cánovas, G. V., (2013). Disinfection of selected vegetables under nonthermal treatments: Chlorine, acid citric, ultraviolet light and ozone. *Food Control, 29*(1), 82–90.
5. Brummell, D. A., & Toivonen, P. M., (2018). *Handbook of Vegetables and Vegetable Processing* (2nd edn., p. 1108). London, U.K.: Wiley Blackwell Publishing.
6. Cantwell, M., Suslow, T., & Lettuce, C., (2002). *Recommendations for Maintaining Postharvest Quality* (p. 10). Department of vegetable crops, University of California, Davis: CA95616.
7. Catunescu, G. M., Rotar, I., Vidican, R., & Rotar, A. M., (2017). Effect of cold storage on antioxidants from minimally processed herbs. *Scientific Bulletin, Series F. Biotechnologies, 21,* 121–126.
8. Cozzolino, M., Delcanale, P., Montali, C., Tognolini, M., Giorgio, C., Corrado, M., & Viappiani, C., (2019). Enhanced photosensitizing properties of protein bound curcumin. *Life Sciences, 233,* 116710–116719.
9. Dhemre, J. K., Shete, M. B., Kad, V. P., & Kotecha, P. M., (2017). Performance of packaging on shelf-life and quality of fenugreek at different storage conditions in Kharif season. *Advances in Research, 2017,* 1–12.
10. Esteban, R., Buezo, J., Becerril, J. M., & García-Plazaola, J. I., (2019). Modified atmosphere packaging and dark/light refrigerated storage in green leafy vegetables have an impact on nutritional value. *Plant Foods for Human Nutrition, 74*(1), 99–106.
11. Goodburn, C., & Wallace, C. A., (2013). The microbiological efficacy of decontamination methodologies for fresh produce: A review. *Food Control, 32*(2), 418–427.
12. Gracheva, A. Y., Zav'yalov, M. A., & Ilyukhina, N. V., (2016). Enhancement of efficiency of storage and processing of food raw materials using radiation technologies. *Physics of Atomic Nuclei, 79*(14), 1682–1687.
13. Hariprasad, P., Durivadivel, P., Snigdha, M., & Venkateswaran, G., (2013). Natural occurrence of aflatoxin in green leafy vegetables. *Food Chemistry, 138*(2/3), 1908–1913.
14. Ioannou, I., (2013). Prevention of enzymatic browning in fruit and vegetables. *European Scientific Journal, 9*(30), 310–341.
15. Islam, M. Z., Lee, Y. T., Mele, M. A., & Choi, I. L., (2019). Effect of modified atmosphere packaging on quality and shelf-life of baby leaf lettuce. *Quality Assurance and Safety of Crops and Foods,* 1–8.
16. Junge, K., Weimar, M., & Blanton, D., (1986). *Precooling Methods for Commercial Vegetable Producers.* Economic Staff Paper Series 24. http://lib.dr.iastate.edu/econ_las_staffpapers/24 (accessed on 18 January 2021).

17. Kaczmarek, M., Avery, S. V., & Singleton, I., (2019). Microbes associated with fresh produce: Sources, types and methods to reduce spoilage and contamination. *Advances in Applied Microbiology, 107,* 29–82.

18. Kasım, M. U., & Kasım, R., (2017). Yellowing of fresh-cut spinach (*Spinacia oleracea L.*) leaves delayed by UV-B applications. *Information Processing in Agriculture, 4*(3), 214–219.

19. Kaur, G., Singla, N., & Singh, A., (2019). Effect of vacuum drying on nutrient retention of some commonly consumed herbs. *Studies on Ethno-Medicine, 13*(02), 62–70.

20. Liberty, J. T., & Okonkwo, W. I., (2013). Evaporative cooling: Postharvest technology for fruits and vegetables preservation. *International Journal of Scientific and Engineering Research, 4*(8), 2257–2266.

21. Llana-Ruíz-Cabello, M., Puerto, M., & Pichardo, S., (2019). Preservation of phytosterol and PUFA during ready-to-eat lettuce shelf-life in active bio-package. *Food Packaging and Shelf-Life, 22,* 100410.

22. López-Gómez, A., Ros-Chumillas, M., & Antolinos, V., (2019). Fresh culinary herbs decontamination with essential oil vapors applied under vacuum conditions. *Postharvest Biology and Technology, 156,* 110942.

23. López-Gómez, A., Soto-Jover, S., & Ros-Chumillas, M., (2016). New technology for enhancement of the food safety of minimally processed fruits and vegetables. In: *VIII International Postharvest Symposium: Enhancing Supply Chain and Consumer Benefits-Ethical and Technological* (No. 1194, pp. 545–552).

24. Matthews, K. R., (2014). Leafy vegetables. In: Matthews, K. R., (ed.), *The Produce Contamination Problem* (pp. 187–206). New York, USA: Academic Press.

25. Mishra, P. K., Singh, P., Prakash, B., & Kedia, A., (2013). Assessing essential oil components as plant-based preservatives against fungi that deteriorate herbal raw materials. *International Biodeterioration and Biodegradation, 80,* 16–21.

26. Niemira, B. A., (2018). Safety and quality of irradiated fruits and vegetables. In: Rosenthal, A., Deliza, R., Welti-Chanes, J., & Barbosa-Cánovas, G. V., (eds.), *Fruit Preservation* (pp. 399–416). New York: Springer.

27. O'Hare, T. J., Able, A. J., Wong, L. S., Prasad, A., & McLauchlan, R., (2001). Fresh-cut Asian vegetables-pak choi as a model leafy vegetable. *Postharvest Handling of Fresh Vegetables, 105,* 113–115.

28. Ouzounidou, G., Papadopoulou, K. K., Asfi, M., Mirtziou, I., & Gaitis, F., (2013). Efficacy of different chemicals on shelf-life extension of parsley stored at two temperatures. *International Journal of Food Science and Technology, 48*(8), 1610–1617.

29. Özer, C., & Yıldırım, H. K., (2019). Some special properties of fermented products with cabbage origin: Pickled cabbage, sauerkraut, and *kimchi. Turkish Journal of Agriculture-Food Science and Technology, 7*(3), 490–497.

30. Pacaphol, K., Seraypheap, K., & Aht-Ong, D., (2019). Development and application of nanofibrillated cellulose coating for shelf-life extension of fresh-cut vegetable during postharvest storage. *Carbohydrate Polymers, 224,* Article ID: 115167.

31. Paskeviciute, E., Zudyte, B., & Luksiene, Z., (2019). Innovative nonthermal technologies: Chlorophyllin and visible light significantly reduce microbial load on basil. *Food Technology and Biotechnology, 57*(1), 126–132.

32. Possas, A., Benítez, F. J., Savran, D., Brotóns, N. J., Rodríguez, P. J., & Posada-Izquierdo, G. D., (2018). Quantitative tools and procedures for shelf-life determination in minimally processed fruits and vegetables. In: Fernando, P. R., Panagiotis, S., &

Vasilis, V., (eds.), *Quantitative Methods for Food Safety and Quality in the Vegetable Industry* (pp. 223–254). New York: Springer.

33. Prakash, A., Baskaran, R., Paramasivam, N., & Vadivel, V., (2018). Essential oil based nanoemulsions to improve the microbial quality of minimally processed fruits and vegetables: A review. *Food Research International, 111*, 509–523.

34. Prakash, B., Praefcke, G. J., Renault, L., Wittinghofer, A., & Herrmann, C., (2000). Structure of human guanylate-binding protein 1 representing a unique class of GTP-binding proteins. *Nature, 403*(6769), 567–573.

35. Romanazzi, G., Feliziani, E., Baños, S. B., & Sivakumar, D., (2017). Shelf-life extension of fresh fruit and vegetables by chitosan treatment. *Critical Reviews in Food Science and Nutrition, 57*(3), 579–601.

36. Saini, R. K., Shang, X. M., Ko, E. Y., Choi, J. H., & Keum, Y. S., (2016). Stability of carotenoids and tocopherols in ready-to-eat baby-leaf lettuce and salad rocket during low-temperature storage. *International Journal of Food Sciences and Nutrition, 67*(5), 489–495.

37. Saltveit, M. E., (2003). Is it possible to find an optimal controlled atmosphere? *Postharvest Biology and Technology, 27*(1), 3–13.

38. Santos, J., Herrero, M., Mendiola, J. A., & Oliva-Teles, M. T., (2014). Fresh-cut aromatic herbs: Nutritional quality stability during shelf-life. *LWT-Food Science and Technology, 59*(1), 101–107.

39. Sanyal, B., Murali, T. S., & Variyar, P. S., (2016). Radiation processing: An effective quality control tool for hygienization and extending shelf-life of a herbal formulation, *Amritamehari Churnam*. *Journal of Radiation Research and Applied Sciences, 9*(1), 86–95.

40. Sargent, S. A., Ritenour, M. A., Brecht, J. K., & Bartz, J. A., (2000). *Handling, Cooling and Sanitation Techniques for Maintaining Postharvest Quality* (Vol. 18, pp. 97–108). University of Florida Cooperative Extension Service, Institute of Food and Agriculture Sciences, EDIS.

41. Selma, M. V., Luna, M. C., Martínez-Sánchez, A., Tudela, J. A., Beltrán, D., Baixauli, C., & Gil, M. I., (2012). Sensory quality, bioactive constituents and microbiological quality of green and red fresh-cut lettuces (*Lactuca sativa* L.) are influenced by soil and soilless agricultural production systems. *Postharvest Biology and Technology, 63*(1), 16–24.

42. Shetty, A. A., Magadum, S., & Managanvi, K., (2013). Vegetables as sources of antioxidants. *Journal of Food and Nutritional Disorders, 2*(1), 2–13.

43. Singh, B. K., Ramakrishna, Y., & Ngachan, S. V., (2014). Spiny coriander (*Eryngium foetiduml.*): A commonly used, neglected spicing-culinary herb of Mizoram, India. *Genetic Resources and Crop Evolution, 61*(6), 1085–1090.

44. Singh, R., Giri, S. K., & Kotwaliwale, N., (2014). Shelf-life enhancement of green bell pepper (*Capsicum annuum* L.) under active modified atmosphere storage. *Food Packaging and Shelf-Life, 1*(2), 101–112.

45. Singh, V., Hedayetullah, M., Zaman, P., & Meher, J., (2014). Postharvest technology of fruits and vegetables: An overview. *Journal of Postharvest Technology, 2*(2), 124–135.

46. Smetanska, I., Hunaefi, D., & Barbosa-Cánovas, G. V., (2013). Nonthermal technologies to extend the shelf-life of fresh-cut fruits and vegetables. In: Yanniotis, S., Taoukis, P. S., Stoforos, N., & Karathanos, V. T., (eds.), *Advances in Food Process Engineering Research and Applications* (p. 375–413). Boston: Springer.

47. Solórzano-Santos, F., & Miranda-Novales, M. G., (2012). Essential oils from aromatic herbs as antimicrobial agents. *Current Opinion in Biotechnology, 23*(2), 136–141.

48. Srinivas, I., Raju, N. S., & Dhemate, A. S., (2012). Effective storage structures for food grains, fruits, and vegetables. *Reshaping Agriculture and Nutrition Linkages for Food and Nutrition Security, 326*, 306–318.

49. Tao, T., Ding, C., Han, N., Cui, Y., Liu, X., & Zhang, C., (2019). Evaluation of pulsed light for inactivation of foodborne pathogens on fresh-cut lettuce: Effects on quality attributes during storage. *Food Packaging and Shelf-Life, 21*, 100358–100369.

50. Tay, F. R. X., & Teo, S. S., (2019). Interactive effects of storage temperatures and packaging methods on sweet basil. *MOJ Food Process Technol., 7*(1), 16–20.

51. Trilokia, M., Sood, M., Bandral, J. D., Ashraf, S., & Manzoor, M., (2019). Changes in quality of microwave blanched vegetables: A review. *IJCS, 7*(2), 732–737.

52. Ubeed, A., Muhsien, H., Humphreys, B. A., & Wanless, E. J., (2018). *Effect of Hydrogen Sulphide on the Postharvest Metabolism of the Green Leafy Vegetables, Pak Choy (Brassica rapa. subsp. Chinensis)* (p. 210). Doctoral dissertation, University of Newcastle.

53. Valdés, A., Mellinas, A. C., Ramos, M., Burgos, N., Jiménez, A., & Garrigós, M. D. C., (2015). Use of herbs, spices and their bioactive compounds in active food packaging. *RSC Advances, 5*(50), 40324–40335.

54. Vega-Gálvez, A., Giovagnoli, C., & Pérez-Won, M., (2012). Application of high hydrostatic pressure to aloe vera (*Aloe barbadensis miller*) gel: Microbial inactivation and evaluation of quality parameters. *Innovative Food Science and Emerging Technologies, 13*, 57–63.

55. Vitas, J., Vukmanović, S., & Čakarević, J., (2019). Fermentation of six medicinal herbs: Chemical profile and biological activity. *Chemical Industry and Chemical Engineering Quarterly, 2019*, 34.

56. Wang, B., Wilson, M. D., & Huynh, N. K., (2018). Shelf-life extension of fresh basil, coriander, mint, and parsley. In: *International Forum on Horticultural Product Quality* (Vol. 1245, pp. 139–144).

57. Watson, A. J., Treadwell, D., Sargent, A. S., Brecht, K. J., & Pelletier, W., (2015). *Postharvest Storage, Packaging, and Handling of Specialty Crops: A Guide for Florida Small Farm Producers*. IFAS Extension University of Florida.

58. Yam, K. L., & Lee, D. S., (2012). *Emerging Food Packaging Technologies: Principles and Practice*. Elsevier.

59. Zhang, L., Shen, H., Xu, J., & Xu, J. D., (2018). UPLC-QTOF-MS/MS-guided isolation and purification of sulfur-containing derivatives from sulfur-fumigated edible herbs: A case study on ginseng. *Food Chemistry, 246*, 202–210.

60. Zhao, L., Fan, H., Zhang, M., Chitrakar, B., Bhandari, B., & Wang, B., (2019). Edible flowers: Review of flower processing and extraction of bioactive compounds by novel technologies. *Food Research International, 126*, 108660.

61. Zhu, Z., Geng, Y., & Sun, D. W., (2019). Effects of operation processes and conditions on enhancing performances of vacuum cooling of foods: A review. *Trends in Food Science and Technology, 85*, 67–77.

62. Zhu, Z., Wu, X., Geng, Y., & Sun, D. W., (2018). Effects of modified atmosphere vacuum cooling (MAVC) on the quality of three different leafy cabbages. *LWT- Food Science and Technology, 94*, 190–197.

CHAPTER 7

ADVANCED METHODS OF ENCAPSULATION

PRIYANKA KUNDU and PRERNA GUPTA

ABSTRACT

The encapsulation is a process of enclosing the substances within an inert material to protect from unwanted reactions and degradation. Two advanced techniques of encapsulation are nanoencapsulation and microencapsulation. Microencapsulation can be done by spray drying, spray chilling, spray cooling, extrusion, vacuum encapsulation, electrostatic encapsulation, air suspension. The microencapsulation has applications in the food industry, agriculture, biotechnology, the pharmaceutical industry. This chapter focuses on different techniques of encapsulation and their importance in the food, pharmaceutical, and the agricultural sectors.

7.1 INTRODUCTION

Encapsulation is a process of enclosing one substance into the coating of another substance [51]. This technique has the potential to create a physical barrier between the core material and the outer environment that leads to the targeted delivery of the product and increased shelf-life of core material [34]. The material to be encapsulated is known as core material, whereas the coating material is called wall material, carrier material, or shell [37].

The technique has come into existence 30 years ago when the company's National Cash Register developed "carbonless carbon paper" by application of coacervation technique and at that time encapsulation technique was having very limited applications but in the current scenario, encapsulation is the most advanced technology with wide applications [24]. There are three techniques of encapsulation, such as, microencapsulation, nanoencapsulation, and bioencapsulation with various methods depending on different core and wall materials [35]. Encapsulation has proved to be effective in the agriculture sector for targeted delivery of pesticides, fertilizers under controlled conditions and field sensing systems to observe the environmental stress [4, 12, 13].

Encapsulation is also an inseparable part of the food industry where it is used to encapsulate various food ingredients and bioactive compounds [15]. On the other hand, pharmaceutical industries are also focusing on the controlled release of drugs at the targeted site. In addition, extracting compounds from natural herbs to encapsulate these compounds for the preparation of drugs is important to treat the chorionic diseases like cancer [7, 8].

Research is needed to perform experiments with this technique, such as, blending of two different wall materials to check their effect on the release mechanism of encapsulated material [2, 47]. Extensive studies are required: (1) to study the mechanisms of encapsulation and their effect on our environment to identify their toxicity level; (2) to understand the difference between single-walled encapsulated product and multi-walled encapsulated product [6].

The cost of encapsulated product is also an important factor, which should be kept in mind, and wall-material from natural resources is an option to reduce the cost of encapsulated product [14]. With all these improvements in the future, this technique will become more effective and suitable for the encapsulation of all kinds of materials [16].

The chapter focuses on various aspects related to nanoencapsulation, such as, the need for encapsulation, mechanism of encapsulation, different wall-materials, and core materials.

7.2 OBJECTIVES OF ENCAPSULATION [20, 31, 37]

- Controlled release of core material;
- Formation of functional foods;
- Formation of new drugs from cheap resources like natural herbs;
- Increased crop yield;
- Protection from outer environment;
- Protection of sensitive food components from oxidation and other contaminations;
- Reducing the overuses of fertilizer (nanofertilizers);
- The increased shelf-life of food products;
- To address the food security problem;
- To coat the oily components in the textile industry;

- To mask the taste, odor, flavor;
- To protect the pure form of compounds (Figure 7.1).

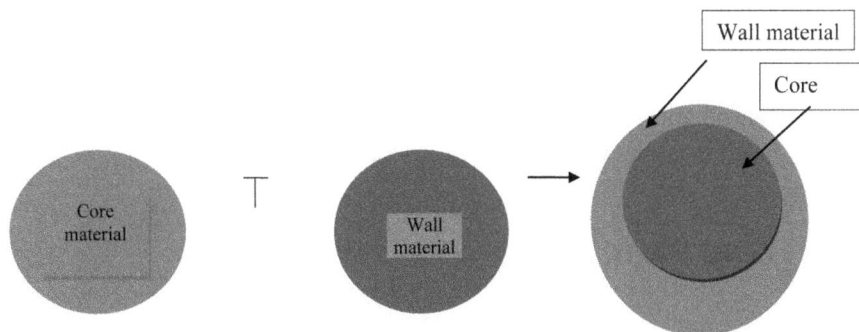

FIGURE 7.1 Mechanism of encapsulation process.

7.3 MICROENCAPSULATION

The natural bioactive compounds from vegetables, fruits, and herbs can be added to various food products to enhance their value so that the pharmaceutical industries can prepare drugs by extracting the anti-cancers, antibacterial, anti-fungal compounds from natural resources (herbs and plants) [21]. However, these compounds are degraded within a very short span of time [32]. Among several techniques, microencapsulation is a technique of enclosing different substances into micro-sized capsules to protect the core material from chemical reactions [40]. This technology was introduced by H.G. Bungenberg de Jong in 1932 [45].

The selection of a microencapsulation method depends on the physical and chemical properties of core material that can be an emulsion, a suspension of solid, a Pickering emulsion or a crystal [7]. The most important step of microencapsulation is to choose a suitable wall material based on its compatibility with the core material (Figure 7.1). The different wall materials are carbohydrates, cellulose, lipids, proteins, polymers, and gum [5].

7.3.1 TYPES OF MICROENCAPSULATION

Microcapsules are particles of 1 to 1000 micron in size. Microcapsules are classified in three types (mononuclear, polynuclear and matrix) based on their morphology [15]:

1. **Mono Nuclear Capsules:** The capsules coated with single wall material and having single core material inside [4, 15].
2. **Poly Nuclear Capsules:** The capsules coated with single wall material and having two or more core material inside [4, 15].
3. **Matrix:** The core material is evenly distributed inside the single wall material. In microencapsulation process, the outer shell for coating can be formed of single-wall material or with the combination of more than one wall material [4, 15].

7.3.2 CORE MATERIAL

The material to be encapsulated is known as core material. It may be in liquid or solid form. The encapsulation of any substance will lead to its stability and controlled delivery at specific sites. Encapsulation of various compounds (Table 7.1) like vitamins, food colorants, drugs, cells, and enzymes can be done [35, 51].

TABLE 7.1 Types of Core Material Along with its Applications

Core Material	Industrial Application	References
Food Sector		
1. Bioactive Compounds: Antioxidants, phenolic compound, polyphenols, carotenoids, probiotic bacteria, vitamins, calcium.	The encapsulation process is used to preserve the sensitive bioactive components and flavors. The encapsulated bioactive components can be used for the preparation of functional foods.	[2, 12, 34, 35, 47, 51]
2. Food Colorants: Anthocyanins, chlorophyll, betacyanin, saffron, turmeric.		
3. Organic acids, Proteins, enzymes.		
4. Sodium bicarbonate.		
5. Citrus oil.		
6. Spices.		
7. Artificial sweeteners		
Agriculture Sector		
Pesticides; fertilizers; agro-chemicals	Microencapsulated fertilizers and pesticides can be used for targeted delivery on specific site to reduce the side effects of pesticides.	[8, 14, 51]

TABLE 7.1   *(Continued)*

Core Material	Industrial Application	References
Pharma Industry		
1. Anticancer and antibacterial drugs.	Microencapsulation is used for its advantage of targeted delivery of drugs and also helps in handling the toxic medicines.	[34, 35, 37, 51]
2. Aspirin.	The outer covering of microcapsule helps in reducing the unpleasant taste and odor of tablets	
3. Bacteria.		
4. Enzymes.		
5. Live bacterial cells.		
6. Live mammalian cells.		
7. Progesterone.		
8. Drugs.		
Cosmetic Sector		
1. Benzoyl peroxide.	Microencapsulation is used to deliver the acids (which are used in cosmetic products such as glycolic acid) on specific site in the skin with this method, we can reduce the harmful effects of acid on the skin.	[4, 37, 43]
2. Glycolic acid.		
3. Linoleic acid (Vitamin F).		
4. Lycopene.		
5. Quercetin.		
6. Rosmarinic acid.		
7. Rutin.		
8. Salicylic acid.		

7.3.2.1   CONTROLLED RELEASE MECHANISM OF CORE MATERIAL

The release mechanism of the core material depends on the compatibility of the core material with the wall material. The important property of the encapsulation method is that the core material should be released at the targeted site on specific time [11]. The controlled release method will increase the effectiveness of the encapsulated product without reacting with other environmental factors. The mechanisms for controlled release are pH, temperature, and use of solvent, degradation, and diffusion [26].

7.3.3 WALL MATERIAL

The wall material should be non-reactive and able to protect the core material from the outer environment, other chemical reactions, and compatible with core material [25, 38]. However, in the food industry, the wall material should be of food-grade based on safety standards [38]. The selection of wall material is the most important step of encapsulation because it will affect the stability and efficiency of encapsulated product [25]. Single material or a combination of two or more wall materials can be used for microencapsulation. Common wall materials can be used in different industries on the basis of product demand (Table 7.2).

TABLE 7.2 Different Wall Materials with Their Applications

Wall Material	Application Sector	References
1. Amylose	Food technology, agriculture and pharmaceutical	[2, 12, 34, 35, 47, 51]
2. Arabic gum		
3. Caseinate		
4. Cellulose		
5. Chitosan		
6. Dextrin		
7. Fatty acids		
8. Fatty alcohols		
9. Gelatine		
10. Gellan		
11. Gluten		
12. Maltodextrin		
13. Pectins		
14. Polydextrin		
15. Polyethylene		
16. Polymers		
17. Polysaccharides		
18. Proteins		
19. Syrups		
20. Wax		
21. Xanthan		

7.3.4 PHYSICAL METHODS OF MICROENCAPSULATION

Microencapsulation process can be performed by different methods. The methods of microencapsulation are classified into three categories: physical methods, chemical methods, and physicochemical methods [11]. The selection of a suitable method of microencapsulation depends on [8, 9, 13, 19, 20]: cost of the process; physical and chemical properties of the core and wall material; required size of microcapsules; and type of core and wall materials. Microencapsulation methods are listed in Figure 7.2.

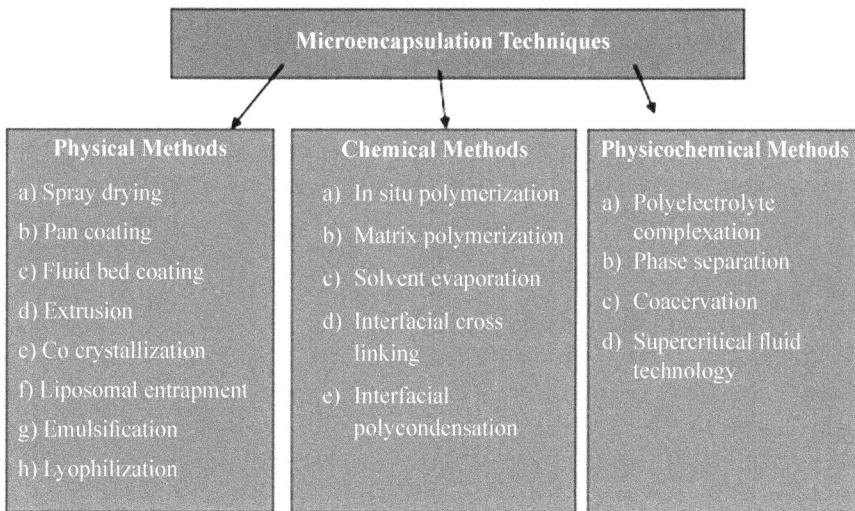

FIGURE 7.2   Different methods of microencapsulation techniques.

7.3.4.1 SPRAY DRYING

The most common physical method of microencapsulation is spray drying that has been used for the encapsulation of pigments, flavors, lipids, drugs, and some acids for cosmetic industries [3, 45, 46]. The microspores formed with this technique are of good quality and low water activity. It is a cost-effective and continuous technique with about 90% applications. However, the disadvantage of this technique is that it is not suitable for thermosensitive components like essential oils due to the high-temperature process [3, 46].

In this process, the first step is the formation of emulsion containing core material and wall material, then the nebulization takes place in a drying chamber with the circulation of hot air, which results into evaporation of water immediately, and the wall material encapsulates the core material [3, 43]. The size of the encapsulated microcapsule will be around 5 micrometers. The size of microcapsules varies depending on the variety of droplets sizes during spraying [38]. For example, lipids have been successfully encapsulated in potato starches, with spray drying [46].

7.3.4.2 AIR-SUSPENSION COATING

This technique provides flexibility and better control of the encapsulated material [46]. The coating of the material is achieved with the solution through suspension in the upward moving air stream. A perforated plate having various patterns of holes inside and outside a cylindrical insert is used for support during a suspension of core material [46].

To fluidize the settling particles only, sufficient air is allowed to rise through the outer annular space. The rising air that flows inside the cylinder causes the rapid rise of the particle. The same cycle repeats when on the top the air stream diverges and slows, the particles settle back onto the outer bed and move downward [34, 51]. This cycle repeats many times within a short period of this method. With the help of the air suspension method, different forms of coatings can be applied on core material, such as aqueous solution, solvent solutions, dispersions, emulsions. The microcapsules can be of micron or submicron in size [45].

7.3.4.3 SPRAY COOLING

In this method, cold air injections are used to solidify the core material. In the first step, the mixture of the core material and wall material is prepared for the atomization by cooled air. The nebulization of this mixture is done by an atomizer, which is followed by passing through the chamber in which airflows [26]. The lower air temperature leads to solidification of the wall material that encapsulates the core material. It is a cost-effective technique due to the use of low temperature compared to other techniques [46]. However, the capacity of encapsulation is low.

The technique is suitable for the encapsulation of frozen liquids, vitamins, minerals, and especially heat-sensitive compounds [3]. The core material is released after the melting of wall material. Spray chilling is also a similar technique. In spray chilling, the temperature will be from 34 to 42°C and on the other hand for spray cooling temperature is higher [3, 45].

7.3.4.4 EXTRUSION

In extrusion method of microencapsulation, the solution of polymer and core material is dropped into the gelling bath. The material can be dropped with different dropping tools, such as spraying nozzle, pipette, or vibrating nozzle. This method is based on the encapsulating the core material in the sodium alginate solution. In addition, dropwise extrusion of core material takes place in this method [26]. The encapsulated material is then broken into small pieces and followed by vacuum drying. Vitamin C and colorants are mainly encapsulated with this technique, which leads to their increased shelf-life due to an impermeable barrier against oxygen. The disadvantage of this technique is the formation of large-size particles, which results in the limited use of this technique in industries [46]. *L. acidophilus* has been encapsulated in a calcium alginate gel through extrusion technique, which leads to the increased survival rate of *L. acidophilus* [45].

7.3.4.5 COACERVATION

In this technique, the polymer deposition is used around the core material by changing the physicochemical properties of the medium, such as, the temperature, pH [46]. The process includes three steps: (1) the liquid manufacturing phase; (2) core material phase; and (3) coating material phase [23]. When a single macromolecule is used, then it is called simple coacervation. However, if more than one macromolecule of opposite charges is used, then it is entitled as a complex coacervation. In this process, we do not require any organic solvent and high temperature [45]. Flavor oils are encapsulated with this method. This technique will take place in limited ranges of pH, colloid concentrations.

7.3.4.6 PAN COATING

Pharmaceutical industries are using this technique for coating the drugs. In this process, the core material is dropped in a pan after that coating material is applied. The microcapsules are of >600 microns in size [23].

7.3.4.7 LYOPHILIZATION

This technique involves dehydration of frozen material under a vacuum sublimation process, which results in the evaporation of water without subjecting the material to high temperature. The major disadvantage of this method is its high processing cost [23].

7.3.4.8 VACUUM AND FREEZE-DRYING

This technique is similar to the drying process. However, it requires low temperature (just above the solvent's freezing point). The disadvantage of this technique is its longer processing time [9].

7.3.4.9 EMULSIFICATION

In the emulsification technique, the organic solvent is used for the dispersion of core material and wall material. The prepared dispersion is subjected to emulsification in water or oil [32]. For the encapsulation, the evaporation of the organic solvent is achieved by continuous stirring those results in the formation of polymer globules in which the core is encapsulated. This technique can be used to encapsulate minerals, enzymes, and microorganisms [52].

7.3.4.10 FLUIDIZED BED TECHNOLOGY

In fluidized bed technology, the wall material evenly coats the core material. In addition, in this technique, all types of wall materials can be used, e.g., proteins, emulsifiers, polysaccharides, fats, powder coatings, enteric coating, etc., [52]. For the proper protection of core material, this technique can be used even after the spray drying of microcapsules. Here waxes, fats, and emulsifiers can be used as wall material [42]. In this technique, a chamber with controlled temperature and humidity high air velocity is used to atomize the wall material. The size of the microcapsules will be in the range of 50 to 500 microns [23].

7.3.4.11 LIPOSOMAL ENTRAPMENT

A lipid bilayer, which entraps more than one liquid compartment, is called a liposome. This technique is used for encapsulating the enzymes, hormones, and vitamins [32]. As this technique includes lipid bilayer, it will lead to more protection to the core material. The outer layer of liposomes is formed of phospholipids [31].

7.3.4.12   CO-CRYSTALLIZATION

In this technique, the core material is encapsulated in the sucrose matrix. To avoid crystallization, the concentrated sucrose syrup is maintained at high temperature [49]. With vigorous mechanical agitation, the core material is added to the syrup. Continuous agitation is required to extend the crystallization until the agglomerates are discharged from the vessel. The encapsulated products are then dried to the desired moisture [45].

7.3.5   CHEMICAL METHODS OF MICROENCAPSULATION

7.3.5.1   SOLVENT EVAPORATION

This technique is mainly used for the encapsulation of drugs in pharmaceutical industry. In this technique, the coating material is used in liquid form and the core material is dissolved in it with continuous agitation to get the encapsulated product of perfect size [49].

7.3.5.2   INTERFACIAL CROSS-LINKING

In this method, Schotten Baumann reaction takes place between acid chloride and hydrogen atom containing the compound. In this reaction, two polymeric reactants meet and they form a thin wall at the microcapsule [31].

7.3.5.3   IN SITU POLYMERIZATION

In this process, the direct polymerization of a single monomer is performed on the surface of the core material [53].

7.4   APPLICATIONS OF MICROENCAPSULATION [24, 34, 50]

- Controlled release of drugs at the targeted sites.

- Encapsulation of pancreatic cells to treat diabetes.
- Immobilization of enzyme or microorganisms.
- The encapsulation of different acids for cosmetic industries.
- To convert liquid into dried forms to make the handling easy.
- To create a barrier between the core material and outer environment.
- To encapsulate the sensitive food ingredients.
- To increase the food value of functional foods.
- To increase the shelf-life of food products.
- To increase the stability of drugs.
- To mask the flavor, odor of the food products.
- To microencapsulate the hepatic cells to treat the liver diseases.
- To protect the food compounds from oxidation and other chemical reactions (Figure 7.3).

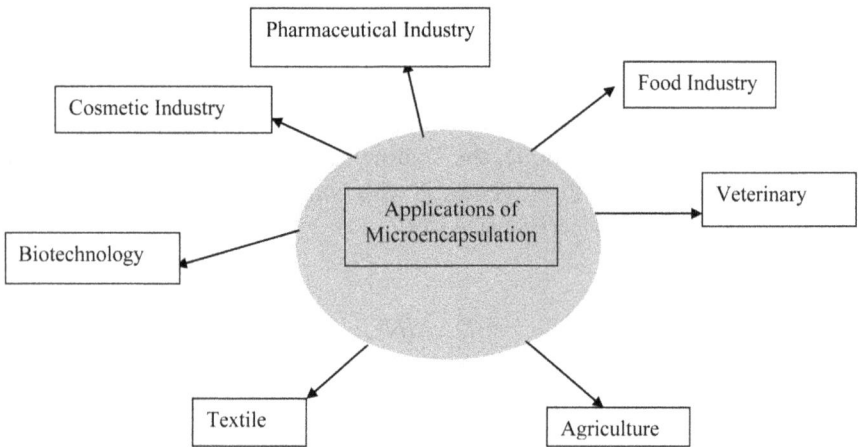

FIGURE 7.3 Applications of microencapsulation technique in different sectors.

7.5 NANOENCAPSULATION

Nanotechnology involves entrapping the particles into nanosize capsules in the range from 1 to 100 nm [36]. In nanoencapsulation, the wall material is called nanocarrier. Nanoencapsulation can be used to encapsulate hydrophilic and lipophilic compounds [18].

Nanoencapsulation is a new opportunity in all sectors for the formulation of different new trends in agriculture and food sectors. The use of nanofertilizer, nanopesticides in agriculture is an innovative step to increase the crop yield by targeted delivery of these agrochemicals [17, 36]. Nanoencapsulation of bioactive compounds, antioxidants, and phenols are examples of the application of nanotechnology in the food sector [41]. Researchers need: (1) to explore the nanoencapsulation techniques to know the effect of wall materials on the nanoparticles; (2) to check whether the nanoparticle is enough to provide the desired results for the prepared product [39].

7.5.1 NANOENCAPSULATION IN AGRICULTURE SECTOR

Nanofertilizer will act on the targeted sites only [41]. Therefore, the new advanced technology nanoencapsulation can provide effective solutions to all issues [1]. To address these issues, farmers can work on-site specific crop management with nanoencapsulation of pesticides and fertilizers. The environmental pollution can be encountered with the help of nanosensors. These improvements in agriculture will be beneficial in resolving the problem of food safety and security [5]. Advantages of nanoencapsulation are:

- Genetically modified plants;
- Improve the seed germination potential;
- Increase the crop yield;
- Protecting the crop from many diseases;
- Sensing environmental stress;
- Site-specific delivery of agrochemicals;
- The enhancement of soil quality;
- The quality of crop is improving.

The wall material of nanoencapsulation is similar to microencapsulation. Only the size distribution of encapsulated material is different [10, 28].

7.6 BIOENCAPSULATION

Bioencapsulation is a technique of encapsulating the particles with the coating of natural polymers, such as, pectin, and agarose. Among all polymers, PEG, and alginate are commonly used polymers. This technique is mostly used in tissue engineering, cell encapsulation and to address the immune rejection of transplanted cells and other body organs [22]. The encapsulation of cells helps in keeping the two immune systems separate and protects the encapsulated cell from host immune response [30]. The preservation of tissue before transplantation is important and sensitive step so that the application of bioencapsulation in cell cryopreservation makes this sensitive step very easy [44].

7.6.1 METHODS OF BIOENCAPSULATION

7.6.1.1 ELECTROSTATIC SPRAY

In this method, the polymer solution droplets are formed on the top of the nozzle, which are then sprayed on the gelling bath. The small micro-capsules are formed due to the electrostatic force between the nozzle and gelling bath [22].

7.6.1.2 MICROFLUIDIC NOZZLE

In this method, monodispersed microcapsules are formed. It is a common method of encapsulation [22].

7.6.1.3 VIBRATIONAL NOZZLE

This method is used in industries for high scale encapsulation. In addition, it is used when the viscosity of the encapsulation solution is low [30].

7.6.1.4 LAMINAR JET BREAKUP

This technique is not commonly used due to its disadvantage of microcap-sules of uneven sizes [22, 30, 44].

7.6.2 APPLICATIONS OF BIOENCAPSULATION [22, 44]

- Bioprinting;
- Bone and tissue engineering;
- Cardiac tissue engineering;
- Cell and tissue cryopreservation;
- Lung tissue engineering;
- Pancreatic and hepatic tissue engineering (Table 7.3).

TABLE 7.3   The Compounds that have been Encapsulated Successfully

Encapsulated Product	References
Encapsulation of anthocyanins	[38, 51]
Encapsulation of chlorophylls	[45]
Encapsulation of anticancer drug	[47]
Encapsulation of ascorbic acid	[4, 13, 15, 34, 37, 51]
Encapsulation of bioactive components	[3]
Encapsulation of biological agents	[32]
Encapsulation of calcium	[21]
Encapsulation of *Camu Camu* juice	[23]
Encapsulation of colorants	[2, 6]
Encapsulation of enzymes	[35, 37]
Encapsulation of essential oils	[45]
Encapsulation of fertilizers	[7, 15, 20, 37]
Encapsulation of hepatic cells to treat liver disease	[47]
Encapsulation of lactobacilli in colon disease	[47]
Encapsulation of microbes for CVD	[2, 47]
Encapsulation of microorganism	[23, 40]
Encapsulation of omega-3 fatty acids	[14]
Encapsulation of pancreatic cells to treat diabetes	[47]
Encapsulation of pesticides	[46]
Encapsulation of polyphenols	[40]
Encapsulation of probiotics	[3]
Encapsulation of protein hydrolysate	[21]
Encapsulation of vitamins	[14]
Microencapsulation of synthetic astaxanthin	[15, 38, 51]

7.7   SUMMARY

The food industry, pharmaceutical industry, cosmetic industry, and agriculture sector are now focusing on encapsulated products, such as fertilizers, pesticides, bioactive compounds (to prepare functional foods), drugs, and other sensitive food products. In this chapter, the authors have discussed different methods of encapsulation with their advantages and disadvantages, and industrial applications of encapsulation.

KEYWORDS

- chorionic diseases
- core material
- encapsulation
- food industry
- microencapsulation
- pharmaceutical industries

REFERENCES

1. Abbas, K. A., Saleh, A. M., Mohamed, A., & Mohd Azhan, N., (2009). The recent advances in the nanotechnology and its applications in food processing: A review. *Journal of Food, Agriculture and Environment, 7*(3/4), 14–17.
2. Agnihotri, N., Mishra, R., & Goda, C., (2012). Microencapsulation: A novel approach in drug delivery: A review. *Indo Global Journal of Pharmaceutical Sciences, 2*(1), 1–20.
3. Aissa, A. F., & Bianchi, M. L. P., (2012). Comparative study of β-carotene and microencapsulated β-carotene: Evaluation of their genotoxic and antigenotoxic effects. *Food and Chemical Toxicology, 50*(5), 1418–1424.
4. Anandaraman, S., & Reineccius, G. A., (1986). Stability of encapsulated orange peel oil. *Journal of Food Technology, 40*(11), 88–93.
5. Anandharamakrishnan, C., (2014). *Techniques for Nanoencapsulation of Food Ingredients* (p. 89). New York: Springer Briefs in Food, Heath, and Nutrition.
6. Arimoto, M., Ichikawa, H., & Fukumori, Y., (2014). Microencapsulation of water-soluble macromolecules with acrylic terpolymers by the Wurster coating process for colon-specific drug delivery. *Powder Technology, 141*, 177–186.
7. Aziz, A., Rabani, F., Jai, J., Raslan, R., & Subuki, I., (2015). Microencapsulation of essential oils application in textile: A review. *Advanced Materials Research, 1113*, 346–351.
8. Bansode, S. S., Banarjee, S., & Gaikwad, D. D., (2010). Microencapsulation: A review. *International Journal of Pharmaceutical Science Review and Research, 1*(2), 38–43.
9. Champagne, C. P., & Fustier, P., (2007). Microencapsulation for the improved delivery of bioactive compounds into foods. *Current Opinion in Biotechnology, 18*(2), 184–190.
10. Chaudhry, Q., Scotter, M., & Blackburn, J., (2008). Applications and implications of nanotechnologies for the food sector. *Food Additives and Contaminants, 25*(3), 241–258.
11. Costa, S. S., Machado, B. A. S., & Martin, A. R., (2015). Drying by spray drying in the food industry: Micro-encapsulation, process parameters and main carriers. *African Journal of Food Science, 9*(9), 462–470.
12. De, Z. T. J., (1995). Food ingredient encapsulation. In: *Encapsulation and Controlled Release of Food Ingredients* (pp. 113–131). Philadelphia, USA: American Chemical Society Symposium Series 590.

13. Dewettinck, K., & Huyghebaert, A., (1999). Fluidized bed coating in food technology. *Trends in Food Science and Technology, 10,* 163–168.

14. Dias, M. I., Ferreira, I. C., & Barreiro, M. F., (2015). Microencapsulation of bioactives for food applications. *Food Function, 6,* 1035–1052.

15. Dziezak, J. D., (1988). Microencapsulation and encapsulated ingredients. *Journal of Food Technology, 42*(4), 136–151.

16. Ersus, S., & Yurdagel, U., (2007). Microencapsulation of anthocyanin pigments of black carrot (*Daucus carota* L.) by spray drier. *Journal of Food Engineering, 80,* 805–812.

17. Fathi, M., Martín, A., & McClements, D. J., (2014). Nanoencapsulation of food ingredients using carbohydrate-based delivery systems. *Trends Food Science Technology, 39,* 18–39.

18. Ghaderi-Ghahfarokhi, M., & Barzegar, M., (2016). Nanoencapsulation approach to improve antimicrobial and antioxidant activity of thyme essential oil in beef burgers during refrigerated storage. *Journal of Food and Bioprocess Technology, 9,* 1187–1201.

19. Gharsallaoui, A., Roudaut, G., Chambin, O., Voilley, A., & Saurel, R., (2007). Applications of spray-drying in microencapsulation of food ingredients: An overview. *International Food Research Journal, 40,* 1107–1121.

20. Gibbs, B. F., Kermasha, S., Alli, I., & Mulligan, C. N., (1999). Encapsulation in the food industry: A review. *International Journal of Food Science and Nutrition, 50,* 213–224.

21. Giunchedi, P., & Conte, U., (1995). Spray-drying as a preparation method of microparticulate drug delivery systems: An overview. *International Journal of Pharma, 5,* 276–290.

22. Gurruchaga, H., Saenz, D. B. L., & Ciriza, J., (2015). Advances in cell encapsulation technology and its application in drug delivery. *Expert Opinion Drug Delivery, 12*(8), 1251–1267.

23. Hussain, S. A., & Abdullah, N., (2018). Review on micro-encapsulation with chitosan for pharmaceuticals applications. *MOJ Current Research and Reviews, 1*(2), 77–84.

24. Hussein, A. M. S., & Kamil, M. M., (2017). Influence of nanoencapsulation on chemical composition, antioxidant activity, and thermal stability of rosemary essential oil. *American Journal of Food Technology, 12,* 170–177.

25. Jafari, S. M., Assadpoor, E., He, Y., & Bhandari, B., (2008). Encapsulation efficiency of food flavors and oils during spray drying. *Journal of Drying Technology, 26*(7), 816–835.

26. Jeyakumari, A., & Zynudheen, A. A., (2016). Microencapsulation of bioactive food ingredients and controlled release: Review. *Journal of Food Processing and Technology, 2*(6), 214–224.

27. Jimenez, M., Garcia, H., & Beristain, C., (2004). Spray drying microencapsulation and oxidative stability of conjugated linoleic acid. *European Food Research and Technology, 219,* 588–592.

28. John, R. P., Tyagi, R. D., & Brar, S. K., (2011). Bio-encapsulation of microbial cells for targeted agricultural delivery. *Critical Reviews in Biotechnology, 31*(3), 211–226.

29. Junyaprasert, V. B., Mitrevej, A., Sinchaipanid, N., Broome, P., & Wurster, D. E., (2001). Effect of process variables on the microencapsulation of vitamin a palmitate by gelatin-acacia coacervation. *Journal of Drug Development and Industrial Pharmacy, 27*(6), 561–566.

30. Kang, A., Park, J., & Ju, J., (2014). Cell encapsulation via microtechnologies. *Biomaterials, 35*(9), 2651–2663.

31. Kirby, C. J., Whittle, C. J., Rigby, N., Coxon, D. T., & Law, B. A., (1991). Stabilization of ascorbic acid by microencapsulation in liposomes. *International Journal of Food Science and Technology, 26*(5), 437–449.

32. Kumar, S., Nakka, S., & Rajabalaya, R., (2011). Microencapsulation techniques and its practices. *International Journal of Pharmaceutical Science Technology, 6*, 1–23.

33. Loksuwan, J., (2007). Characteristics of microencapsulated β-carotene formed by spray drying with modified tapioca starch, native tapioca starch, and maltodextrin. *Food Hydrocolloids, 21*, 928–935.

34. Madene, A., Jacquot, M., Scher, J., & Desobry, S., (2006). Flavor encapsulation and controlled release: A review. *International Journal of Food Science and Technology, 41*(1), 1–21.

35. Manojlovic, V., Rajic, N., & Djonlagic, J., (2008). Application of electrostatic extrusion flavor encapsulation and controlled release. *Journal of Sensors, 8*, 1488–1496.

36. Maryam, I., Huzaifa, U., Hindatu, H., & Zubaida, S., (2015). Nanoencapsulation of essential oils with enhanced antimicrobial activity: A new way of combating antimicrobial resistance. *Journal of Pharmacognosy Phytochemistry, 4*, 165–170.

37. Mellema, M., Van, B. A. J., Boer, B., Von, H. J., & Visser, A., (2006). Wax encapsulation of water-soluble compounds for application in foods. *Journal of Microencapsulation, 23*, 729–740.

38. Nunes, I. L., & Mercadante, A. Z., (2007). Encapsulation of lycopene using spray-drying and molecular inclusion processes. *Journal of Food Science and Technology, 50*(5), 893–900.

39. Paredes, A. J., Asensio, C. M., & Llabot, J. M., (2016). Nanoencapsulation in the food industry: Manufacture, applications and characterization. *Journal of Food Bioengineering and Nanoprocessing, 1*(1), 36–44.

40. Peanparkdee, M., & Iwamoto, S., (2016). Microencapsulation of bioactive compounds from mulberry (*Morus alba L.*) leaf extracts by protein-polysaccharide interactions. *International Journal of Food Science and Technology, 51*, 531–536.

41. Rigo, L. A., & Silva, C. R., (2015). Nanoencapsulation of rice bran oil increases its protective effects against UVB radiation-induced skin injury in mice. *European Journal of Pharmaceutics and Biopharmaceutics, 93*, 11–17.

42. Rodriguez-Huezo, M., & Pedroza-Islas, R., (2004). Microencapsulation by spray drying of multiple emulsions containing carotenoids. *Journal of Food Science, 69*, 351–359.

43. Saénz, C., Tapia, S., Chávez, J., & Robert, P., (2009). Microencapsulation by spray drying of bioactive compounds from cactus pear (*Opuntia ficus-indica*). *Food Chemistry, 114*(2), 616–622.

44. Schoebitz, M., López, M. D., & Roldán, A., (2013). Bioencapsulation of microbial inoculants for better soil-plant fertilization: Review. *Agronomy for Sustainable Development, 33*(4), 751–765.

45. Shah, N. P., & Ravula, R. R., (2000). Microencapsulation of probiotic bacteria and their survival in frozen fermented dairy desserts. *The Australian Journal of Dairy Technology, 55*(3), 139–144.

46. Silva, P. T. D., & Fries, L. L. M., (2014). Microencapsulation: Concepts, mechanisms, methods, and some applications in food technology. *Journal of Food Technology, 44*(7), 1304–1311.

47. Tomaro-Duchesneau, C., & Saha, S., (2013). Microencapsulation for the therapeutic delivery of drugs, live mammalian and bacterial cells, and other biopharmaceutics:

Current status and future directions. *Journal of Pharmaceutics, 2013*, 9. article ID: 103527.

48. Trindade, M. A., & Grosso, C. R. F., (2000). The stability of ascorbic acid microencapsulated in granules of rice starch and in gum Arabic. *Journal of Microencapsulation, 17*(2), 169–176.

49. Uddin, M. S., Hawlader, M. N. A., & Zhu, H. J., (2001). Microencapsulation of ascorbic acid: Effect of process variables on product characteristics. *Journal of Microencapsulation, 18*(2), 199–209.

50. Wilson, N., & Shah, N. P., (2007). Microencapsulation of vitamins: A review. *Asian Food Journal, 14*(1), 1–14.

51. Wischke, C., & Schwendeman, S. P., (2008). Principles of encapsulating hydrophobic drugs in PLA/PLGA, microparticles. *International Journal of Pharmaceutics, 364*(2), 298–327.

52. Yeo, Y., Baek, N., & Park, K., (2001). Microencapsulation methods for delivery of protein drugs. *Biotechnology and Bioprocess Engineering, 6*, 213–230.

53. Zuidam, N. J., & Shimoni, E., (2010). Overview of microencapsulates for use in food products or processes and methods to make them. In: *Encapsulation Technologies for Active Food Ingredients and Food Processing* (pp. 3–29). Dordrecht, The Netherlands: Springer.

CHAPTER 8

OZONATION: POTENTIAL APPLICATIONS IN OILSEEDS STORAGE

GURJEET KAUR, GAGANDEEP KAUR SIDHU, JASHANDEEP SINGH, and PREETI BIRWAL

ABSTRACT

Oilseeds (groundnut, rapeseed, safflower, sesame, and soybean) need proper care and handling during storage. Several techniques have been adopted to reduce the deterioration of oilseeds, such as the surrounding environmental conditions, controlling gaseous composition, modified atmosphere packaging (MAP), vacuum packaging, and ozonation. Ozone alone or in combination with other treatments has a promising role in maintaining the stability of food crops. The factors related to ozonation to enhance the shelf-life of oilseeds are discussed in this chapter.

8.1 INTRODUCTION

The oilseeds (such as rapeseed, mustard, groundnut, sunflower, soybean, sesame, nigger, castor, and linseed) are a rich source of food, energy, feed, and commerce. It is the second-largest food commodity, which is valued at about 5% of GNP (gross national product) in India. Most of the oilseeds consist of 15–30% of protein and 20–50% of fats. The oil that is extracted from the oilseeds is vastly used for cooking, such as groundnut oil, sunflower oil, sesame oil, etc. Vegetable oil is a good source of Vitamin E and fat.

This chapter elaborates on factors affecting the storability of oilseeds and techniques adopted to overcome issues on quality deterioration during storage. This chapter also explores the potential of ozonation technology in the context of maintaining the shelf-life of oilseeds.

8.2 FACTORS AFFECTING STORAGE OF OILSEEDS

The storage of seeds depends on physicochemical factors, such as relative humidity (RH) of the surrounding air, temperature, moisture content, and chemical composition of the seeds. The quality of the seeds depends on vigor and viability. The vigor consists of those properties, which determine the potential for uniform development of normal seedlings, while the viability refers to the degree to which seeds contain the enzymes that are metabolically active for the reactions involved in the germination and growth of seedlings.

The type of packaging material and storage structure also affects the quality of seeds for storing for a longer period. The quality deterioration may include the oxidation of seed reserves (carbohydrates, lipids, proteins, vitamins, and aroma components), off-flavor, mold growth, insect attack, etc. Therefore, the moisture content of the seeds should be in an appropriate range before storing it for a longer period.

8.3 TRADITIONAL/CONVENTIONAL METHODS OF OILSEED STORAGE

The conventional method of storing the oilseeds is to store them in bags or bulk. The storage of the seeds at the farm level is done in jute bags but most of the oilseeds are kept for sowing purpose and rest of the harvested one are sold in the market in less than 2 months [56]. The controlled atmosphere (CA) is required not only from harvesting to storage but also during storage for the longer storage. The soybean seeds stored in 500 gauge polyethylene bags at 10% moisture content for 12 months were better than in gunny bags [11]. The groundnut seeds stored in the gunny bags lost their viability to nil after 9 months of storage, while there was retention in the viability of the seeds that were stored in the polyethylene bags [42].

The effect of jute bags, PE lined jute bags, metal bins, and heaping the groundnut pods over a storage period of 6 months was examined for insect and fungal infection, oil content, and protein content. The satisfactory results were observed in the polyethylene lined jute bags and metal bins with minimum effect on the parameters under evaluation [44]. Another research suggested that the untreated groundnut seeds retain their viability for 9 months and 15 months in cloth bags and polyethylene bags, respectively [20], while the soybean seeds showed satisfactory germination up to 4 months of storage in cloth bags [47].

8.4   INNOVATIVE METHODS OF OILSEED STORAGE

In order to maintain low oxygen content inside the polyethylene bags, the vacuum packaging and gas flushing have been used to prevent the oxidation of seeds. Although gas flushing or vacuum packaging is used to extend the shelf-life of the commodity, yet aerobic spoilage can occur due to the residual oxygen left in the headspace of the packaged product. This can be overtaken by reducing the moisture content as revealed in one study that storing the groundnut seeds under sealed vacuum packages at low moisture content (1.7–3.5%) at 35°C resulted in viable storage of 288 weeks [45].

In another study, groundnut seeds were packaged in gunny bags, high-density polyethylene (HDPE) bags and vacuum-packed bags and were stored under ambient (25 ± 2C) and cold storage (4 ± 1°C) conditions for 18 months. The results revealed that the seeds packed in vacuum bags maintained better seed quality with respect to germination percent, seedling vigor index, seedling dry weight, and root length compared to seeds stored in polyethylene and gunny bags [36].

In case of oilseeds, few studies have been reported on the effect of ozonation, temperature, and CO_2. The combined effect of these parameters on seeds quality of rapeseed and barley resulted in the decrease in growth and production [9]. The impact of ozone on food quality could be positive or negative depending on crop, specific quality parameter, etc., [54].

8.4.1   OZONATION

Pure ozone (O_3) gas is pale blue and the liquid is blue color, having pungent and characteristic aroma [22]. The density of ozone gas is 2.14 g of liquid at 0°C and 101.3 kPa [57]. Ozone is more stable in the gaseous than in the aqueous phase [51]. Stability of ozone in aqueous state is influenced by its concentration, pH, UV radiation exposure, the existence of radical scavengers [32, 53], temperature [46] and the presence of organic matter and metal ions.

8.4.2   METHODS OF PRODUCTION OF OZONE

Ozone in gaseous form is produced in small amounts from lighting or energetic UV radiations acting on oxygen naturally in the stratosphere [26].

Ozone production starts once a diatomic oxygen molecule splits. This splitting action requires a huge amount of energy. The formed free radical oxygen can freely react with other diatomic oxygen and lead to the formation of a triatomic ozone molecule, O_3 [6, 18].

However, a small fraction of ozone is also produced in the troposphere due to released end products from forest fires, vehicle exhausts, industrial emissions, and volcano eruptions, such as hydrocarbons, oxygen, and nitrogen. The produced gas is not stable and readily dissociates in the atmosphere [31].

For industrial applications, ozone production occurs in a closed system [18]. Two widely used methods for artificial generation of ozone are Ultraviolet (UV) radiation and corona discharge. At commercial levels, the most common method used is the corona discharge method [6, 13, 18, 22], because of its relatively low cost and high ozone production efficiency compared to systems with UV radiation.

8.4.2.1 WORKING PRINCIPLE OF OZONATION

The system consists of a filter for purifying ambient air, a fan for conveying purified air, two electrodes to provide corona discharge process, a steel shielding, and an outlet for generated ozone. Corona or UV systems trigger the occurrence of free radical oxygen, which is the primary step of ozone production. UV lamps give ozone when light of a wavelength within a range of 185 to 254 nm strikes on an oxygen molecule. In corona discharge, high voltage of ozone generator produce arcs (Figure 8.1).

Ambient air is extracted in the generator through filters. The purified air is then conveyed in the cavity by using fans. As air comes in contact with the discharge of the system, they split the oxygen molecule from the ambient air supply into atoms, which later on join to form ozone. The generated ozone is then transferred to the container having food samples. For proper treatment, a diffuser must be used to ensure uniform supply of ozone to each and every part of the food sample.

In the electrochemical method for ozone production, electrolysis of water leads to separate water into hydrogen and oxygen atoms. Then from this gas and water mixture, hydrogen molecules are collected and remaining oxygen atoms join and produce ozone and oxygen. The concentration of produced ozone is three to four times higher (10–18%) compared to ozone from the corona discharge method [31].

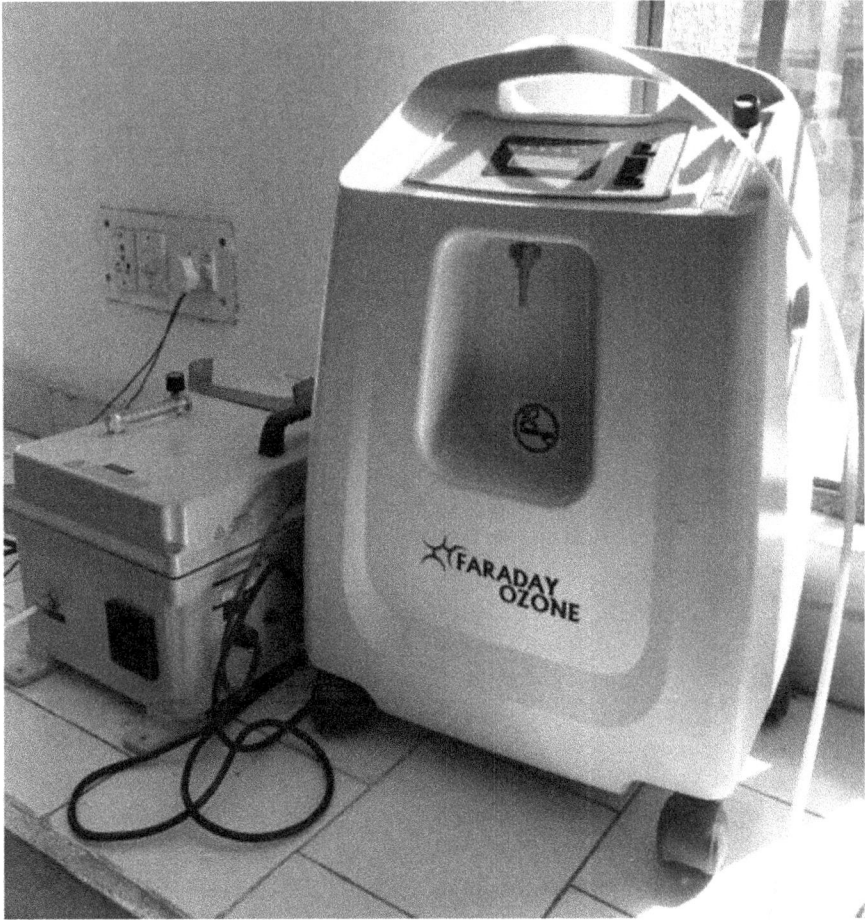

FIGURE 8.1 Ozone generator.

8.4.3 DESTRUCTIVE MECHANISM OF OZONE TREATMENT

Ozone acts as a potential oxidizing agent and eliminates disease-causing microorganisms due to its antimicrobial properties without leaving any residue in the treated food. Microorganisms can be inactivated by treating foods with low concentrations of ozone for the limited period [33].

As microorganisms are exposed to ozone, destruction occurs due to the oxidation of cellular constituents. The main focus of ozonation is the surface of bacterial cell. Bacterial destruction results from two basic mechanisms: (1) oxidation of sulphydryl groups and amino acids of enzymes, peptides, and proteins to smaller peptides; (2) oxidation of polyunsaturated fatty acids to acid peroxides. As unsaturated lipids present in bacterial cell membrane are destructed by ozone, the inner cellular material gets oozed out. Unsaturated lipids are particularly prone to the ozone attack [32]. Ozonation causes effective destruction and damage of genetic material, which ultimately leads to the death of bacterial cell [22]. It also harms the intracellular enzymes and lipopolysaccharides coating of Gram-negative bacteria [58].

8.5 APPLICATIONS OF OZONATION

Ozone is efficient for disinfection in water treatment. This holds second position after fluorine in the oxidizing power [13, 18]. There are other applications of ozone in the industry, such as, sanitation of equipments in food plant and treatment of wastewater. In a research study, ozone was used in a CIP (cleaning in place) system of a wine industry [19]. The treatments were ozonized water at $28 \pm 1°C$, hot water, peracetic acid, and caustic soda solution with peracetic acid. Observations postulated that ozonized water was more efficient than the single use of peracetic acid and the combination of soda and peracetic acid. Ozone can be applied by using various techniques as follows:

- In the gaseous state: it can be discharged at low levels in the cold stores, coolers used for meat product storage and directly in food containers;
- In the liquid state: dissolved ozone in water can reduce the microbial load on fruits, vegetables, meat, and poultry by washing;
- In solid-state: ozone sterilized ice cubes can be used in the storage of perishable products.

8.5.1 APPLICATIONS OF OZONE FOR SHELF-LIFE EXTENSION OF FOOD PRODUCTS

Ozone has an oxidation-reduction potential of 2.07 V [5], i.e., it is the strongest oxidant applicable in food applications [22]. Ozone is one of the non-thermal technologies for food preservation, which leads to give safe foods without any harm to the quality of foods. In addition, ozone application in food products leaves no residues in treated foods, as its decomposition rate is very high [21]. Table 8.1 shows the current applications of ozone in food industries.

TABLE 8.1 Recent Applications of Ozone in Food Processing

Product	Dose	Effect	References
Cantaloupe	15.008 mg m^{-3} for 1 h	Decline in respiration rate, microbial growth, high levels of firmness, pectin, etc., in sarcocarp and exocarp	[7]
Chestnut	800 and 1,600 ppm for 30 min	Reduction of 1.6 logs in mesophilic bacteria and 1.0 in fungi	[55]
Coconut water	0.075 to 0.37 mg/mL	Fully inactivation of peroxidase activity without affecting phenolic content	[43]
Grape tomatoes	800 and 1600 ppm	Reduction in Salmonella populations below 0.6 log CFU/fruit	[14]
Papaya	1.5 ppm ozone for 4 h	Anthracnose severity reduced resulting in extending shelf-life up to 7 days	[40]
Satsuma mandarin	2.5 µg L^{-1} for 24 h	Postharvest disease control and shelf-life extension	[61]
Smoke sharks	1.3 mg L^{-1} for 5, 10, 15, and 20 min	Sensory quality was improved. Additionally, urea, and microbe count were decreased. Optimal results were found for treatment with 15 minutes due to 3 log cycles decrease in microbial population	[52]
Strawberry	0.1 ppm for 1–4 min	2 min ozone treatment cause 21% decay in physiological loss in weight, firmness increased to 19%, 46% in contrast to control during storage of 14 days	[39]
Summer squash	0.2, 0.4 and 0.6 ppm 30 min	Controlled decay at 0.6 ppm, total bacteria inactivated to 3.92 log CFU/g and yeast and mold to 3.65 log CFU/g	[27]
Whole and skim milk	28 mgL^{-1} for 5, 10, 15 min	Based upon time and composition of milk, ozone elevated reduction in microbial load of *Pseudomonas*	[37]
Winter jujube	1.5, 2, 2.5, and 3 mg/L ozone in water for 5 min	Color change was limited, firmness, ascorbic acid, titratable acidity, total soluble solids were maintained to a large extent	[35]

8.5.1.1 FRUITS AND VEGETABLES

Ozone shows promising results for surface treatment for sanitization of fruits and vegetables and for processing water in food manufacturing units [16, 23, 38]. For this purpose, ozone can be used in two ways: either in gaseous form or liquid form (ozonated water). Various studies have been documented for ozone application in tomatoes, carrots, apples, kiwifruit, pears, broccoli, cucumber, berries, etc., [15, 17, 48, 59].

8.5.1.2   MEAT PRODUCTS

The temperature, pH, humidity, RH, and microbial activities affect the meat quality. For chicken meat, levels of *Escherichia coli* bacteria at <10 CFU/g and total mesophyll aerobic bacteria <10^7 CFU/g are permissible. Any deviations from these quality acceptable limits indicate that the chicken meat has been deteriorated [29]. Ozone treatment is a promising application in processing and storage of meat. Ozone treatment also increases water holding capacity and cooking yield of turkey samples, which improve the technological characteristics [2]. It was concluded that with ozone concentrations of 3 ppm in the water for the time duration of 10 min and stored at 2–7°C with ozonation could enhance the shelf-life of chicken meat up to 10th day [29].

8.5.1.3   CEREALS

Ozone is being used as a fumigant in storage of cereals. Flours made from soft and hard wheat showed no negative effect of ozone tempering at a dose of 11.5 mg ozone/L on physicochemical and rheological properties. Ozone treatment also lowered the microbial count to significant levels [24]. Ozone also affects the mechanical characteristics of cereal grains. The treatment of wheat grains before milling resulted in 10–20% decline in energy required for breaking during milling without any effect on yield of wheat flour [12].

8.5.1.4   SPICES

To avoid microbial contamination in herbal and aromatic plants, ozone has been successfully applied. Detoxification of aflatoxin B$_1$ in red pepper influenced by ozone treatment at different concentrations was evaluated. The reductions in levels of aflatoxin B$_1$ in flaked and chopped red peppers were 80% and 93%, respectively [25]. In another study, treatment of thyme (*Thymus vulgaris*), lemon verbena (*Aloysia triphyla syn. Lippia citriodora*), oregano (*Origanum vulgare ssp. hirtum*), mountain tea (*Sideritis raeseri*), and chamomile (*Matricaria chamomilla*) with gaseous ozone was used as an alternative method of disinfection [30].

8.6   USE OF COMBINED PRESERVATION TECHNIQUE

In the case of ozone, foods can be treated first with ozone and then packaged and/or stored, or foods processed using several methods, and packaged in material treated with ozone. Ozone due to its versatile nature can affect the properties of packaging films.

As the polymers are treated with ozone, it degrades polymer chains by forming oxygen-containing functional groups. It leads to alterations in physical (surface tension), mechanical (tensile strength) and barrier properties (water vapor transmission rate, oxygen transmission rate) of packaging films. Polarity and surface tension are most common. Mechanical properties of polyethylene films were changed to a large extent when treated with ozone at higher temperature. However, barrier properties of both polyethylene and polyamide polymers were altered to significant levels when exposed to ozone. Ozone treatment showed significant increase adhesion properties due to an increase in surface tension and hydrophobicity of polyethylene (PE), polypropylene (PP), and polyethylene terephthalate (PET) macromolecules [41].

In a study to evaluate variation in diffusion rates of common plastic films having different ventilation areas, the ozone diffusion followed the trend, such as: high-density polyethylene > polypropylene > low-density polyethylene (LDPE), with non-significant differences. Ozone diffusion was increased at a moderate rate with the increase in ventilation area [28].

8.6.1   OZONE TREATMENT WITH MODIFIED ATMOSPHERE PACKAGING (MAP)

In modified atmosphere packaging (MAP) of foods, concentration of O_2 and CO_2 is modified to inhibit microbial outbreak and extend the shelf-life by avoiding decay and physiological changes in foods [1, 49, 60]. In a study, semi-dried buckwheat noodle was treated with aqueous ozone (2.21 ppm), which postulated 47% decline of initial total plate counts. Furthermore, the samples in MAP with 70% CO_2 was able to arrest the microbial growth. This combination of ozone and MAP retained the overall quality during the storage period of 9 days [3].

Quality evaluation of asparagus was done using chlorinated and ozonated water to lower the microbial load on fresh green asparagus and packaged in MAP for storage of 23 days at 4°C. No significant differences in the amount of Escherichia coli contamination among washing methods were found [49]. The postharvest quality of chilies delays in senescence and firmness during storage was better retained by ozone (30 mg) treatment compared to chlorine-treatment and untreated samples. Additional positive effect on quality was shown by passive modified atmospheric packaging and prolonged shelf-life of chilies stored at (8 ± 1°C) [8].

8.6.2 OZONE AND CONTROLLED ATMOSPHERE (CA) STORAGE

Ozone at low concentration along with adequate low temperature can maintain product quality throughout the transportation. To study the growth of *Listeria monocytogenes*, blueberries were stored at 4 or 12°C under different CA conditions (5% O_2, 15% CO_2, 80% N_2) or ozone gas (O_3) 4 ppm at 4°C or 2.5 ppm at 12°C for 10 days inside a plastic clamshell. The 3 and 2-log reductions were observed in case of ozone compared to air treatment at 4 and 12°C, respectively [10].

In addition, survival of *Salmonella enteritidis* in cherry tomatoes was analyzed during passive MAP, CA, and the results were compared with samples stored in the air at 7 and 22°C. It was found that ozone can be used as a sanitizer for the surface as gaseous ozone treatment has a bactericidal effect on *S. enteritidis*. Ozone gas treatment at 10 mg/l was found to be effective [11].

8.6.3 PACKAGING WITH OZONE GENERATION AND MONITORING

Under the in-package treatments, the ozone is generated inside the package with the advent of electrodes called PK-1 system. This was capable of reducing *E. coli* O157:H7 on prepackaged, ready-to-eat (RTE) spinach leaves and its reduction rate was increased with increased storage period. Greater reductions of the pathogen were recorded when spinach was stored in oxygen gas as opposed to air. Ozone production levels were higher in the initial stages when produced by using oxygen compared to air and then decreased with time [34]. A sensor device senses ozone and UV radiation energy inside of the storage space. The attached sensor transmits information to the processor [21].

8.6.4 OTHER METHODS

SAMRO-AG, Switzerland, a plant with a ventafresh technology, is being used to clean, disinfect, and storage of potatoes. Raw potatoes are first washed with ozone water and then potatoes move through a tunnel, where UV radiation is applied to continue disinfection and remove ozone residues. Then these potatoes are stored in a clean warehouse under high humidity conditions and exposure of ozone at low concentrations. The process is suited to all root vegetables and also to fruits, such as apples, pears, tomatoes [50].

8.6.5   POTENTIAL APPLICATIONS OF OZONE IN OILSEEDS

Various studies have been documented for ozone application in fruits, vegetables, grains, spices, and meat and poultry products. Its application in oilseeds is still not vast. Oilseeds are prone to deterioration in quality due to outbreak of mycotoxins. Ozone prevents microbial contamination due to oxidation of amino acids, glycoproteins, etc., and cellular material present in the membrane. Due to its potential for controlling fungi, ozonation can be considered as a promising technique in handling of oilseeds.

8.6.5.1   ADVANTAGES

Various benefits of ozone technique are cost-effective, eco-friendly, non-residual effect, substitute to chlorine, highly efficient than chlorine to control microbial growth, easy to apply.

8.6.5.2   LIMITATIONS

Although it is considered as a safe technology, yet exposure to ozone for a long time can cause irritation in the throat and nose, headache, difficulty in breathing, and can cause corrosion of equipments at high doses.

8.7   SUMMARY

Common oilseeds are mainly mustard, rapeseed, groundnut, soybean, sunflower, sesame, castor, and linseed. Vigor and viability of seeds define the quality of oilseeds. RH of the surrounding air, temperature, moisture content of the seeds, chemical composition of the seeds, packaging material, and storage structure affect the long-term storability of oilseeds. The ozonation helps in extending the shelf-life of food products and has potential in the postharvest handling of oilseeds.

KEYWORDS

- gross national product
- modified atmosphere packaging
- oilseeds
- ozonation
- ozone gas
- shelf-life

REFERENCES

1. Alexander, E. M., & Brandão, T. R., (2011). Modeling microbial load reduction in foods due to ozone impact. *Procedia Food Science, 1*, 836–841.
2. Ayranci, U. G., Ozunlu, O., Ergezer, H., & Karaca, H., (2019). Effects of ozone treatment on microbiological quality and physicochemical properties of turkey breast meat. *Ozone: Science and Engineering, 2019*, 1–9.
3. Bai, Y. P., Guo, X. N., Zhu, K. X., & Zhou, H. M., (2017). Shelf-life extension of semi-dried buckwheat noodles by the combination of aqueous ozone treatment and modified atmosphere packaging. *Food Chemistry, 237*, 553–560.
4. Bhattacharya, K., & Raha, S., (2002). Deteriorative changes of maize, groundnut and soybean seeds by fungi in storage. *Mycopathologia, 155*, 135–141.
5. Brady, J. E., & Humiston, G. E., (1978). *General Chemistry Principles and Structure* (2nd edn., p. 314). New York, John Wiley & Sons.
6. Bocci, V., (2006). Is it true that ozone is always toxic? The end of a dogma. *Toxicology and Applied Pharmacology, 216*(3), 493–504.
7. Chen, C., Zhang, H., Zhang, X., Dong, C., Xue, W., & Xu, W., (2020). The effect of different doses of ozone treatments on the postharvest quality and biodiversity of cantaloupes. *Postharvest Biology and Technology, 163*, 1–10.
8. Chitravathi, K., Chauhan, O. P., Raju, P. S., & Madhukar, N., (2015). Efficacy of aqueous ozone and chlorine in combination with passive modified atmosphere packaging on the postharvest shelf-life extension of green chilies (*Capsicum annuum L.*). *Food Bioprocessing Technology, 8*, 1386–1392.
9. Clausen, S. K., Frenck, G., & Linden, L. G., (2011). Effects of single and multifactor treatments with elevated temperature, CO_2 and ozone on oilseed (Rape and Barley). *Journal of Agronomy and Crop Science, 197*(6), 442–453.
10. Concha-meyer, A., Eifert, J., Williams, R., Marcy, J., & Welbaum, G., (2014). Survival of *Listeria monocytogenes* on fresh blueberries (*Vaccinium corymbosum*) stored under controlled atmosphere and ozone. *Journal of Food Protection, 77*(5), 832–836.
11. Dadlani, M., & Veena, S., (2006). Effect of packaging on vigor and viability of soybean (*Glycine max* L. Merrill) seed during ambient storage. *Seed Research, 31*(1), 27–32.

12. Desvignes, C., & Chaurand, M., (2008). Changes in common wheat grain milling behavior and tissue mechanical properties following ozone treatment. *Journal of Cereal Science, 47*(2), 245–251.

13. Duguet, J. P., (2004). Basic concepts of industrial engineering for the design of new ozonation processes. *Ozone News, 32*(6), 15–19.

14. Fan, X., Sokorai, K. J. B., & Gurtler, J. B., (2020). Advanced oxidation process for the inactivation of salmonella typhimurium on tomatoes by combination of gaseous ozone and aerosolized hydrogen peroxide. *International Journal of Food Microbiology, 312*, 1–6.

15. Fonteles, T., Do Nascimento, R., Rodrigues, S., & Fernandes, F., (2018). Effects of ozone pretreatment on drying kinetics and quality of granny smith apple dried in a fluidized bed dryer. In: *21st International Drying Symposium Proceedings* (pp. 789–794).

16. Garcia, A., Mount, J. R., & Davidson, P. M., (2003). Ozone and chlorine treatment of minimally processed lettuce. *Journal of Food Science, 68*, 2747–2751.

17. Goffi, V., Magri, A., Botondi, R., & Petriccione, M., (2019). Response of antioxidant system to postharvest ozone treatment in Soreli kiwifruit. *Journal of Science of Food and Agriculture, 100*, 961–968.

18. Goncalves, A. A., (2009). Ozone: An emerging technology for the seafood industry. *Brazilian Archives of Biology and Technology, 52*(6), 1527–1539.

19. Guillen, A. C., Kechinsk, C. P., & VitorManfroi, V., (2010). The use of ozone in a CIP system in the wine industry. *Ozone: Science and Engineering, 32*(5), 355–360.

20. Gupta, A., & Aneja, K. R., (2004). Seed deterioration in soybean varieties during storage: Physiological attributes. *Seed Research, 32*, 26–32.

21. Gutman, J., & Raton, B., (2006). *Self-Monitoring Ozone Containing Packaging System for Sanitizing Application* (p. 21). US Patent Application No. 11/226,123 Pub. No.: US206/0840A1.

22. Guzel-Seydim, Z. B., Greene, A. K., & Seydim, A. C., (2004). Use of ozone in the food industry. *Lebensm-Wiss-U- Technology, 37*, 453–460.

23. Han, Y., Floros, J. D., Linton, R. H., Nielsen, S. S., & Nelson, P. E., (2002). Response surface modeling for the inactivation of *E. coli* O157:H7 on green peppers (*Capsicum annuum*) by ozone gas treatment. *Journal of Food Science, 67*, 1188–1193.

24. Ibanoglu, S., (2001). Influence of tempering with ozonated water on the selected properties of wheat flour. *Journal of Food Engineering, 48*, 345–350.

25. Inan, F., Pala, M., & Doymaz, I., (2007). Use of ozone in detoxification of aflatoxin b1 in red pepper. *Journal of Stored Products Research, 43*, 425–429.

26. Jakob, S. J., & Hansen, F., (2005). New chemical and biochemical hurdles. *Emerging Technologies for Food Technology, 2005*, 387–418.

27. Kannaujia, P. K., Asrey, R., Singh, A. K., Varghese, E., & Bhatia, K., (2019). Influence of ozone treatment on postharvest quality of stored summer squash. *Indian Journal of Horticulture, 76*(2), 350–354.

28. Karaca, H., & Smilanick, J. L., (2011). The influence of plastic composition and ventilation area on ozone diffusion through some food packaging materials. *Postharvest Biology and Technology, 62*, 85–88.

29. Karamah, E. F., Adi, S. Z., & Wajdi, N., (2019). Effect of ozone exposure time and ozonated water replacement to control the quality of chicken meat. *IOP Conf. Series 1295: Journal of Physics.* Article ID: 012068; doi: 10.1088/1742-6596/1295/1/012068.

30. Kazi, M., Parlapani, F. F., Boziaris, I. S., Vellios, E. K., & Lykas, C., (2017). Effect of ozone on the microbiological status of five dried aromatic plants. *Journal of Science, Food and Agriculture, 98*(4), 1369–1373.

31. Khadre, M. A., & Yousef, A. E., (2001). Decontamination of a multilaminated aseptic food packaging material and stainless steel by ozone. *J. Food Safety, 21*(1), 1–13.
32. Kim, J. G., (1998). *Ozone as an Antimicrobial Agent in Minimally Processed Foods* (p. 219). PhD thesis; The Ohio State University, Columbus, OH.
33. Kim, J. G., Yousef, A. E., & Dave, S., (1999). Application of ozone for enhancing the microbiological safety and quality of foods: A review. *J. Food Protect, 62*(9), 1071–1087.
34. Klockow, P. A., & Keener, K. M., (2009). Safety and quality assessment of packaged spinach treated with a novel ozone-generation system. *LWT – Food Science and Technologies, 42,* 1047–1053.
35. Li, H., Xiong, Z., Gui, D., & Li, X., (2019). Effect of aqueous ozone on quality and shelf-life of Chinese winter jujube. *Journal of Food Processing and Preservation, 43,* 1–8.
36. Meena, M. K., Chetti, M. B., & Nawalagatti, C. M. (2017). Influence of different packaging materials and storage conditions on the seed quality parameters of groundnut (*Arachis hypogaea L.*). *International Journal of Pure and Applied Bioscience, 5*(1), 933–941.
37. Munhõs, M. C., Navarro, R. S., Nunez, S. C., Kozusny-Andreani, D. I., & Baptista, A., (2019). Reduction of *Pseudomonas* inoculated into whole milk and skin milk by ozonation. *XXVI Brazilian Congress on Biomedical Engineering, 70,* 837–840.
38. Muthukumarappan, K., Halaweish, F., & Naidu, A. S., (2000). *Ozone: Natural Food Antimicrobial Systems* (pp. 783–800). Boca Raton- FL: CRC Press.
39. Nayak, S. L., Sethi, S., Sharma, R. R., Sharma, R. M., Singh, S., & Singh, D., (2020). Aqueous ozone controls decay and maintains quality attributes of strawberry (*Fragaria 3 ananassa Duch.*). *Journal of Food Science and Technology, 57*(1), 319–326.
40. Neto, O. P. S., & Pinto, E. V. S., (2019). Ozone slows down anthracnose and increases shelf-life of papaya fruits. *Brazilian Magazine of Fruit Culture, 41*(5), 1–12.
41. Ozen, B. F., Mauer, L. J., & Floros, J. D., (2002). Effects of ozone exposure on the structural, mechanical, and barrier properties of select plastic packaging films. *Packaging Technology and Science, 15,* 301–311.
42. Patra, A. K., Tripathy, S. K., & Samui, R. C., (2000). Effect of drying and storage methods on seed quality of summer groundnut (*Arachis hypogaea L.*). *Seed Research, 28*(1), 32–35.
43. Porto, E., Filho, E. G. A., & Silva, L. M. A., (2020). Ozone and plasma processing effect on green coconut water. *Food Research International, 131,* 1–9.
44. Radadia, L. B., Patel, N. C., & Chauhan, P. M., (1992). Studies on storage characteristics on groundnut. *Bulletin of Grain Technology, 30,* 231–235.
45. Sastry, D. V., Upadhyaya, H. D., & Gowda, C. L., (2007). Survival of groundnut seeds under different storage conditions. *SAT E-Journal, 5*(1), 1–3.
46. Sease, W. S., (1976). ozone mass transfer and contact system. *Second International Symposium on Ozone Technology* (pp. 1–14). International Ozone Association, Vienna, VA.
47. Singh, K. K., & Dadlani, M., (2003). Effect of packaging on vigor and viability of soybean (*Glycine max L. Merrill*) seed during ambient storage. *Seed Research, 31*(1), 27–32.
48. Skog, L. J., & Chu, C. L., (2001). Effect of ozone on qualities of fruits and vegetables in cold storage. *Canadian Journal of Plant Science, 81*(4), 773–778.
49. Sothornvit, R., & Kiatchanapaibul, P., (2009). Quality and shelf-life of washed fresh-cut asparagus in modified atmosphere packaging. *LWT-Food Science and Technology, 42,* 1484–1490.

50. Steffen, H., & Zumstein, R. G., (2010). Fruits and vegetables disinfection at SAMRO, Ltd. Using hygienic packaging by means of ozone and UV radiation ozone: *Science and Engineering, 32,* 144–149.

51. Stumm, W., (1958). Ozone as a disinfectant for water and sewage. *Journal of the Boston Society of Civil Engineers, 46,* 68.

52. Suryaningsih, W., Supriyono, Hariono, B., & Kurnianto, M. F., (2020). Improving the quality of smoked shark meat with ozone water technique. *IOP Conf. Series: Earth and Environmental Science, 411,* 1–8.

53. Tomiyasu, H., Fukutomi, H., & Gordon, G., (1985). Kinetics and mechanisms of ozone decomposition in basic aqueous solution. *Inorganic Chemistry, 24,* 2962–2966.

54. Vandermeiren, K., & De Bock, M., (2012). Ozone effects on yield quality of spring oilseed rape and broccoli. *Atmospheric Environment, 47,* 76–83.

55. Vettraino, A. M., & Vinciguerra, V., (2020). Gaseous ozone as a suitable solution for post-harvest chestnut storage: Evaluation of quality parameter trends. *Food and Bioprocess Technology, 13,* 187–193.

56. Vijay, S., & Ansari, S. U., (1991). Farmers level survey on insects and mites on stored groundnut in Andhra Pradesh. *Bulletin of Grain Technology, 29,* 14–21.

57. Wojtowicz, J. A., (1998). Ozone. In: Kirk, R. E., & Othmer, D. E., (eds.), *Encyclopedia of Chemical Technology* (4th edn., Vol. 17, pp. 953–994). New York: Wiley-Interscience.

58. Yousef, A. E., Kim, J. G., & Dave, S., (1999). Application of ozone for enhancing the microbiological safety and quality of foods: Review. *Journal of Food Protection, 62,* 1071–1087.

59. Wang, L., Fan, X., Sokorai, K., & Sites, J., (2019). Quality deterioration of grape tomato fruit during storage after treatments with gaseous ozone at conditions that significantly reduced populations of *salmonella* on stem scar and smooth surface. *Food Control, 103,* 9–20.

60. Zhang, M., Meng, X., Bhandari, B., Fang, Z., & Chen, H., (2015). Recent application of modified atmosphere packaging (MAP) in fresh and fresh-cut foods. *Food Reviews International, 31*(2), 172–193.

61. Zhu, X., Jiang, J., Yin, C., Li, G., Jiang, Y., & Shan, Y., (2019). Effect of ozone treatment on flavonoid accumulation of Satsuma mandarin (*Citrus unshiu* Marc.) during ambient storage. *Biomolecules, 9,* 1–12.

CHAPTER 9

ROLE OF VACUUM TECHNOLOGY IN FOOD PRESERVATION

ASWIN S. WARRIER

ABSTRACT

The vacuum (sub-baric or low-pressure) technology finds numerous roles in processing, handling, packaging, etc., in the food industry. Sub-baric or low-pressure conditions are valuable in the processing of products, which are reactive to heat or oxygen. This chapter discusses the role of vacuum technology in vacuum evaporation, vacuum drying, vacuum frying, vacuum deaeration, vacuum impregnation, vacuum cooling, *sous vide* cooking, vacuum packaging, etc.

9.1 INTRODUCTION

The vacuum has several useful applications ranging from simple pumps to sophisticated laboratory instruments; and in processing, handling, packaging of foods. Although the SI unit of pressure is Newtons per square meter or Pascal (Pa), yet the vacuum is usually expressed in torrs or sometimes in millimeters of mercury (mmHg) below the atmospheric pressure.

The use of vacuum is imperative in the manufacture of several foods, as it permits processes, which cannot be normally carried out under atmospheric conditions. The physics and philosophy of vacuum, methods of creating and maintaining vacuum, other uses of vacuum, outer space, etc., are all beyond the scope of this chapter.

This chapter focuses on vacuum evaporation, vacuum drying, vacuum frying, vacuum deaeration, vacuum impregnation, vacuum cooling; *sous vide* cooking, vacuum packaging of foods.

9.2 APPLICATIONS OF VACUUM TECHNOLOGY

9.2.1 VACUUM EVAPORATION

Evaporation is the process of transition of a liquid to gas, when its vapor pressure increases beyond the surrounding pressure. Usually, this happens as a result of heating in case of many liquids at atmospheric pressure. However, evaporation can also be induced by bringing down the pressure of a closed region below the vapor pressure of the liquid. Thus, by adjusting the pressure level in a closed vessel, we can manipulate the temperature at which evaporation will take place. In the food industry, this technique has been in use for more than a century to evaporate moisture from the food product, thereby concentrating it and enhancing its shelf-life by arresting microbial growth and enzyme activity while minimizing loss of flavor and nutrition.

Vacuum evaporation or bringing down the boiling point is used in concentration of heat-labile foods (such as milk, egg white, fruit juices, etc.). A glance at steam tables will give an idea about the pressure-temperature relations in evaporation. For example, by bringing down the pressure to 38.6 kPa, the boiling temperature comes down to 75°C, while it further drops to 50°C if the pressure further lowers to 12.4 kPa [52].

The equipments for evaporation mainly consist of the heat exchanger with provision for vapor separation, condenser, and vacuum pump. Vacuum pan is the most basic equipment used for evaporation, which is in fact, a closed chamber with a steam jacket and a pressure reduction system. The Howard's vacuum pan is still popular among many sugar manufacturers [34]. In the case of continuous systems, the heat exchanger may be of plate type, tubular or scraped-surface, the most common being the falling-film type shell and tube evaporators for milk or fruit juices. In addition, multiple-effect evaporators and vapor recompression systems are employed to enhance the steam economy. Plate heat exchangers offer the advantage of providing large heat transfer areas while occupying limited space in comparison with the tubular heat exchangers. They have more spacing between the plates and larger openings to accommodate the phase change and can be used for concentrating milk and fruit juices, etc., [10].

Scraped surface evaporators are best suited for viscous products, such as, purees. Centritherm® is a type of evaporator marketed by the Australian company Flavortech, in which centrifugal force is used to spread the product as a thin film on one side of a rotating cone. On the other side, steam condenses, thereby heating the cone and the product. The whole process takes place in a vacuum, and the company claims that evaporation takes place at 35°C [32].

9.2.2 VACUUM DRYING

Drying is a prominent unit operation in food processing. Most of the direct contact type driers use hot air as the heating medium, which makes drying of heat and oxygen-sensitive foods a difficult task. A popular solution for this difficulty is the use of vacuum. Difficulty in adapting to a continuous process, long drying times, and higher costs limit the use of vacuum drying to high-valued foods.

A vacuum tray dryer is the most prevalent equipment for vacuum drying, mainly in laboratory and pilot-scale applications. Unlike atmospheric tray dryers, where the dominant mode of heat transfer is convection between hot air and the food, heat transfer takes place mainly by conduction and radiation in vacuum tray dryers [73]. Other options for vacuum drying are microwave vacuum drying and freeze-drying.

Microwave processing is an innovative technology that works on the dielectric heating principles. The fluctuating electromagnetic field created by the microwaves cause the polar molecules in the food to vibrate and reorient repeatedly (2450 million times in a second), causing the food to heat up from inside. The ability to heat rapidly and volumetrically, without causing issues like case hardening makes microwave an attractive option for drying. Microwave vacuum drying is a novel process that combines the advantages of microwave drying and vacuum drying.

For freeze-drying a food, at first it is frozen and its temperature and pressure is brought down to a level below its triple point to ensure that melting will not take place. Some amount of heat is then slowly provided to the food so that sublimation of moisture takes place directly from solid to vapor phase. This step ensures the removal of most of the moisture contained in food. A secondary desorption step, by providing some more heat follows, which ensures that the freeze-drying is complete, after which the vacuum is released. Though not always a vacuum drying process, yet spray freeze-drying combines spray drying with freeze-drying, where atomized particles are frozen and then subjected to sublimation at low temperature and pressure [45].

Castro-Albarran et al. [15] dried human milk using spray drying and freeze-drying to study the retention of immunoglobulins, and they reported that freeze-drying at 30°C showed maximum retention. Santos et al. [81] reviewed the use of spray drying and freeze-drying in the manufacture of yogurt powder. Ambros et al. [2] studied the efficacy of using microwave vacuum drying for preserving lactic cultures including probiotics. They observed better viability of the cultures.

Numerous studies related to the use of vacuum for drying fruits and vegetables have been reported. Figiel and Michalska [29] have extensively reviewed several works that used vacuum and microwaves for drying fruits and vegetables, while Fan et al. [27] have reviewed microwave-assisted freeze-drying. Vacuum drying have been successfully evaluated for drying of mango slices [75], lemon slices [91], Pears [53], blueberries [96], Plum [62] and tomato [70], carrot slices [17], Okra snacks [47], pumpkin slices [37, 65] and button mushroom slices [80].

Ma et al. [57] used ultrasound to improve the effect of vacuum drying of liquid whole egg and reported enhancement in mass transfer and drying rate. Wang et al. [93] used pulsed-spouted microwave vacuum drying for samples of a puffed product containing varying proportions of duck egg white. They reported that the variation in egg white quantity significantly affected the expansion of the product.

Aykin-Dincer et al. [5] studied the kinetics, moisture sorption characteristics, and quality of beef slices dried under vacuum. Baslar et al. [8] studied the drying kinetics of salmon and trout fillets processed using a combination of ultrasound and vacuum. They reported that the time required for drying can be significantly reduced by this combination. Cantalejo et al. [14] evaluated a combination of ozone and with freeze-drying for processing of chicken meat. They claimed that this combination was able to double the shelf-life of chicken meat in comparison with normal freeze-drying.

Monteiro et al. [64] made some modifications to a domestic microwave oven, to make it function as a microwave vacuum dryer with turntable. They made trials using several foods at different power levels and reported that this inexpensive equipment gave excellent results and hence, has scope for laboratory level research. Tsuruta et al. [90] studied the shrinkage of foods during microwave vacuum drying and observed that the drying time can be reduced and the evaporation rate was increased by providing a gas flow inside the chamber. They conducted mathematical simulations and compared with experimental results of shrinkage and movement of moisture during microwave vacuum drying.

9.2.3 VACUUM FRYING

In frying, food item is dipped in a heated edible oil or fat, which acts as a medium for heat and mass transfer at the same time. Various physicochemical changes take place in the food during frying, helping it to develop its characteristic appearance, flavor, and texture. However, some of these changes are not very attractive to health-conscious consumers. The major chemical changes seen in oils are oxidation, hydrolysis, and polymerization [22]. Furthermore, depending on factors, such as, type, and quality of oil used, temperature, and time taken for frying, whether the oil is reused, etc., the frying can even lead to adverse health effects. The health concerns associated with fried foods, and the benefits vacuum frying can offer over the conventional frying method have been described by Dueik and Bouchon [26].

Andres-Bello et al. [4] have written an extensive review on vacuum frying in which they have explained the different equipments available, effects of various pretreatments like blanching, drying, osmotic dehydration and coating, effects of frying conditions, physicochemical changes, along with mathematical modeling of the vacuum frying process.

Vacuum frying is usually done at pressures <50 torr, and offers the advantages of increased nutrient retention, reduced oil absorption and reduced oxidation. Moreover, the chances of formation of carcinogenic substances like acrylamide are also limited [66]. Batch type vacuum frying equipment generally consists of an enclosed frying pan, where vacuum can be maintained, vacuum pump, condenser, and an oil separation mechanism. A continuous vacuum fryer consists of frying pan fixed inside a vacuum chamber with airlock facilities so that the vacuum is not broken due to the conveyance of food to and from the chamber.

Yang [94] was one of the early researchers to have patented equipment for frying and separating absorbed oil from fries under vacuum. Gupta et al. [42] reported the design of a vertical vacuum dryer, which is more productive according to them than the horizontal design due to increased utilization of available frying space and mass distribution.

Potato chips are one of the most popular fried snacks across the globe. Therefore, a large number of published works related to vacuum frying refers to fried potato products: potato chips or crisps and French fries. The study by Garayo and Moreira [35] is the earliest work on vacuum frying of potato chips, in which they found that the frying temperature and the vacuum pressure significantly affected the drying rate and oil uptake.

Granda et al. [39] reported that the acrylamide production during vacuum frying of potato chips was 94% less in comparison with normal frying. Moreira et al. [67] observed that although the oil absorption was less in vacuum frying of potato chips, the oil from the surface tends to enter the open pores of the food once the vacuum is released. Hence, they incorporated an oil separation mechanism based on centrifugation with the fryer and reported a better quality product with less oil content. Su et al. [88]

studied a microwave vacuum fryer for making potato chips and compared it with conventional vacuum frying. They observed that potato chips made by microwave vacuum frying absorbed less oil, had faster moisture evaporation rate, was crispier, and had better color in comparison with vacuum drying.

Belkova et al. [9] investigated the effect of vacuum frying process on the oil used for making potato crisps. They reported that the acrylamide production was less by 98%; the formation of products of Maillard reaction like alkylprazines was decreased, along with reduced oxidation. In addition to potato, vacuum frying of apple, apricot, banana, carrot, fish, jackfruit, mushrooms, etc., have also been studied [25].

9.2.4 VACUUM DEAERATION

Deaeration is the process of removing gases, dissolved or free, like oxygen from the liquid. The presence of air in liquid foods can make processes like pumping complicated, affect the quality of the product by inducing oxidation, or affect the heat transfer rate. Furthermore, deaeration is helpful in enhancing filling accuracy and reducing packaging volume. There are several complications associated with oxygen content in food, and hence the industry uses several methods to overcome this issue. Garcia-Torres et al. [36] have reviewed in detail how dissolved oxygen affects fruit juices and how this can be removed. Though the review is on fruit juices, many of the chemical processes that take place are similar in other foods also, and the equipments used also work on similar principles.

The application of vacuum for removing air has been in practice for several decades. The basic system consists of a vacuum chamber, having an outlet at the bottom for deaerated product along with another outlet on top, connected to the vacuum pump for removing the gases and other volatile components. When the space above the food surface is at a lower pressure, any enclosed gaseous particles within the food will try to escape out, leading to deaeration. Currently, continuous deaerators are commercially available, which may be centrifugal, spray or film type.

Sharp et al. [85] proposed vacuum deaeration for reducing oxidation and protecting vitamin C in the pasteurized milk. Greenbank and Wright [41] studied the effect of deaeration of raw milk done in a vacuum pan on the development of tallowy flavor in dried milk. The noted that the redox potential was substantially reduced, and the formation of tallowy flavor was brought under control by deaeration.

Anderson et al. [3] found that flavor retention during storage of concentrated sweetened cream was enhanced by vacuum deaeration. Loesecke et al. [56] proposed vacuum deaeration along with flash pasteurization as a solution to preserve flavor in orange juice. Marshall [60] reported the removal of oxygen from apple juice using a vacuum deaerator. Kanner et al. [49] found a reduction in accumulation of furfural in stored orange juice by employing vacuum deaeration. Jordan et al. [48] observed that a significant portion of volatile aroma compounds was lost during vacuum deaeration of orange juice.

9.2.5 *VACUUM IMPREGNATION*

Vacuum impregnation is the process of filling the pores of a solid food product with a solution of choice (for obtaining some specific benefit or functionality), with the help of low pressure. Unlike osmotic dehydration in which the mass transfer takes place by the concentration gradient, mass transfer in vacuum impregnation is driven by the pressure difference induced by applied vacuum [76]. This is a rapid process and can endow the food with several advantages, such as nutrition, shelf-life, or flavor while causing minimal damage to the product structure.

The scope of vacuum impregnation was extended to food processing in the early 1940s when a few patents were granted in this regard. However, it took some more time to catch the interest of the research community, and the majority of works in this area have been conducted in the past 25 years. Still vacuum impregnation is a trendy topic and research studies are going on and expanding the scope of this technology.

The food to be treated by vacuum impregnation is initially immersed in the external solution in a closed chamber. The pressure level in the chamber is then brought down with the help of a vacuum pump, and the low-pressure condition is maintained for some specified holding time. During this period, the air present in the capillaries and other available spaces within the food matrix is removed. Subsequently, the vacuum is released and the chamber is brought to atmospheric pressure and held for some time. As a consequence, there will be a pressure difference on either side of the food surface, which rapidly drive the surrounding liquid to enter the pores, which were previously evacuated.

Fito et al. [30] proposed vacuum deaeration as a pretreatment to drying of fruits, and studied the effect on physical and transport properties, along with changes in the drying process. They observed that vacuum impregnation affects the density, thermal conductivity, diffusivity, dielectric properties, etc.

Fito et al. [31] investigated the efficacy of vacuum impregnation to incorporate physiological active components to fruits and vegetables for making functional foods. Chiralt et al. [21] successfully experimented with the application of vacuum impregnation in salting of meat, fish, and Manchego type cheese. Gras et al. [40] studied the effect of this technique on carrot, beetroot, zucchini, eggplant, mushroom, etc. Mujica-Paz et al. [68] studied its effect on apple, banana, mango, papaya, melon, etc. In both these studies, how the variation in porosities affected the process was ascertained. Betoret et al. [12] vacuum impregnated apple pieces using apple juice spiked with probiotic organisms to study the application of this method in preparation of probiotic dry fruit.

Zhao and Xie [95] reviewed practical applications of vacuum impregnation in fruits and vegetables processing, where they have highlighted applications like pretreatment to drying and freezing, fortification, and development of minimally processed foods. Panarese et al. [72] studied mechanisms of mass transfer in vacuum impregnation process using a microscopic method and image analysis to evaluate gas outflow and liquid inflow during the process. Recently, Sharanabasava et al. [84] reported on using vacuum impregnation for soaking balls of Gulab Jamun with sugar syrup.

9.2.6 VACUUM COOLING

Vacuum cooling is the process of fast cooling of foods based on the principle of evaporative cooling. As this technique involves extracting water out of the food and evaporating it, this is applicable to foods having high moisture content and a porous structure. Vacuum cooling has been in use since the 1950s for cooling leafy vegetables [92]. Unlike conventional refrigeration systems, this system does not require any refrigerant circulation mechanism and offers better cooling efficiency and faster cooling. Moreover, the cooling obtained is almost uniform and volumetric, without any temperature gradient. The operational cost involved in cooling is also less. However, moisture content, porosity, surface characteristics, etc., affect the efficiency of cooling, and this method is not suitable for all food products.

The equipment used for vacuum cooling primarily consists of a vacuum chamber and a vacuum pump. When the food product to be processed is kept in the vacuum chamber and the pressure is lowered, its boiling point drops due to reduction in vapor pressure needed for vaporization. This leads to evaporation of moisture particles in the food. When the water, either on the surface or inside the pores, evaporates, it absorbs the necessary latent heat for changing phase from its surroundings, i.e., the food itself. This leads to temperature of food being reduced without using any refrigerant.

Most of the reported research in this area has been done using batch systems. Switzerland based Aston Foods introduced a continuous vacuum cooling system named "Continua" in 2012 [71]. This patented system, which uses a food conveyor passing through a cooling chamber, extends the benefits of vacuum cooling to large-scale industries.

Cheyney et al. [20] reported applying vacuum cooling to wrapped lettuce while studying the effect of packaging film on the cooling effect. Burton et al. [13] compared cooling of mushrooms using vacuum with conventional method and reported that vacuum cooled mushrooms had better color, while it also showed increased weight loss. Houska et al. [44] developed a mathematical model for forecasting temperatures of liquid foods during vacuum cooling, and verified it with a process vessel. McDonald et al. [61] cooled cooked beef using vacuum cooling and compared its quality with conventional cooling. They observed that very little time was taken for vacuum cooling, and yield was reduced by more than 100% due to weight loss. They also found that the microbial quality was better for vacuum cooled product.

Sun and Hu [89] performed mathematical modeling of heat and mass transfer during vacuuming of porous foods using computational fluid dynamics (CFD) simulation. They verified the accuracy of their model in a laboratory-scale vacuum cooler. To deal with the weight loss during vacuum cooling, Cheng, and Sun [19] combined vacuum cooling with water immersion cooling and studied its feasibility on cooked pork ham. They observed that the weight loss was less than vacuum cooling alone and the time taken was less than air blast cooling.

Cheng and Hsueh [18] studied the pre-cooling of cabbage using vacuum cooling and observed that releasing the pressure using chilled air could be used to control the sudden rise of temperature when the product is brought back to atmospheric pressure. Li et al. [54] used vacuum cooling for lettuce to study its effect on *Escherichia coli* O157:H7 and concluded that this technology is not very advantageous against the microorganism. Schmidt et al. [82] combined cooking with vacuum cooling of chicken in a single vessel, and compared different cooking techniques like vapor cooking, water cooking, and water immersion cooking in the integrated system.

9.2.7 VACUUM PACKAGING

Vacuum packaging involves placing the food product in the package, flushing the air out of the package with the help of a vacuum pump, and then sealing it. This ensures that the quality of food is maintained and shelf-life is ensured for a longer period. Creating vacuum is also an important intermediate step in modified atmosphere packaging (MAP), where the evacuated space is filled with another gas. The major objective of vacuum packaging is to remove oxygen from the headspace of the food, thereby controlling oxidative deterioration, growth of aerobic bacteria and loss of volatile components and moisture. However, this technology is not without limitations, the major one being its inability to control anaerobic microorganisms like *Clostridium botulinum.*

The idea of preservation of foods using vacuum was conceived in the 17th century [23]. However, the commercial use of this technology started only in the late 1940s [74]. Vacuum packaging is usually done in flexible or semi-rigid packages made of materials like PVDC (polyvinylidene chloride) or EVOH (ethylene vinyl alcohol), to preserve mostly dry or solid foods and sometimes even liquid foods.

The vacuum packaging can be batch type or continuous, and may be of nozzle vacuumizing or chamber vacuumizing like thermoforming or vacuum skin packaging type [74]. In nozzle vacuumizing, a nozzle attached to the vacuum pump is inserted into the package to create the vacuum. In chamber vacuumizing, the product placed in a flexible package is kept in a vacuum chamber. Thermoforming type of equipment is a kind of form-fill-seal machine in which the packaging material is thermoformed into required shape; food is loaded in it and then sealed under vacuum. Vacuum skin packaging is similar to thermoforming, however, the heat and vacuum provided during sealing ensures that the packaging material is tightly packed (skin-packed) around the food.

Seideman and Durland [83] reviewed vacuum packaging of beef. Recently, Stella et al. [86] have reviewed the influence of vacuum skin packaging on quality of raw beef. The major characteristics of vacuum packaging beef included inhibition of spoilage bacteria, prevention of color and odor development, and survival of lactobacilli. Depending on the residual oxygen content in the package, the color of meat will be purple or brown due to the presence of deoxymyoglobin or metmyoglobin, respectively, and will not be regular red, which is formed due to oxymyoglobin [58].

Barros-Velazquez et al. [7] have reported that the growth of anaerobes and lactobacilli could also be controlled using vacuum skin packaging. Vacuum packaging has been reported for several meat products, such as: chicken [51], turkey [1], goat [77], pork [16], lamb [11], venison [28], and camel [59]. Gorris and Peppelenbos [38] have reviewed the application of vacuum packaging and MAP to respiring foods, such as fruits, and vegetables, starting from early works by S. Burg and including moderate vacuum packaging, MAP, microbial aspects, safe packaging, etc.

Li et al. [55] extensively reviewed several low-pressure preservation strategies, including vacuum packaging for fruits and vegetables. Min et al. [63] compared several packaging technologies including vacuum packaging on the flavor retention of milk powder. Several researchers have studied the effects of vacuum packaging on different types of cheese [46, 50].

9.2.8   SOUS VIDE COOKING

Sous vide is a French word meaning 'under vacuum.' This term is commonly used term by food scientists and chefs in the early 1990s [6]. In sous vide cooking; a food product hermetically sealed in a package is heated at a low temperature for a long period. This process is also referred to as cryovacking, cryovac being a pioneering brand name in vacuum packaging. The recommended time-temperature combination for this process is 70°C for 2 minutes or proportional in case of low shelf-life foods, to ensure 6 log reduction of *Lysteria monocytogenes*; or 90°C for 10 minutes or proportional in case of foods that need to be stored for more than 10 days, to ensure 6 log reduction of *Clostridium botulinum* spores [87].

Nowadays, sous vide cooking is used by several restaurants, even at low temperatures, mostly for minimally processed foods. Along with the quality and safety, sous vide cooking also ensures efficient heat transfer, retention of nutrients and flavors, prevention of recontamination, oxidation, and moisture loss, etc. The possibility of perfectly controlling the temperature and doneness makes this method attractive to chefs.

For sous vide cooking, the food product is initially vacuum-packaged in heatable pouch, and the package is then placed in a temperature-controlled water bath providing mild heating for a longer time than that required for normal temperature. As vacuum packaging is done, air will not act as a barrier between the heating medium and the food, thus providing an even heating. According to Baldwin [6], sous vide cooking can be of two types, namely, cook-serve, and cook-chill. Cook-serve method is used when the food is to be immediately consumed. In case of cook-chill, the food after cooking is rapidly cooled, stored at a low temperature for few days, and then reheated before serving.

Most of the reported research refers to the cooking of meat, fish, or vegetables. Tenderization of meat is a common application of this technique. Hong et al. [43] cooked chicken breast sous vide, and studied the changes in its quality during refrigerated storage. They suggested this as an attractive option for cooking. Naveena et al. [69] observed that sous vide cooking reduced lipid oxidation and growth of microorganisms in chicken sausages during storage, and enhanced its shelf-life.

Rondanelli et al. [78] applied sous vide cooking to cereals and legumes, and demonstrated increased mineral retention. The mineral retention as a result of sous vides cooking was further confirmed in Brassica vegetables by Florkiewicz and Berski [33].

Stringer and Metris [87] developed mathematical models for predicting the growth or reduction in populations of microorganisms in *sous vide* foods cooked at different temperatures. They also highlighted the limitations in extending the models for conventional cooking to sous vides cooking. Though sous vide cooking is a method used to obtain tender and juicy meat, this technique is not effective in producing browning or roasted flavor. Hence, the chefs follow a two-stage process in which sous vide cooking is combined with oven roasting. Ruiz-Carrascal et al. [79] studied the effects and changes in quality with this combination and suggested the use of reducing sugars during sous vide cooking to induce maillard browning.

9.3 SUMMARY

The vacuum technology in food processing (also known as sub baric processing) is used for numerous applications in the food industry, including but not limited to vacuum evaporation, vacuum drying, vacuum frying, vacuum deaeration, vacuum impregnation, vacuum cooling, sous vide cooking, vacuum packaging, etc. This chapter intends to give a brief overview of these technologies.

KEYWORDS

- food industry
- low-pressure processing
- modified atmosphere packaging
- sub-baric processing
- vacuum evaporation
- vacuum processing

REFERENCES

1. Ahn, D. U., Ajuyah, A., Wolfe, F. H., & Sim, J. S., (1993). Oxygen availability affects prooxidant catalyzed lipid oxidation of cooked turkey patties. *Journal of Food Science, 58*(2), 278–282.

2. Ambros, S., Foerst, P., & Kulozik, U., (2018). Temperature-controlled microwave-vacuum drying of lactic acid bacteria: Impact of drying conditions on process and product characteristics. *Journal of Food Engineering, 224*, 80–87.

3. Anderson, H., Bell, R., & Tittsler, R., (1960). The effect of deaeration on the flavor stability of concentrated sweetened cream. *Journal of Dairy Science, 43*(6), 230–241.

4. Andrés-Bello, A., García-Segovia, P., & Martínez-Monzó, J., (2011). Vacuum frying: An alternative to obtain high quality dried products. *Food Engineering Reviews, 3*(2), 63–70.

5. Aykın-Dinçer, E., & Erbaş, M., (2018). Drying kinetics, adsorption isotherms, and quality characteristics of vacuum-dried beef slices with different salt contents. *Meat Science, 145*, 114–120.

6. Baldwin, D. E., (2012). Sous vide cooking: A review. *International Journal of Gastronomy and Food Science, 1*(1), 15–30.

7. Barros-Velazquez, J., Carreira, L., Franco, C., Vazquez, B. I., Fente, C., & Cepeda, A., (2003). Microbiological and physicochemical properties of fresh retail cuts of beef packaged under an advanced vacuum skin system. *Journal of Food Protection, 66*(11), 2085–2092.

8. Başlar, M., Kılıçlı, M., & Yalınkılıç, B., (2015). Dehydration kinetics of salmon and trout fillets using ultrasonic vacuum drying as a novel technique. *Ultrasonics Sonochemistry, 27*, 495–502.

9. Belkova, B., Hradecky, J., Hurkova, K., Forstova, V., Vaclavik, L., & Hajslova, J., (2018). Impact of vacuum frying on quality of potato crisps and frying oil. *Food Chemistry, 241*, 51–59.

10. Berk, Z., (2016). *Citrus Fruit Processing* (p. 330). London: Academic Press.

11. Berruga, M. I., Vergara, H., & Gallego, L., (2005). Influence of packaging conditions on microbial and lipid oxidation in lamb meat. *Small Ruminant Research, 57*(2–3), 257–264.

12. Betoret, N., Puente, L., & Dıaz, M. J., (2003). Development of probiotic-enriched dried fruits by vacuum impregnation. *Journal of Food Engineering, 56*(2/3), 273–277.

13. Burton, K. S., Frost, C. E., & Atkey, P. T., (1987). Effect of vacuum cooling on mushroom browning. *International Journal of Food Science and Technology, 22*(6), 599–606.

14. Cantalejo, M. J., Zouaghi, F., & Pérez-Arnedo, I., (2016). Combined effects of ozone and freeze-drying on the shelf-life of broiler chicken meat. *LWT-Food Science and Technology, 68*, 400–407.

15. Castro-Albarrán, J., & Aguilar-Uscanga, B. R., (2016). Spray and freeze-drying of human milk on the retention of immunoglobulins (IgA, IgG, IgM). *Drying Technology, 34*(15), 1801–1809.

16. Cayuela, J. M., Gil, M. D., Bañón, S., & Garrido, M. D., (2004). Effect of vacuum and modified atmosphere packaging on the quality of pork loin. *European Food Research and Technology, 219*(4), 316–320.

17. Chen, Z. G., Guo, X. Y., & Wu, T., (2016). A novel dehydration technique for carrot slices implementing ultrasound and vacuum drying methods. *Ultrasonics Sonochemistry, 30*, 28–34.

18. Cheng, H. P., & Hsueh, C. F., (2007). Multi-Stage vacuum cooling process of cabbage. *Journal of Food Engineering, 79*(1), 37–46.

19. Cheng, Q., & Sun, D. W., (2006). Feasibility assessment of vacuum cooling of cooked pork ham with water compared to that without water and with air blast cooling. *International Journal of Food Science and Technology, 41*(8), 938–945.

20. Cheyney, C., Kasmire, R., & Morris, L., (1979). Vacuum cooling wrapped lettuce. *California Agriculture, 33*(10), 18, 19.

21. Chiralt, A., Fito, P., & Barat, J. M., (2001). Use of vacuum impregnation in food salting process. *Journal of Food Engineering, 49*(2/3), 141–151.

22. Choe, E., & Min, D. B., (2007). Chemistry of deep-fat frying oils. *Journal of Food Science, 72*(5), R77–R86.

23. Cowell, N. D., (1998). The contributions of Robert Boyle and Denis Papin to food preservation. *Transactions of the Newcomen Society, 70*(1), 123–133.

24. Derossi, A., De Pilli, T., & Severini, C., (2012). The application of vacuum impregnation techniques in food industry. *Scientific, Health and Social Aspects of the Food Industry,* 25–56.

25. Diamante, L. M., Shi, S., Hellmann, A., & Busch, J., (2015). Vacuum frying foods: Products, process and optimization. *International Food Research Journal, 22*(1).

26. Dueik, V., & Bouchon, P., (2011). Development of healthy low-fat snacks: Understanding the mechanisms of quality changes during atmospheric and vacuum frying. *Food Reviews International, 27*(4), 408–432.

27. Fan, K., Zhang, M., & Mujumdar, A. S., (2019). Recent developments in high efficient freeze-drying of fruits and vegetables assisted by microwave: A review. *Critical Reviews in Food Science and Nutrition, 59*(8), 1357–1366.

28. Farouk, M. M., & Freke, C., (2008). Packaging and storage effects on the functional properties of frozen venison. *Journal of Muscle Foods, 19*(3), 275–287.

29. Figiel, A., & Michalska, A., (2017). Overall quality of fruits and vegetables products affected by the drying processes with the assistance of vacuum-microwaves. *International Journal of Molecular Sciences, 18*(1), 71–82.

30. Fito, P., Chiralt, A., & Barat, J. M., (2001). Vacuum impregnation for development of new dehydrated products. *Journal of Food Engineering, 49*(4), 297–302.

31. Fito, P., Chiralt, A., & Betoret, N., (2001). Vacuum impregnation and osmotic dehydration in matrix engineering: Application in functional fresh food development. *Journal of Food Engineering, 49*(2/3), 175–183.

32. Flavourtech. (2020). *Centritherm® Evaporator.* https://flavourtech.com/products/centritherm-evaporator/ (accessed on 19 January 2021).

33. Florkiewicz, A., & Berski, W., (2018). Application of sous vide method as an alternative to traditional vegetable cooking to maximize the retention of minerals. *Journal of Food Processing and Preservation, 42*(2), e13508.

34. Galloway, J. H., (2005). *The Sugar Cane Industry: An Historical Geography from its Origins to 1914* (Vol. 12, p. 284). Cambridge: Cambridge University Press.

35. Garayo, J., & Moreira, R., (2002). Vacuum frying of potato chips. *Journal of Food Engineering, 55*(2), 181–191.

36. García-Torres, R., & Ponagandla, N. R., (2009). Effects of dissolved oxygen in fruit juices and methods of removal. *Comprehensive Reviews in Food Science and Food Safety, 8*(4), 409–423.

37. Ghaboos, S. H. H., & Ardabili, S. M. S., (2016). Combined infrared-vacuum drying of pumpkin slices. *Journal of Food Science and Technology, 53*(5), 2380–2388.

38. Gorris, L. G. M., & Peppelenbos, H. W., (1992). Modified atmosphere and vacuum packaging to extend the shelf-life of respiring food products. *Hort. Technology, 2*(3), 303–309.

39. Granda, C., Moreira, R. G., & Tichy, S. E., (2004). Reduction of acrylamide formation in potato chips by low-temperature vacuum frying. *Journal of Food Science, 69*(8), E405–E411.

40. Gras, M., Vidal-Brotóns, N., Betoret, A., & Fito, P., (2002). The response of some vegetables to vacuum impregnation. *Innovative Food Science and Emerging Technologies, 3*(3), 263–269.

41. Greenbank, G. R., & Wright, P. A., (1951). The deaeration of raw whole milk before heat treatment as a factor in retarding the development of the tallowy flavor in its dried product. *Journal of Dairy Science, 34*(8), 815–818.

42. Gupta, A., Choudhari, A., Kadaka, T., & Rayar, P., (2019). Design and analysis of vertical vacuum fryer. In: Vasudevan, H., Kottur, V. K. N., & Raina, A. A., (eds.), *Proceedings of International Conference on Intelligent Manufacturing and Automation* (p. 720). Singapore: Springer.

43. Hong, G. E., Kim, J. H., Ahn, S. J., & Lee, C. H., (2015). Changes in meat quality characteristics of the sous-vide cooked chicken breast during refrigerated storage. *Korean Journal for Food Science of Animal Resources, 35*(6), 757–765.

44. Houska, M., Podloucký, S., Zitný, R., & Gree, R., (1996). Mathematical model of the vacuum cooling of liquids. *Journal of Food Engineering, 29*(3/4), 339–348.

45. Ishwarya, S. P., Anandharamakrishnan, C., & Stapley, A. G., (2015). Spray-Freeze-drying: A novel process for the drying of foods and bioproducts. *Trends in Food Science and Technology, 41*(2), 161–181.

46. Jalilzadeh, A., Tunçtürk, Y., & Hesari, J., (2015). Extension shelf-life of cheese: A review. *International Journal of Dairy Science, 10*(2), 44–60.

47. Jiang, N., Liu, C., Li, D., Zhang, Z., Liu, C., Wang, D., Niu, L., & Zhang, M., (2017). Evaluation of freeze-drying combined with microwave vacuum drying for functional okra snacks: Antioxidant properties, sensory quality, and energy consumption. *LWT-Food Science and Technology, 82*, 216–226.

48. Jordan, M. J., Goodner, K. L., & Laencina, J., (2003). Deaeration and pasteurization effects on the orange juice aromatic fraction. *LWT-Food Science and Technology, 36*(4), 391–396.

49. Kanner, J., Harel, S., Fishbein, Y., & Shalom, P., (1981). Furfural accumulation in stored orange juice concentrates. *Journal of Agricultural and Food Chemistry, 29*(5), 948–949.

50. Khoshgozaran, S., Azizi, M. H., & Bagheripoor-Fallah, N., (2012). Evaluating the effect of modified atmosphere packaging on cheese characteristics: A review. *Dairy Science and Technology, 92*(1), 1–24.

51. Kim, J. Y., & Song, K. B., (2004). Effect of vacuum packaging on the microbiological profile of chilled chicken during storage. *Journal of Applied Biological Chemistry, 47*(1), 35–37.

52. Koretsky, M. D., (2012). *Engineering and Chemical Thermodynamics* (2nd edn., p. 704). New York-USA: John Wiley & Sons.

53. Lech, K., Siudek, M., Michalska, A., & Figiel, A., (2016). Drying kinetics and textural properties of pears dehydrated by combined method with application of vacuum-microwaves. *Zeszyty Problemowe Postępów Nauk Rolniczych, 11*, 20–31.

54. Li, H., Tajkarimi, M., & Osburn, B. I., (2008). Impact of vacuum cooling on *Escherichia coli* O157: H7 infiltration into lettuce tissue. *Applied Environmental Microbiology, 74*(10), 3138–3142.

55. Li, L., Zhang, M., Adhikari, B., & Gao, Z., (2017). Recent advances in pressure modification-based preservation technologies applied to fresh fruits and vegetables. *Food Reviews International, 33*(5), 538–559.

56. Loesecke, H. W. V., Mottern, H. H., & Pulley, G. N., (1934). Preservation of orange juice by deaeration and flash pasteurization. *Industrial and Engineering Chemistry, 26*(7), 771–773.

57. Ma, Y., Zhu, W., Bai, X., Luo, L., Huang, J., Yu, B., & Li, N., (2018). Drying characteristics and kinetic model of liquid whole egg during ultrasound-reinforced vacuum drying. *Shipin Kexue/Food Science, 39*(3), 142–149.

58. Mancini, R. A., & Hunt, M., (2005). Current research in meat color. *Meat Science, 71*(1), 100–121.

59. Maqsood, S., Al Haddad, N. A., & Mudgil, P., (2016). Vacuum packaging as an effective strategy to retard off-odor development, microbial spoilage, protein degradation and retain sensory quality of camel meat. *LWT-Food Science and Technology, 72*, 55–62.

60. Marshall, C. R., (1951). Oxidation in apple juice, II: Some observations on deaeration. *Journal of the Science of Food and Agriculture, 2*(7), 321–327.

61. McDonald, K., Sun, D. W., & Kenny, T., (2000). Comparison of the quality of cooked beef products cooled by vacuum cooling and by conventional cooling. *LWT-Food Science and Technology, 33*(1), 21–29.

62. Michalska, A., Wojdyło, A., Lech, K., Łysiak, G. P., & Figiel, A., (2016). Physicochemical properties of whole fruit plum powders obtained using different drying technologies. *Food Chemistry, 207*, 223–232.

63. Min, D. B., Lee, S. H., Lindamood, J. B., Chang, K. S., & Reineccius, G. A., (1989). Effects of packaging conditions on the flavor stability of dry whole milk. *Journal of Food Science, 54*(5), 1222–1224.

64. Monteiro, R. L., Carciofi, B. A. M., Marsaioli, Jr. A., & Laurindo, J. B., (2015). How to make a microwave vacuum dryer with turntable. *Journal of Food Engineering, 166*, 276–284.

65. Monteiro, R. L., Link, J. V., Tribuzi, G., Carciofi, B. A., & Laurindo, J. B., (2018). Microwave vacuum drying and multi-flash drying of pumpkin slices. *Journal of Food Engineering, 232*, 1–10.

66. Moreira, R. G., (2014). Vacuum frying versus conventional frying: An overview. *European Journal of Lipid Science and Technology, 116*(6), 723–734.

67. Moreira, R. G., Da Silva, P. F., & Gomes, C., (2009). The Effect of a de-oiling mechanism on the production of high quality vacuum fried potato chips. *Journal of Food Engineering, 92*(3), 297–304.

68. Mujica-Paz, H., & Valdez-Fragoso, A., (2003). Impregnation properties of some fruits at vacuum pressure. *Journal of Food Engineering, 56*(4), 307–314.

69. Naveena, B. M., & Khansole, P. S., (2017). Effect of Sous vide processing on physicochemical, ultrastructural, microbial and sensory changes in vacuum packaged chicken sausages. *Food Science and Technology International, 23*(1), 75–85.

70. Orikasa, T., Koide, S., & Sugawara, H., (2018). Applicability of vacuum microwave drying for tomato fruit based on evaluations of energy cost, color, functional components, and sensory qualities. *Journal of Food Processing and Preservation, 42*(6), e13625.

71. Packaging Insights, (2012). *Aston Foods Unveils Vacuum Cooling for Baked Goods.* https://www.packaginginsights.com/news/Aston-Foods-Unveils-Vacuum-Cooling-For-Baked-Goods.html (accessed on 19 January 2021).

72. Panarese, V., Dejmek, P., Rocculi, P., & Galindo, F. G., (2013). Microscopic studies providing insight into the mechanisms of mass transfer in vacuum impregnation. *Innovative Food Science and Emerging Technologies, 18*, 169–176.

73. Parikh, D. M., (2015). Vacuum drying: Basics and application. *Chem. Eng., 122*(4), 48–54.

74. Perdue, R., (2010). Vacuum packaging. In: Yam, K. L., (ed.), *Encyclopedia of Packaging Technology* (3rd edn., pp. 1259–1264). USA: John Wiley & Sons.

75. Pu, Y. Y., & Sun, D. W., (2017). Combined hot-air and microwave-vacuum drying for improving drying uniformity of mango slices based on hyperspectral imaging visualization of moisture content distribution. *Biosystems Engineering, 156*, 108–119.

76. Radziejewska-Kubzdela, E., Biegańska-Marecik, R., & Kidoń, M., (2014). Applicability of vacuum impregnation to modify physicochemical, sensory, and nutritive characteristics of plant origin products: A review. *International Journal of Molecular Sciences, 15*(9), 16577–16610.

77. Rajkumar, V., Agnihotri, M. K., & Sharma, N., (2004). Quality and shelf-life of vacuum and aerobic packed chevon patties under refrigeration. *Asian-Australasian Journal of Animal Sciences, 17*(4), 548–553.

78. Rondanelli, M., Daglia, M., & Meneghini, S., (2017). Nutritional advantages of sous vide cooking compared to boiling on cereals and legumes: Determination of ashes and metals content in ready-to-eat products. *Food Science and Nutrition, 5*(3), 827–833.

79. Ruiz-Carrascal, J., & Roldan, M., (2019). Sous-vide cooking of meat: A maillarized approach. *International Journal of Gastronomy and Food Science, 2019*. Article ID: 100138.

80. Salehi, F., Kashaninejad, M., & Jafarianlari, A., (2017). Drying kinetics and characteristics of combined infrared-vacuum drying of button mushroom slices. *Heat and Mass Transfer, 53*(5), 1751–1759.

81. Santos, G. D., Nogueira, R. I., & Rosenthal, A., (2018). Powdered yoghurt produced by spray drying and freeze-drying: A review. *Brazilian Journal of Food Technology, 21*.

82. Schmidt, F. C., Aragão, G. M. F., & Laurindo, J. B., (2010). Integrated cooking and vacuum cooling of chicken breast cuts in a single vessel. *Journal of Food Engineering, 100*(2), 219–224.

83. Seideman, S. C., & Durland, P. R., (1983). Vacuum packaging of fresh beef: A review. *Journal of Food Quality, 6*(1), 29–47.

84. Sharanabasava, Ravindra, M. R., (2019). Evaluation of vacuum impregnation as a novel approach for soaking of fried Gulab jamun balls. *Journal of Food Science and Technology, 56*(5), 2764–2770.

85. Sharp, P. F., Guthrie, E. S., & Hand, D. B., (1941). New method of retarding oxidized flavor and preserving vitamin C-deaeration. *Journal of Dairy Science, 24*, 252, 253.

86. Stella, S., Bernardi, C., & Tirloni, E., (2018). Influence of skin packaging on raw beef quality: A review. *Journal of Food Quality*. https://doi.org/10.1155/2018/7464578.

87. Stringer, S. C., & Metris, A., (2018). Predicting bacterial behavior in sous vide food. *International Journal of Gastronomy and Food Science, 13*, 117–128.

88. Su, Y., Zhang, M., Zhang, W., Adhikari, B., & Yang, Z., (2016). Application of novel microwave-assisted vacuum frying to reduce the oil uptake and improve the quality of potato chips. *LWT- Food Science and Technology, 73*, 490–497.

89. Sun, D. W., & Hu, Z., (2003). CFD simulation of coupled heat and mass transfer through porous foods during vacuum cooling process. *International Journal of Refrigeration, 26*(1), 19–27.

90. Tsuruta, T., Tanigawa, H., & Sashi, H., (2015). Study on shrinkage deformation of food in microwave-vacuum drying. *Drying Technology, 33*(15/16), 1830–1836.
91. Wang, J., Law, C. L., Nema, P. K., Zhao, J. H., Liu, Z. L., Deng, L. Z., Gao, Z. J., & Xiao, H. W., (2018). Pulsed vacuum drying enhances drying kinetics and quality of lemon slices. *Journal of Food Engineering, 224*, 129–138.
92. Wang, L., & Sun, D. W., (2001). Rapid cooling of porous and moisture foods by using vacuum cooling technology. *Trends in Food Science and Technology, 12*(5, 6), 174–184.
93. Wang, T., Zhang, M., Fang, Z., & Liu, Y., (2016). Effect of processing parameters on the pulsed-spouted microwave vacuum drying of puffed salted duck egg white/starch products. *Drying Technology, 34*(2), 206–214.
94. Yang, C. S., (1989). U.S. Patent No. 4,873,920 (p. 21). Washington, DC: U.S. Patent and Trademark Office.
95. Zhao, Y., & Xie, J., (2004). Practical applications of vacuum impregnation in fruit and vegetable processing. *Trends in Food Science and Technology, 15*(9), 434–451.
96. Zielinska, M., Sadowski, P., & Błaszczak, W., (2016). Combined hot air convective drying and microwave vacuum drying of blueberries (*Vaccinium corymbosum L.*): Drying kinetics and quality characteristics. *Drying Technology, 34*(6), 665–684.

PART III
Food Processing Techniques for Product Formulation

CHAPTER 10

ADVANCED ENCAPSULATION METHODOLOGIES FOR HERBAL FOOD PRODUCTS

SADHNA MISHRA, ARVIND KUMAR, SHIKHA PANDHI, and DINESH CHANDRA RAI

ABSTRACT

The bioactive constituents from herbs and plants have countless significant functional groups, such as oxygenated compounds like aromatics, alcohols, aldehydes, phenols, sesquiterpene alcohol, esters, lactones, oxides, ethers; and they also bear monoterpenes, sesquiterpenes, and aromatics. All bioactive compounds are extremely susceptible to the environmental and chemical factors, which are responsible for the degradation and inactivation of the compounds. Encapsulation technology provides an effective approach for the protection for these phytoactive compounds. In encapsulation, the inner phase is a phytoactive herbal compound, and the outer phase is a second core material phase. In encapsulation, the coating of tiny bioactive particles can be done by homo or heterogeneous matrix. This chapter focuses on nanoprecipitation, electrospinning, solvent diffusion, dialysis, emulsification, and solvent evaporation methods of encapsulation. By using these strategies for stabilization of food constituents, it is possible to enhance the nutraceutical properties of food products.

10.1 INTRODUCTION

Since ancient times, the herbs have been used as remedial, pungent, and as flavoring agents [43, 47]. Approximately 25% of the drugs with different curative properties are extracted from herbal sources [51]. Around 21,000 plant species are used for remedial purposes. Apart from medicinal uses of the bioactive constituents extracted from medicinal plants, the other utilities of these bioconstituents include food preservation [61].

Medicinal plants with higher antioxidant activity can prevent many disorders, including degenerative and chronic diseases [4]. However, most of the population has always preferred herbal products compared to synthetic drugs. Therefore, the identification, stabilization, and characterization of bioactive compounds from herbal sources have significant research studies [61]. Only a few evaluation studies are available on the fitness as a whole herbal extract [44].

Herbal extracts are used for the value addition in food and medicines sectors because these herbal bioactives are an excellent source of aromas and have many medicinal properties, such as: antipyretic, anti-mutagenic, anti-cancerous, antifungal, antiviral, antibacterial, antioxidant, anti-chemotactic, and anti-inflammatory, etc., [29, 46, 51]. The phytoactive constituents from the herbal origin are also very notorious, such as phytoactives, active agents, bioactives, etc. All bioactive compounds (or phytochemicals or bioactives or phytoactives) from herbal plants, such as polyphenols, vitamins, proteins, minerals, and isothiocyanates, etc., have great potential and other functional properties, which have impacted the production of health-promoting foodstuffs [17, 18]. Properties of the polyphenols, such as bioactivity, bioavailability, and constancy are contingent on the food preservation. The compounds with remedial activities cannot be used in high concentrations because of their harmful and toxic effects.

Oral administration of certain phytoactive compounds can also cause health problems due to slow bioavailability inside the gut and inadequate gastrointestinal (GI) habitation period. Many of the deficiencies can be alleviated by the utilization of the encapsulated bioactive compounds as compared to unencapsulated bioactive compounds due to poor bioavailability. On the other hand, due to unsteadiness of the active compounds under different conditions throughout the food processing and stowage or even the effect of the hydrogen content [pH], enzymes, and presence of some other mineral compounds, which are in GIT, restrict their potential activity for health promotion [5]. In addition, the solubility of the lipids is poor; hence the bioactive substances express fewer bioactivities in a hydrophobic environment.

According to the recent data, the increment has been noticed in the expansion of deliverance of the encapsulated phytoactive components with novel nutraceutical properties [21]. Encapsulation means the protection by covering and entirely envelopment of the desired bioactive compound or whole herbal extract by using a feasible wall material [62].

The controlled discharge of desired encapsulated bioactive constituents under defined circumstances is possible [38]. The suitable delivery of the desired compounds is possible by applying the encapsulation technology. Encapsulation helps to hide the nasty taste at the time of eating that includes harshness of polyphenols, pungent taste, and also to protect other biocompounds with high antioxidant activity. It is an efficient process for the improvement of the bioavailability of bioactive compounds [40].

Formation of the particle with a defined diameter by entrapping the substance inside the other substance is carried out by encapsulation. The desired bioactive compounds, which have to be encapsulated, are known as heart phase, vigorous agent, core material or cargo phase. The materials, which are used for the encapsulation of the core material, are termed as external phase, wall material, coating material or carrier compound, etc. The wall materials, which are used for the encapsulation of herbal food products, must-have characteristics, such as the material can stabilize the core particles by their surroundings, and it must be of food-grade. Consequently, the protective machinery is necessary for the protection of bioactive constituents up to the stretch of consumption, which enables the deliverance of desired molecules in accordance with the metabolic pathway of the organisms [8]. Encapsulation has extensive applications for the design of the functional food and other food ingredients, and it also presents momentous issues in businesses, such as nutraceutical, pharmacological, and cosmetics [38]. Therefore, there are many encapsulation technologies, which are used for the encapsulation of herbal food products.

This chapter focuses on the advanced encapsulation technologies, which are used for the encapsulation of bioactive compounds with various remedial properties.

10.2 AIMS OF THE ENCAPSULATION

Encapsulation technology is differentiated from each other based on the bioactive compounds, which have to be encapsulated. The assortment of accurate encapsulation progression is very crucial for the design of various food formulations with specific physiognomies [2]. In the organized and stable encapsulation process, the external material or the wall material entraps the active substances (core material/desired bioactive compound). In the food industry, the encapsulation technology is a very efficient process to enhance the deliverance of desired bioactive compounds, which are encapsulated, including enzymes, vitamins, polyphenols, and probiotics [64]. These additive effective biocompounds are used because of their protective nature for the preservation, consistency, color, and essence belongings.

Additive compounds are prone to chemical, environmental, and other factors; therefore, an effective encapsulation approach provides protection. In the food industry, encapsulated ingredients have many benefits. The main aim of the encapsulation is to preserve all properties of the phytoactive compound including the prevention of disagreeable interactions of the foodstuffs and throughout the stowage and processing (Figure 10.1). Hence, the bioactive compounds can prevent inactivation of their activities by oxidation/hydrolysis and increase their bioavailability during delivery at the required spot through encapsulation [30]. Consequently, the functionality of the bioactive agent can be maintained.

Encapsulation technology maintains a barrier between the activity of the bioactive compound and their surroundings; and by this, the flavor and taste differentiation and façade of the undesirable taste of polyphenols is conceivable. Decrease in evaporation rate and slow-down of the degradation of the volatile biocompounds is also possible by encapsulation technology. Encapsulation of the functional agent also aims to prevent the reactions between the bioactive agent and other molecules, such as water, and oxygen present the foodstuffs. Beyond the overhead, the other food processing processes including metabolite production and fermentation technology is also used encapsulation for the immobilization of enzymes and living cells like bifidobacterium. It is a high demand to invent an appropriate approach, which will deliver the highest-level of productivity and superior quality of the final foodstuffs at the same time. An efficient treatment for cancer is possible by double encapsulation method to ensure the targeted deliverance and prolonged release of the encapsulated bioagent.

Encapsulation is also applied for some other physical modifications in the original bioconstituents [38], such as:

- Easiness during the handling;
- For providing equal dispersion and satisfactory concentration of the bioactive agent;
- To prevent reaction of the biocompounds between each other by a separation.

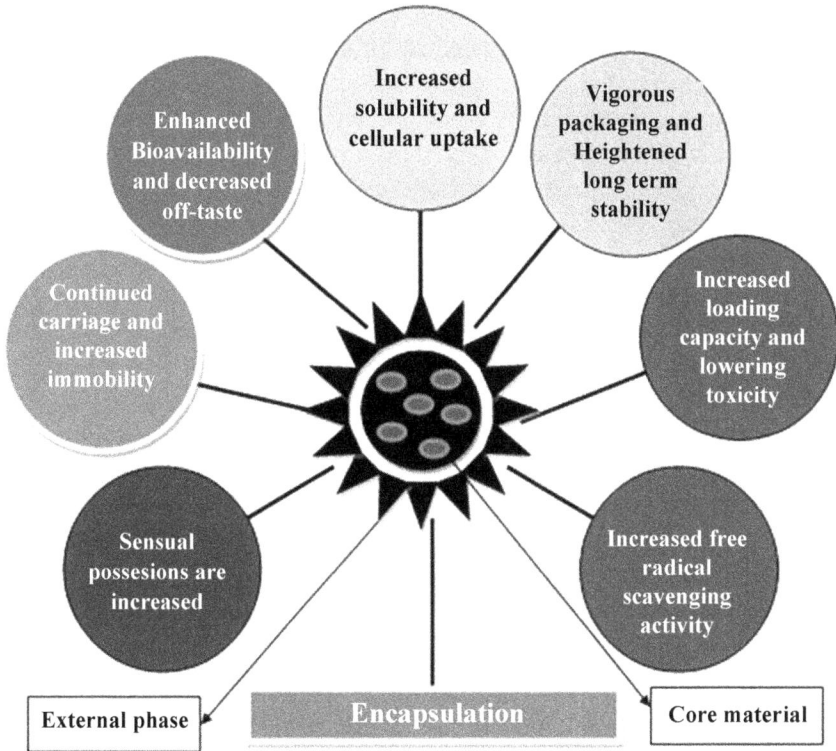

FIGURE 10.1 The encapsulation technology for bioactive agents: Improvements in the bioactivates of the encapsulated particle at the center).

10.3 EXTERNAL PHASES FOR THE ENCAPSULATION

Biomolecules are a major group of the compounds in the food sector for the encapsulation purpose. Beyond the natural belonging of the external material, which is used to cover the bioactive agent during encapsulation for protection of the desired bioactive compound from physical, chemical, and ecological circumstances throughout the stowage and processing, it ensures prevention of the reaction between bioactives and encapsulating material, and provide virtuous rheological physiognomies if it is used in high concentration and informal work aptitude.

Many of the fiber and other chemicals from the bacterial source with antibacterial activity are synthetic-based encapsulating materials, which are not eco-friendly. Polysaccharides are widely used biocompounds among other encapsulating materials for encapsulation of foodstuffs. The polydextrose, amylose, amylopectin, dextrin, cellulose, and their derivatives, maltodextrin, and syrups fall under the category of starch and their derivatives are frequently used. The encapsulating bioagent from protein and lipid in nature are likewise used appropriately for stable encapsulation. Many of the wall materials in the food sector are derived from plant extracts and exudates, and these include pectin, galactomannans, soluble soybean polysaccharides, gum Arabic, and gum Karaya.

Carrageenan and alginates as wall materials are from marine origin and are frequently used. Some of the external phases used as an encapsulating material including gellan, xanthan caseins, gelatin gluten, and chitosan are animal-based and microbial polysaccharides. The other suitable wall materials of lipid in nature include waxes (such as bee, *Candellia camauba* wax), fatty alcohols, fatty acids, glycerids, and phospholipids. Other materials for the encapsulation include: paraffin, polyvinylpyrrolidone (PVP), and shellac [65].

In the encapsulation process, the compounds, which are selected to encapsulate the bioactives, are adsorbed on the periphery of the particles [34]. Recently, many of the external phases (such as albumin, gelatin, etc.), have been prepared from natural polymer. Instead of indigenous remedial effects, the deliverance of the desired compounds, including vaccines, enzymes, living cells to their defined targeted organ is possible by the formation of nanoparticles and microparticles [22].

Properties of the polymer, including their ultimate demand and poisonousness stand as criteria for their selection. The ecological polymer for the encapsulation must be nontoxic and stable throughout the metabolism and whole circulation [37]. Other biodegradable polymers for the encapsulation of the bioactives comprise of poly (caprolactone, lactide, lactic acid, glycolic acid, and lactic-co-glycolic acid (PLGA)). Many of the remedial molecules composed of cisplatin, insulin, progesterone, paclitaxel, haloperidol, tamoxifen, dexamethasone, tyrphostin, 9-nitrocamptothecin, and estradiol are encapsulated effectively by using several organic and artificial polymers [23].

10.4 ENCAPSULATION TECHNOLOGIES

The encapsulation technology-based drying is used for the encapsulation of molecules, which are often found in liquid form. Hence, technologies (such as liposome entrapment, extrusion coating, spray drying, fluidized bed coating, spray-chilling, and spray-cooling) are accessible to encapsulate such bioactives [70].

In the food sector, the spray drying method is commonly used for the encapsulation of food additives. Since the spray drying technique has properties, such as flexibility, continuity, and economical operation, therefore it is frequently used to encapsulate the herbal bioactive constituents in the food sector. Because of the relatively high temperature throughout the whole desiccating procedure, which can cause the degradation and inactivation of the encapsulated compound, decreased encapsulation steadiness, and efficiency of the encapsulated element [29].

Encapsulation technologies for the phytoactive compounds are classified into two main groups:

- On the basis of performed polymer dispersion, which is physico-chemical representative based process; and
- On the basis of polymerization of the monomers, which are Chemistry linked progressions.

Monomer polymerization includes mini-emulsion, microemulsion, radical polymerization, and interfacial polymerization methods. The reformed polymer comprises of nanoprecipitation, solvent diffusion, dialysis, and solvent evaporation methods [2]. These techniques are used to develop several kinds of transporters, such as phytosomes, niosomes, liposomes, etc., [27, 63]. The emulsification technique is also efficient to encapsulate the bioactive constituents, which can be dissolved in aqueous solution [21].

Hydrophobic characteristics of the molecule to be encapsulated are very efficient for the selection of suitable encapsulation [45]. Some more expensive carriers (such as phytosomes, liposomes) and molecular inclusion in cyclodextrins offer specific topographies to the bioactive agents. The inner pocket of the cyclodextrins is lipophilic in which the phytoactive compound is entrapped reversibly in an aqueous environment and is of uniform size. The hydrophobic and hydrophilic interactions between the water molecule and phospholipid are main reacting mechanisms during the formation of the liposomes [21].

In food and pharmaceutical industries, electrospinning is a feasible route for the encapsulation of the bioactives. Electrospinning technique is a multipurpose, forthright, and superficial for the production of the nanofibers employing porosity. Nonattendance of the temperature is the main beneficial function of the electrospinning and it is also very efficient for the enhancement of the encapsulation efficacy and maintenance of the structure of the encapsulated bioactive on storage and processing. An electrospun fiber with bioactive compound as core can retain boosted functionality and stability [28].

The encapsulating polymers (such as proteins and polysaccharides), which are food-grade, attain extensive attentiveness because they are sustainable, biocompatible, nontoxic, and biodegradable. Conversely, several polymers are restricted for the electrospinning even though they are food grade [15, 30]. The selection criteria and other additional conditions of biocompatible, food-grade polymers for the formation of delivery carriers in the foodstuffs are extensively being evaluated [6, 11, 39, 41].

10.4.1   NANOPRECIPITATION TECHNIQUE

The formation of the nanocarriers by using nanoprecipitation method was principally developed by Hatem Fessi [12].

In the nano-precipitation, the organic phase contains the mixture of the film creating constituents like a polymer with aqueous annexable solvent plus active agent while the aqueous phase contains water plus stabilizer. It means that the encapsulation of the bioactive compounds by nanoprecipitation technique requires two annexable phases [33, 37]. The particle formed by the present technique comprises of three-step process, such as nucleation, growth, and aggregation [38].

The hydrophobic bioactive compounds cannot be encapsulated by nanoprecipitation method. Some of the bioactive agents with hydrophilic nature can be encapsulated by adopting some modifications in the present methodology [7]. Polymers (such as polylactide-co-glicolide (PLGA), poly-e-caprolactone (PCL), lactide) are biocompatible and eco-friendly, so that they are frequently used for the nanoprecipitation method. Acetone is frequently used as a solvent for polymer in this method. Ethanol and some other organic solvents are used for the suspension of bioactive substance and oil. The polysorbate-80 and poloxamer-188 and buffer solution are used as stabilizer and non-solvent, respectively.

10.4.2   EMULSIFICATION PROCESS

For the formation of the emulsion, at least two immiscible liquors are required in which one is a continuous phase while another is a dispersed phase, such as O/W emulsion or W/O emulsion. For the stabilization of the emulsion, an emulsifier (or a surfactant) can also be used. The emulsification method, amount, and nature of the emulsifier play main role during the particle formation and their size, and conversely, the effectiveness of the encapsulation depends on the particle.

The desired phytoactive compound is implanted in the continuous phase through the process of emulsification; consequently, the active agent is fully protected from their inactivation and thermos-oxidation. In addition to these properties, the overall physical status of the core compound is contingent on the droplet size, disseminated segment, suspension viscosity, dispersal of bioactives in the continuous phase, and conditions during processing, etc.

Homogenizer and ultrasonicator equipments are frequently used for nanoemulsion formation. For the encapsulation of the extract from *Aloe vera* by using tragacanth gum as a wall material to produce a natural wound healing product by microemulsion method, whereas the extract of leaves of olive is encapsulated within a soybean oil by the nanoemulsion technique with improved antioxidant actions [13, 34].

10.4.3   SOLVENT EVAPORATION TECHNIQUE

The formation of nanocarriers primarily includes the mixing of the aqueous phase with the solution of organic polymer. The postponement is developed after the fraternization of the oil with bioactive agent in a non-solvent with an organic solution of the polymer. After suspension formation, the solvent is vaporized, entrapment of the bioactive agent within the polymer has occurred as the nanoparticles are formed. Solvent evaporation is also a simple technique. The formation of the polynuclear microparticles with 1.38 µm in diameter and 38% encapsulation efficiency was achieved by the encapsulation of seed extract of grapes with polylactide as a wall material by using two combined methods (one is solvent evaporation and the other is a double emulsion) [66, 67]. The extract of Pink pepper was obtained with antioxidant, anticancer, and anti-inflammatory activities due to encapsulation with polylactic polymer by emulsification followed by solvent extraction [1].

10.4.4   EMULSION DIFFUSION TECHNIQUE

The emulsion diffusion method was used by Quintanar-Guerrero and Fessi for the formation of nanoparticles by using polylactide [43]. Emulsion diffusion technique is very effective for the formation of nanoparticle of bioactive compounds, which are lipophilic and hydrophilic in nature. In addition, the phytoactive compounds with hydrophobic properties are commonly encapsulated by emulsion diffusion technique.

Three phases (such as dilution, aqueous, and organic phases) are required in the emulsion diffusion method [48]. The carbon-based phase (organic) is comprised of an organic solvent partially annexable in water, bioactive compound; and polymer plus oil requires saturation with water [32].

The operational conditions throughout this method including temperature and volume of the water (which is used for the dilution), polymer concentration, and stirring rate during emulsification, the proportion of the phases and concentration of stabilizer (surfactant) can disturb the particle size of developed nanocarriers [36]. The equipment-based processes, which are used for the emulsification (such as sonication and homogenization), also affect the particle size. It was noticed that sonication is very crucial compared with the homogenization [32].

The organic phase plays the role of solvent; however organic phase also contains the oil solvent or active ingredient solvent. The aqueous dispersion with stabilizing agent forms the aqueous phase by the employment of the solvent saturated water, whereas the dispersed aqueous phase of the surfactant forms the dilution phase. Various polymers, such as PLA, PCL, and eudragit® are commonly used for nanoparticle formation by this method [35].

10.4.5   REVERSE PHASE EVAPORATION TECHNIQUE

Nanocarriers (such as unilamellar and oligolamellar liposomes) by using acetazolamide are prepared by using this method [16]. The lipid constituents (such as phosphatidylcholine or phosphatidylethanolamine) are used either alone or in combination with the charge agents inducing diacetyl phosphate or stearyl amine; and chloroform and methanol is used as a solvent.

10.4.6   DOUBLE EMULSION TECHNIQUE

The double emulsion is a multifaceted process and is named after suspension within suspension. Frequently, these emulsions are categorized into two different groups: oil-water-oil (O-W-O) and water-oil-water (W-O-W). The double emulsion technique requires a two-step process to develop the nanoparticles, and the droplet size of the particles are frequently polydispersed. In this technique, the aqueous phase is dispersed into the organic solvent, which is non-miscible; the primary emulsion is prepared by the high-shear homogenization or by applying short-time low-power sonication. After this step of the principal emulsion, it is disseminated into additional phase, which contains hydrophilic emulsifier; and after that, sonication or homogenization step is conducted repeatedly the same conditions.

If the sonication process is used for the emulsion formation, then it must be for few minutes by providing less voltage otherwise particle breaking will occur in the primary emulsion. Sometimes, the small particles formed after the double emulsion process generally contain more than one particle, whereas occasionally the large particle contains approximately 50–100 small particles [52]. Deliverance profile and stability of the carrier particle can be significantly enhanced by altering the amount of the encapsulated material inside the particle and use of surfactant or stabilizer.

Bioavailability, versatility, and biocompatibility are some of the advantages of double emulsions possess. However, the double emulsion process presents several issues, such as formulation misfortune, bulkiness, and exposure to the chemical and physical degradation. To overcome the steadiness difficulties of double emulsions, efforts like altering the amount of stabilizer, primary emulsion, interfacial complexation, polymerization gelling, steric stabilization, and addition of excipients are tried [37].

The construction of relatively assorted, size sensitive, thermodynamically instable and process complexity of the nanoparticle formation is few disadvantages of this process. The double emulsion technique has the advantage to encapsulate the bioactive agent with hydrophilic and lipophilic properties. The evaporation rate, relative phase proportion, polymer concentration, molecular weight, plus nature of polymer, homogenizer, or sonication speed are also the influencing factors for the nanoparticles specificity formed by double emulsion method [20, 59]. The most advanced encapsulation techniques (such as emulsion diffusion, microemulsion, nanoprecipitation, and high-pressure homogenization methods) do not require organic solvents throughout the encapsulation process [19].

10.4.7   IONIC GELATION METHOD

Alginate and chitosan are the charged polymers, which are encapsulated by ionic gelation method. The formation of the carrier particles occurs by the interaction of the oppositely charged molecules with a cross-linking property. For the encapsulation of the turmeric oil, sodium alginate aqueous solution was used as an external phase tracked through the gelification of chitosan with calcium chloride by subsequent solvent evaporation [26].

10.4.8 SOLID LIPID NANOPARTICLES (SLNs)

The pasteurization viability, squat toxicity, scaling up, and the increased biocompatibility are some main advantages of solid lipid nanoparticles (SLNs) [8, 59]. Along with the protection of the solid content, the SLNs are also responsible for the deliverance of the encapsulated bioactives (Figure 10.2). The organizational resemblance of these nanocarriers with fatty acids of the skin epidermis layer and the auspicious transdermal deliverance of the encapsulated bioactives occurs [41, 50].

On the basis of recent data, the SLNs can boost the skin moisturization by preventing the water loss from the skin. The properties like the dissemination of trivial sized particle make easy penetration of encapsulated bioactive agent [36].

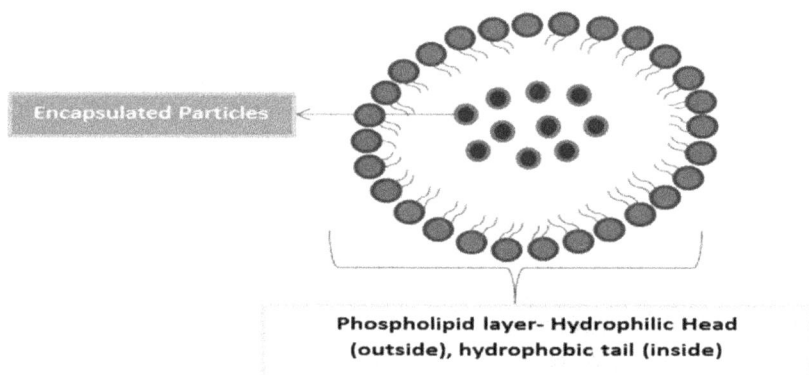

Phospholipid layer- Hydrophilic Head (outside), hydrophobic tail (inside)

FIGURE 10.2 Schematic diagram of the solid lipid nanoparticle.

10.4.9 PHYTOSOME AND LIPOSOME

The anhydrous co-solvent lyophilization, solvent evaporation, anti-solvent precipitation, and precipitation process are frequently used for the formation of phytosomes [3, 24, 42, 43, 47, 56, 58, 68, 69].

The production of Phytosomes occurs by cross-linkage between the equal number of phospholipids (such as phosphatidylethanolamine or phosphatidylcholine, phosphatidylserine) and phytoactive compounds (such as polyphenols) in a solvent (ethyl acetate or dioxane). The production of the Phytosomes involves blending of the phospholipid, biomaterial, and inorganic solvent till the creation of clear solution tracked by evaporation of the solvent and creation of the reedy film (hydration and sonication), respectively (Figure 10.3).

On the other hand, the Liposomes are made up of single or supplementary layers (Figure 10.4) of the used polymer describe as single or numerous washy core partitions [36]. The foremost components used for the formation of the liposomes include phospholipids and cholesterol.

Liposomes are formed as sphere-shaped vesicles and can be classified on the basis of the used polymer, such as natural and/or synthetic phospholipids [25]. The reverse-phase evaporation and Hydration lipid film methods are efficiently used for the production of liposomes [19, 35, 36].

10.4.10 ELECTROSPINNING

The nanofiber can be formed by the process of electrospinning. Because of submicron-nanoscale size with the large surface expanse, the electrospun fibers are responsible for causing vicissitudes to the contiguous surroundings, including all the physical and environmental factors (RH and temperature), and to empower the tunable release of entrapped compounds compared other nanocarriers [60]. In the food sector, the electrospinning technology has moderately unique and less explored potential applications [6, 9, 31, 40, 65]. The successful encapsulation of phytoactive agent within the electrospun fibers is frequently accomplished by mixing the encapsulating agent with the compatible polymer solution followed by electrospinning.

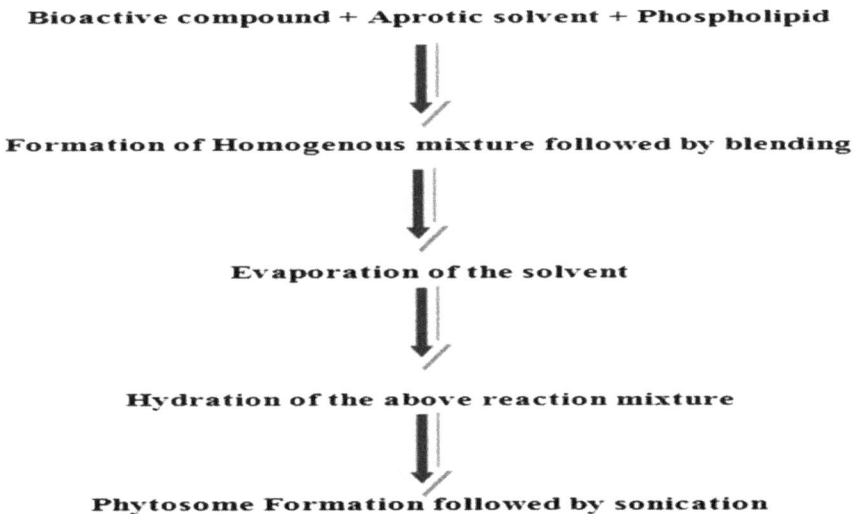

Bioactive compound + Aprotic solvent + Phospholipid

↓

Formation of Homogenous mixture followed by blending

↓

Evaporation of the solvent

↓

Hydration of the above reaction mixture

↓

Phytosome Formation followed by sonication

FIGURE 10.3 Schematic demonstration of the phytosome formation by solvent evaporation and sonication method.

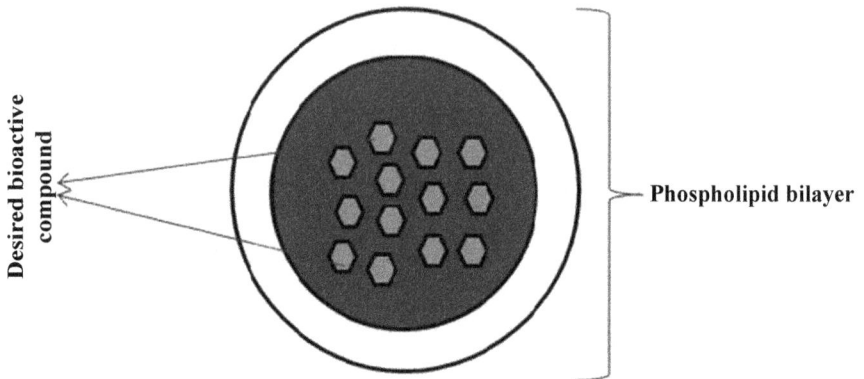

FIGURE 10.4   Schematic diagram of liposome.

10.5   SUMMARY

By using advanced encapsulation techniques, it is possible to enhance the stability, shelf-life, and proper deliverance of the encapsulated compound, which is incorporated into the various kind of foodstuffs. The encapsulation technology is very beneficial for the development of various nutraceutical functional foods.

KEYWORDS

- encapsulation
- flavonoids
- oil-water-oil
- phytoactive compounds
- polyphenols
- sesquiterpene alcohol

REFERENCES

1. Andrade, K. S., Poncelet, D., & Ferreira, S., (2017). Sustainable extraction and encapsulation of pink pepper oil. *Journal of Food Engineering, 204*, 38–45.
2. Armendariz-Barragan, B., Zafar, N., Badri, W., Galindo-Rodriguez, S. A., Kabbaj, D., Fessi, H., & Elaissari, A., (2017). Plant extracts: From encapsulation to application. *Expert Opinion on Drug Delivery, 13*(8), 1165–1175.

3. Awasthi, R., & Kulkarni, J. T. V., (2011). Phytosomes: An approach to increase the bioavailability of plants extracts. *International Journal of Pharmacy and Pharmaceutical Science, 3*(2), 13.

4. Barba, F. J., & Jambrak, A. R., (2017). Innovative technologies for encapsulation of Mediterranean plants extracts. *Trends in Food Science and Technology, 69*, 1–12.

5. Bell, L. N., (2001). Stability testing of nutraceuticals and functional foods. In: Wildman, R. E. C., (ed.), *Handbook of Nutraceuticals and Functional Foods* (pp. 501–516). Boca Raton: CRC Press.

6. Bhushani, J. A., & Anandharamakrishnan, C., (2014). Electrospinning and electrospraying techniques: Potential food-based applications. *Trends in Food Science and Technology, 38*(1), 21–33.

7. Bilati, U., Allemann, E., & Doelker, E., (2005). Development of a nanoprecipitation method intended for the entrapment of hydrophilic drugs into nanoparticles. *European Journal of Pharmaceutical Sciences, 24*(1), 67–75.

8. Castro-Enriquez, D. D., & Montano-Leyva, B., (2019). Stabilization of betalains by encapsulation: A review. *Journal of Food Science and Technology, 2019*, 1–14.

9. Chen, L., Remondetto, G. E., & Subirade, M., (2006). Food protein-based materials as nutraceutical delivery systems. *Trends in Food Science and Technology, 2006*, 272–283.

10. Dolatabadi, J. E. N., Hamishehkar, H., Eskandani, M., & Valizadeh, H., (2014). Formulation, characterization, and cytotoxicity studies of alendronate sodium-loaded solid lipid nanoparticles. *Colloids and Surfaces B: Bio-Interfaces, 117*, 21–28.

11. Fathi, M., Martin, A., & McClements, D. J., (2014). Nanoencapsulation of food ingredients using carbohydrate-based delivery systems. *Trends in Food Science and Technology, 39*(1), 18–39.

12. Fessi, H. P. F. D., & Puisieux, F., (1989). Nanocapsule formation by interfacial polymer deposition following solvent displacement. *International Journal of Pharmaceutics, 55*(1), R1–R4.

13. Galindo-Rodriguez, S. A., Allemann, E., Fessi, H., & Doelker, E., (2005). Polymeric nanoparticles for oral delivery of drugs and vaccines: A critical evaluation of *in vivo* studies. *Critical Review in Therapeutic Drug Carrier Systems, 22*(5), 167–177.

14. Ghayempour, S., Montazer, M., & Rad, M. M., (2016). Encapsulation of aloe vera extract into natural Tragacanth gum as a novel green wound healing product. *International Journal of Biological Macromolecules, 93*, 344–349.

15. Ghorani, B., & Tucker, N., (2015). Fundamentals of electrospinning as a novel delivery vehicle for bioactive compounds in food nanotechnology. *Food Hydrocolloids, 51*, 227–240.

16. Granato, D., Nunes, D. S., & Barba, F. J., (2017). An integrated strategy between food chemistry, biology, nutrition, and statistics in the development of functional foods: A Proposal. *Trends in Food Science and Technology, 62*, 13–22.

17. Guinedi, A. S., Mortada, N. D., Mansour, S., & Hathout, R. M., (2005). Preparation and evaluation of reverse-phase evaporation and multilamellar niosomes as ophthalmic carriers of acetazolamide. *International Journal of Pharmaceutics, 306*(1/2), 71–82.

18. Hashemi, S. M. B., & Khaneghah, A. M., (2017). Fermented sweet lemon juice (*Citrus limetta*) using *Lactobacillus plantarum* LS5: Chemical composition, antioxidant, and antibacterial activities. *Journal of Functional Foods, 38*, 409–414.

19. Heneweer, C., Gendy, S. E. M., & Penate-Medina, O., (2012). Liposomes and inorganic nanoparticles for drug delivery and cancer imaging. *Therapeutic Delivery, 3*, 645–656.

20. Iqbal, M., Zafar, N., Fessi, H., & Elaissari, A., (2015). Double emulsion solvent evaporation techniques used for drug encapsulation. *International Journal of Pharmaceutics, 496*(2), 173–190.

21. Karimi, N., Ghanbarzadeh, B., & Hamishehkar, H., (2015). Phytosome and liposome: The beneficial encapsulation systems in drug delivery and food application. *Applied Food Biotechnology, 2*(3), 17–27.

22. Khoee, S., & Yaghoobian, M., (2009). Investigation into the role of surfactants in controlling particle size of polymeric nanocapsules containing penicillin-G in double emulsion. *European Journal of Medicinal Chemistry, 44*(6), 2392–2399.

23. Kumar, S., Kumar, R., & Khan, A., (2011). Medicinal plant resources: Manifestation and prospects of life-sustaining healthcare system. *Continental Journal of Biological Sciences, 1*, 19–29.

24. Kumari, A., Yadav, S. K., & Yadav, S. C., (2010). Biodegradable polymeric nanoparticles based drug delivery systems. *Colloids Surfaces B Bio-Interfaces, 75*, 1–18.

25. Kusumawati, I., & Yusuf, H., (2011). Phospholipid complex as carrier of *Kaempferia galanga* rhizome extract to improve its analgesic activity. *International Journal of Pharmacy and Pharmaceutical Science, 3*, 44–46.

26. Lertsutthiwong, P., Rojsitthisak, P., & Nimmannit, U., (2009). Preparation of turmeric oil-loaded chitosan-alginate biopolymeric nanocapsules. *Materials Science and Engineering: C, 29*(3), 856–860.

27. Li, Y., Yang, D. J., Chen, S. L., Chen, S. B., & Chan, A. S., (2008). Process parameters and morphology in puerarin, phospholipids, and their complex microparticles generation by supercritical antisolvent precipitation. *International Journal of Pharmaceutics, 359*, 35–45.

28. Lopez-Rubio, A., & Sanchez, E., (2012). Electrospinning is a useful technique for the encapsulation of living bifidobacteria in food hydrocolloids. *Food Hydrocolloids, 28*(1), 159–167.

29. Lv, J., Huang, H., Yu, L., & Whent, M., (2012). Phenolic composition and nutraceutical properties of organic and conventional cinnamon and peppermint. *Food Chemistry, 132*(3), 1442–1450.

30. McClements, D., & Lesmes, U., (2009). Structure-function relationships to guide rational design and fabrication of particulate food delivery systems. *Trends Food Science and Technology, 20*, 448–457.

31. Mendes, A. C., Stephansen, K., & Chronakis, I. S., (2017). Electrospinning of food proteins and polysaccharides. *Food Hydrocolloids, 68*, 53–68.

32. Miladi, K., Ibraheem, D., Iqbal, M., Sfar, S., Fessi, H., & Elaissari, A., (2014). Particles from preformed polymers as carriers for drug delivery. *EXCLI Journal, 13*, 28–37.

33. Miladi, K., Sfar, S., Fessi, H., & Elaissari, A., (2013). Drug carriers in osteoporosis: Preparation, drug encapsulation and applications. *International Journal of Pharmaceutics, 445*(1/2), 181–195.

34. Miladi, K., Sfar, S., Fessi, H., & Elaissari, A., (2016). Nanoprecipitation process: From particle preparation to *in vivo* applications. In: Vauthier, C., & Ponchel, G., (eds.), *Polymer Nanoparticles for Nanomedicines* (pp. 17–53). Switzerland: Springer, Cham, Springer International Publishing.

35. Mohammadi, A., Jafari, S. M., Esfanjani, A. F., & Akhavan, S., (2016). Application of nano-encapsulated olive leaf extract in controlling the oxidative stability of soybean oil. *Food Chemistry, 190*, 513–519.

36. Mora-Huertas, C. E., Fessi, H., & Elaissari, A., (2010). Polymer-based nanocapsules for drug delivery. *International Journal of Pharmaceutics, 385*(1/2), 113–142.

37. Mosaddik, A., Ravinayagam, V., & Elaanthikkal, S., (2018). Development and use of polymeric nanoparticles for the encapsulation and administration of plant extracts. In: *Natural Products as Source of Molecules with Therapeutic Potential* (pp. 391–463). Switzerland: Springer.

38. Nedovic, V., Kalusevic, A., Manojlovic, V., Levic, S., & Bugarski, B., (2011). An overview of encapsulation technologies for food applications. *Procedia Food Science, 1*, 1806–1815.

39. Nieuwland, M., Geerdink, P., & Brier, P., (2013). Food-grade electrospinning of proteins. *Innovative Food Science and Emerging Technologies, 20*, 269–275.

40. Nikmaram, N., Leong, S. Y., & Koubaa, M., (2017). Effect of extrusion on the anti-nutritional factors of food products: An overview. *Food Control, 79*, 62–73.

41. Noruzi, M., (2016). Electrospun nanofibers in agriculture and the food industry: A review. *Journal of the Science of Food and Agriculture, 96*(14), 4663–4678.

42. Pallerla, S. M., & Prabhakar, B., (2013). Review on solid lipid nanoparticles. *International Journal Pharmaceutical Sciences Review and Research, 20*, 196–206.

43. Pathan, R., & Bhandari, U., (2011). Preparation and characterization of embelin phospholipid complex as effective drug delivery tool. *J. Incl. Phenom. Macrocycl. Chem., 69*, 139–147.

44. Paur, I., Carlsen, M. H., Halvorsen, B. L., & Blomhoff, R., (2011). Antioxidants in herbs and spices. In: Benzie, I. F. F., & Wachtel-Galor, S., (eds.), *Herbal Medicine: Biomolecular and Clinical Aspects* (pp. 21–27). Boca Raton - FL: CRC Press.

45. Peng, Q., Gong, T., Zuo, J., Liu, J., Zhao, D., & Zhang, Z., (2008). Enhanced Oral bioavailability of salvianolic acid b by phospholipid complex loaded nanoparticles. *Pharmazie, 63*, 661–666.

46. Petronilho, S., Maraschin, M., Coimbra, M. A., & Rocha, S. M., (2012). *In vitro* and *in vivo* studies of natural products: Challenge for their valuation: Case study of chamomile (*Matricaria recutita* L.). *Industrial Crops and Products, 40*, 1–12.

47. Petrovska, B. B., (2012). Historical review of medicinal plants' usage. *Pharmacognosy Reviews, 6*(11), 1–21.

48. Qin, X., Yang, Y., Fan, T. T., Gong, T., Zhang, X. N., & Huang, Y., (2010). Preparation, characterization and *in vivo* evaluation of bergenin-phospholipid complex. *Acta Pharmacol. Sin., 31*, 127–136.

49. Quintanar-Guerrero, D., Fessi, H., Allemann, E., & Doelker, E., (1996). Influence of stabilizing agents and preparative variables on the formation of poly (D, L-lactic acid) nanoparticles by an emulsification-diffusion technique. *International Journal of Pharmaceutics, 143*(2), 133–141.

50. Ram, D. T., Debnath, S., Babu, M. N., Nath, T. C., & Thejeswi, B., (2012). Review on solid lipid nanoparticles. *Research Journal of Pharmacy and Technology, 5*(11), 2–14.

51. Roby, M. H. H., Sarhan, M. A., Selim, K. A. H., & Khalel, K. I., (2013). Evaluation of antioxidant activity, total phenols and phenolic compounds in thyme (*Thymus vulgaris* L.), sage (*Salvia officinalis* L.), and marjoram (*Origanum majorana* L.) extracts. *Industrial Crops and Products, 43*, 827–831.

52. Sahoo, N., Manchikanti, P., & Dey, S., (2010). Herbal drugs: Standards and regulation. *Fitoterapia, 81*, 462–471.

53. Schuch, A., Deiters, P., Henne, J., Kohler, K., & Schuchmann, H. P., (2013). Production of W/O/W (water-in-oil-in-water) multiple emulsions: Droplet breakup and release of water. *Journal of Colloid and Interface Science, 402*, 157–164.

54. Seju, U., Kumar, A., & Sawant, K. K., (2011). Development and evaluation of olanzapine-loaded PLGA nanoparticles for nose-to-brain delivery: *In vitro* and *in vivo* studies. *Acta Biomaterials, 7*, 4169–4176.

55. Semalty, A., Semalty, M., Singh, D., & Rawat, M. S. M., (2010). Preparation and characterization of phospholipid complexes of naringenin for effective drug delivery. *Journal of Inclusion Phenomena and Macrocyclic Chemistry, 67*, 253–260.

56. Sikarwar, M. S., Sharma, S., Jain, A. K., & Parial, S. D., (2008). Preparation, characterization, and evaluation of marsupsin-phospholipid complex. *AAPS Pharmaceuticals Science and Technology, 9*, 129–137.

57. Singh, D., Rawat, M. S., Semalty, A., & Semalty, M., (2012). Rutin phospholipid complex: An innovative technique in novel drug delivery system- NDDS. *Current Drug Delivery, 9*, 305–314.

58. Singh, D., Rawat, M. S. M., Semalty, A., & Semalty, M., (2013). Chrysophanol-phospholipids complex. *Journal of Thermal Analysis and Calorimetry, 111*(3), 2069–2077.

59. Van, D. V. H., & Vermeersch, M., (2011). PLGA nanoparticles loaded with the antileishmanial saponin β-aescin: Factor influence study and *in vitro* efficacy evaluation. *International Journal of Pharmaceutics, 420*(1), 122–132.

60. Vega-Lugo, A. C., & Lim, L. T., (2009). Controlled release of allyl isothiocyanate using soy protein and poly(lactic acid) electrospun fibers. *Food Research International, 42*(8), 933–940.

61. Vincekovic, M., Viskic, M., & Juric, S., (2017). Innovative technologies for encapsulation of Mediterranean plants extracts. *Trends Food Science and Technology, 69*, 1–12.

62. Vincekovic, M., Viskic, M., Juric, S., & Giacometti, J., (2014). Bioactive components from leaf vegetable products. In: *Studies in Natural Products Chemistry* (Vol. 41, pp. 321–346).

63. Vos, P., Faas, M. M., Spasojevic, M., & Sikkema, J., (2010). Review: Encapsulation for preservation of functionality and targeted delivery of bioactive food components. *International Dairy Journal, 20*, 292–302.

64. Wandrey, C., Bartkowia, A., & Harding, S. E., (2010). Materials for encapsulation. In: Zuidam, N. J., & Nedovic, V., (eds.), *Encapsulation Technologies for Active Food Ingredients and Food Processing* (pp. 31–100). New York: Springer-Verlag.

65. Wen, P., Zong, M. H., Linhardt, R. J., Feng, K., & Wu, H., (2017). Electrospinning: A novel nano-encapsulation approach for bioactive compounds. *Trends in Food Science and Technology, 70*, 56–68.

66. Yourdkhani, M., Leme-Kraus, A. A., Aydin, B., Bedran-Russo, A. K., & White, S. R., (2017). Encapsulation of grape seed extract in polylactide microcapsules for sustained bioactivity and time-dependent release in dental material applications. *Dental Materials, 33*(6), 630–636.

67. Zafar, N., Fessi, H., & Elaissari, A., (2014). Cyclodextrin containing biodegradable particles: From preparation to drug delivery applications. *International Journal of Pharmaceutical, 461*, 351–366.

68. Zaidi, S. M. A., Pathan, S. A., Ahmad, F. J., Surender, S., Jamil, S., & Khar, R. K., (2011). Neuropharmacological evaluation of *Paeonia emodi* root extract phospholipid complex in mice. *Planta Medicine, 77*, 123.

69. Zhang, J., Gao, W., Bai, S. H., Chen, H., Qiang, Q., & Liu, Z., (2011). Glycyrrhizic acid-phospholipid complex: Preparation process optimization and therapeutic and pharmacokinetic evaluation in rats. *American Journal of Pharmaceutical, 30,* 1621–1630.
70. Zuidam, N. J., Jublot, L., Suijker, M. J., Ziere, A., & Smit, G., (2009). Encapsulates to deliver great flavors to food products. In: *XVII International Conference on Bioencapsulation* (p. 20).

CHAPTER 11

ENCAPSULATION OF PROBIOTICS BY ELECTROSPINNING

ADITYA P. SUKUMAR, P. DEVIKRISHNA,
F. MAGDALINE ELJEEVA EMERALD, HEARTWIN A. PUSHPADASS,
and B. SURENDRA NATH

ABSTRACT

The International Dairy Federation (IDF) has raised the availability of a population of probiotic microorganisms from 10^6 CFU/g to at least 10^7 CFU/g in dairy products throughout their shelf-life. Therefore, it is mandatory to protect the probiotic microorganisms from an adverse environment and to maintain their viability. Encapsulation of probiotic microorganisms can ensure the controlled dosage using spray drying, freeze-drying, extrusion, coacervation, and spray chilling, etc. Electrospinning can be an alternative to these methods to encapsulate probiotics in the form of nano-microfibers with desired characteristics for food applications. Moreover, it is a cost-effective and scalable technology, and is carried out at ambient conditions without the risk of thermal damage to the microorganisms. This chapter focuses on various techniques for encapsulation of probiotic microorganisms.

11.1 INTRODUCTION

The probiotics concept was first instituted in the twentieth century by Elie Metchnikoff, and the global market for probiotics and probiotic foods is witnessing a growth of 7%. The world market of probiotics is valued at US$49.4 billion in 2018 and is anticipated to attain US$ 69.3 billion in 2023 [73, 90].

Probiotics deliver many health benefits on the host by enhancing the balance of intestinal microbiota. The most prominent definition of probiotics was given by FAO/WHO as *"live microorganisms that, when administered in adequate amounts, confer a health benefit on the host"* [23]. Active digestion capacity, modulation of gut beneficial microflora, antagonistic activity against pathogens, improved colon integrity, immunomodulation, down-regulated allergic responses, anti-carcinogenic properties, reduction of serum cholesterol, mitigation of lactose intolerance and nutritional enrichment (production of B complex vitamins and absorption of calcium) of probiotics lead to the incorporation of probiotic microorganisms in dairy products, such as yogurt and *dahi* [39]. Commercially used probiotic strains are *L. plantarum, L. acidophilus, L. rhamnosus, L. reuteri, L. casei, E. faecium, S. thermophilus, B. brief, B. bifidum, B. lactis,* and *B. longum* [79].

There are several approaches to enhance the stability of probiotics in foods, such as fortification with probiotic substrates, aseptic packaging, and adjustment of pH, refrigerated storage, and freeze-drying. Encapsulation is an alternative method to protect the delivery of probiotics into the intestines and the GI transit for selective release at the targeted area (colon) [43].

Encapsulation of probiotics to protect them throughout processing and storage, and during their transit in the GI tract is a key area of research in academia and industry [50]. Encapsulation of probiotics is done by coacervation, spray drying, freeze-drying, spray chilling, emulsification, extrusion, etc., [11, 50]. Electrospinning is a simple technique that does not demand extreme conditions of temperature, pressure, use of chemical solvents, but favors the use of a wide variety of food-grade, biocompatible, and biodegradable polymers as encapsulating materials [8, 28, 29].

This chapter focuses on novel techniques for encapsulation of probiotics.

11.2 TECHNIQUES FOR ENCAPSULATION OF PROBIOTICS

Encapsulation of probiotics has provided the scope for microorganisms to be immobilized within semipermeable and biocompatible materials that govern the release of probiotics in the GI tract [17] to protect them against the adverse environments rather than controlled release [98]. Selection of a particular technique requires analysis of probiotic viability, process conditions encountered during food processing; storage conditions prior to consumption of end product, size, and density of encapsulate needed, mechanism of release, and economic constraints [97].

11.2.1 SPRAY DRYING

The major controlling parameters in spray drying of encapsulation (Figure 11.1) are gas flow rate and drying air temperature. Reported inlet air temperatures for probiotic encapsulation ranges from 100 to 170°C and outlet air temperature ranges from 45 to 105°C [79].

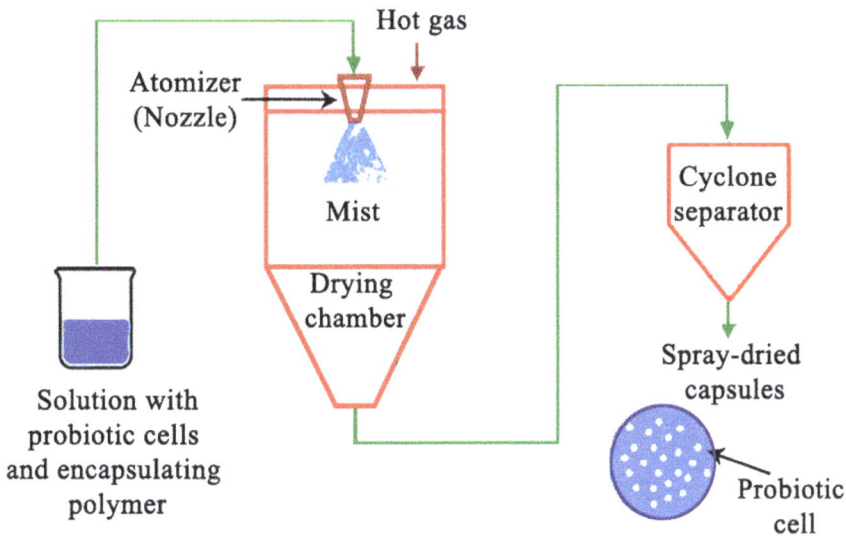

FIGURE 11.1 Encapsulation by spray drying.

Despite many advantages of this method, the high temperatures to evaporate water affect the activity and viability of the probiotic microorganisms in the dried powder. The cellular membrane is most susceptible and the targeted part of cells may cause cellular pores, leading to the expulsion of the intracellular substances [5]. However, proper adjustments and control of inlet and outlet temperatures limit the adverse effects.

11.2.2 SPRAY CHILLING

Possibility of using lipid matrix in spray chilling for probiotic encapsulation [26, 65] is highly favorable as the lipid carrier offers high viability during encapsulation and also during storage by blocking it from high stressors such as water and H^+ ions. The use of lipid carriers allows the probiotics to be released near the favorable site of interest as the lipid matrix is easily digested by lipases in the intestine [25, 65].

L. acidophilus, S. boulardii, and B. bifidum were successfully encapsulated as single and double-layered structures using spray drying and spray chilling methods with hydrogenated palm oil as wall material [6]. Similarly, B. animalis subsp. lactis and L. acidophilus were successfully encapsulated in cocoa butter by this technique [69].

11.2.3 FREEZE DRYING

Direct freeze-drying involves mixing of cryoprotectants in media containing probiotics and then freeze-drying the mixture to remove water [8, 53, 82]. In the indirect method, probiotics are first immobilized within the polymer matrix and then transferred to the freeze drier for lyophilization. Microencapsulation of L. paracasei by spray freeze-drying in trehalose and maltodextrin matrix [84] was successful with high viability (>60%), and trehalose concentration was a critical factor governing the final viability of the encapsulated probiotics.

11.2.4 EMULSIFICATION

Encapsulating probiotics in emulsion droplets greatly enhances the survivability of cells under adverse conditions of the digestive system [32, 34, 46, 75]. Hou et al. reported that L. delbrueckii ssp. bulgaricus encapsulated in emulsion formed by sesame oil showed approximately 10^4 times higher survivability under simulated GI tract conditions as compared to free cells (Figure 11.2).

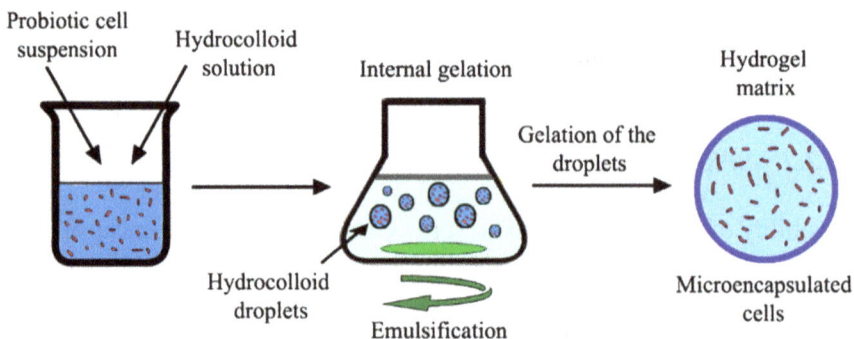

FIGURE 11.2 Encapsulation by emulsification.

Pimentel-González et al. [70] encapsulated *L. rhamnosus* cultured in sweet whey in the inner aqueous phase of a double water-in-oil-in-water emulsion employing whey as an emulsifier. The study concluded that *L. rhamnosus* was protected against simulated GI tract conditions using the double emulsion formulation.

11.2.5 COACERVATION

The microencapsulation of *Bifidobacterium lactis* and *Lactobacillus acidophilus* was achieved by complex coacervation method followed by spouted bed drying, with a combination of pectin and casein as wall materials [16, 40]. These encapsulated probiotics were more viable in acidic media than as free cells [66]. Encapsulation of probiotics by coacervation offers several advantages, such as no requirement of high temperature or use of organic solvents, and cost-effective [91]. Coacervation also allows the incorporation of a large amount of microorganisms in relation to the encapsulant. However, it is a batch process and scale-up is difficult. In addition, further drying of the precipitate is needed to extend its shelf-life, which could be detrimental to the probiotics.

11.2.6 EXTRUSION

Extrusion involves mixing of probiotics with hydrocolloids, followed by extrusion through an orifice as droplets into a solution containing cross-linking agent. As the polymer comes in contact with the cross-linking agent, gelation occurs (Figure 11.3). Controlling factors are: orifice diameter, feed flow rate, solution viscosity, temperature, and the distance between orifice and cross-linking agent [12, 21].

Many polymers can be used to form encapsules by this technique, including alginates, *k*-carrageenan, and whey proteins. Among them, alginate, gellan, and xanthan gums have been considered for encapsulating probiotics [79]. However, this technique is ineffective in producing particles below 500 µm, and requires optimal viscosity of polymer solution and nozzles of relatively large diameter [76, 77].

Shah and Ravula [85] encapsulated *L. acidophilus* and *Bifidobacterium* using calcium alginate in a fermented milk-based frozen dessert. Free and encapsulated microorganisms were added to the product before freezing. The results indicated that encapsulation improved the survivability of probiotic bacteria. Similarly, *L. casei* was encapsulated by extrusion technique with alginates as encapsulating material [22]. Probiotic cell survival study was carried out under simulated gastric GI conditions during processing of fermented milk, and also during storage at 4°C. The results revealed that the alginate matrix had great potential to encapsulate the probiotics and guard them against unfavorable conditions of the GI tract.

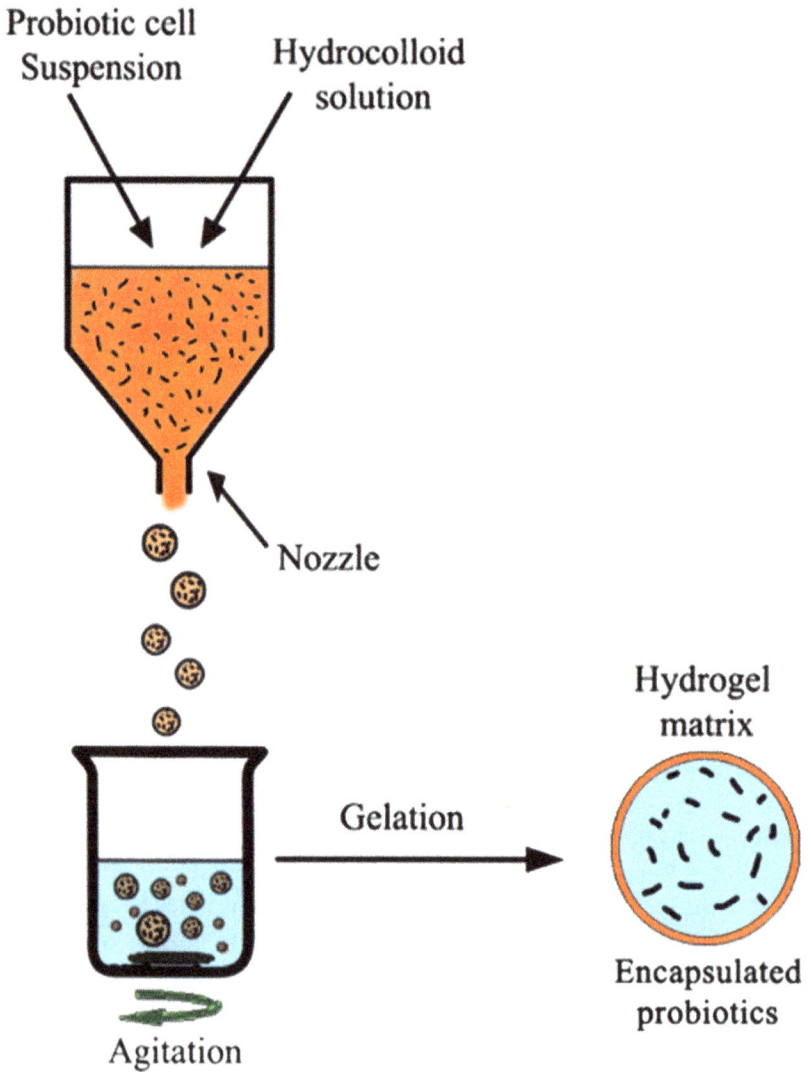

FIGURE 11.3 Encapsulation by extrusion.

11.3 ELECTROSPINNING FOR ENCAPSULATION OF PROBIOTICS

Electrospinning consists of a high voltage source, a capillary tube with spinneret, and a metal collector (Figure 11.4). The high voltage creates an electrically charged liquid jet between electrodes having opposite polarity. When the electrostatic repulsion exceeds surface tension, the hemispherical surface of the droplet at the spinneret tip elongates to form the 'Taylor cone' [36, 52].

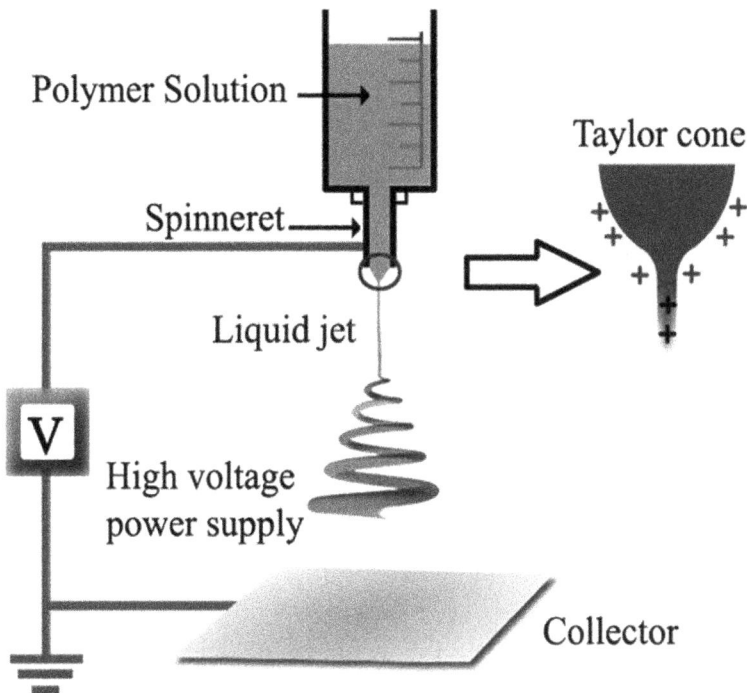

FIGURE 11.4 Electrospinning setup.

Electrospinning completely removes the solvent in the dried fibers, and the protective encapsulant helps to enhance the stability of probiotic microorganisms and extend their shelf life. The electrospraying is based on the variation in the polymer solution concentration. When the polymer concentration or entanglement of molecular chains in the liquid is sufficient, the jet formed in the process does not disperse into droplets but forms into a micro or nanofiber and the process is called electrospinning [38].

Electrospun fibers could be used for encapsulation to maintain the stability and viability of probiotics and other microorganisms during processing and storage, using carbohydrates and proteins as wall materials [13]. Salalha et al. [81] enhanced the viability of *E. coli* and *St. albus* by encapsulating in electrospun nanofibers of polyvinyl alcohol (PVA). The authors postulated that electrospinning opened a new field in developing an alternative to lyophilization to preserve microorganisms.

Keskin et al. [44] converted cyclodextrin into ultrathin nanofibers by electrospinning to encapsulate *Lysinibacillus sp.* for bioremediation in wastewater treatment. The bacteria showed viability for 7 days at 4°C, and it was reported that the nanofibers obtained could be used to remove heavy metals from wastewater. López-Rubio et al. [49] encapsulated *B. animalis ssp. lactis* using electrospinning in PVOH fibers. The encapsulated probiotic bacteria retained viability for 40 days at ambient temperature, and 130 days at refrigerated conditions. In contrast, the viability was decreased for non-encapsulated bacteria.

An attempt to encapsulate *L. acidophilus* in soluble dietary fibers from soybean (okara), oil palm fond and oil palm trunk by electrospinning was reported [28]. The cell viability was 78.6–90%, which was retained at 4°C for 21 days. López-Rubio et al. [51] highlighted the feasibility of electrospinning to encapsulate *B. animalis ssp. lactis* using pullulan and whey protein concentrate (WPC) as encapsulants as they are capable of forming micro-nano capsules. Encapsulation by electrospinning substantially increased the viability of microbial strain, and even at high relative humidity (RH), WPC showed better protection ability than pullulan. Nagy et al. [60] prepared a nanofibrous solid dosage of *L. acidophilus* for curing bacterial vaginosis by electrospinning.

Coghetto et al. [19] applied electrospraying technique to microencapsulate *L. planatarum* using Na-alginate and Na-alginate-citric-pectin matrices as encapsulating materials. The authors demonstrated that the survivability of organism increased in the digestive system and in refrigerated conditions, and the cells remained stable and viable over 6 months at 25°C.

Feng et al. [27] encapsulated *L. planatarum* using electrospinning and studied its effect on viability and thermal stability. Fructooligosaccharides (FOS) were used as encapsulating material. It was observed that the addition of 2.5% FOS improved thermal stability and viability of the probiotic organisms. In another study, the feasibility of electrohydrodynamic atomization to encapsulate *L. casei* using whey protein isolate (WPI) and WPC cross-linked with transglutaminase enzyme was reported [2]. Encapsulation by this method had high efficiency, which subsequently increased the stability of electrosprayed capsules in GI fluids [2].

Hu et al. [35] developed probiotic biofilm integrated electrospun nanofiber membranes using *L. plantarum* as a starter culture for the production of fermented milk. The authors reported that GI resistance and the amount of viable cells increased after 3 h incubation in simulated GI media. The fermentation time was shorter due to the higher viability of microorganisms.

11.3.1   WALL MATERIALS FOR ENCAPSULATION

The interest in the application of electrospun fibers by food industries depends upon the ability of biopolymers to form fibers in micro to nanosize, and encapsulation of probiotics [78]. The typical properties of biopolymers (such as biocompatibility, biodegradability), and physicochemical behavior create a good demand for these products as encapsulating wall material [24, 78]. Carbohydrates (alginate, carrageenan, xanthan, pectin, chitosan, pullulan, fructooligosaccharides, inulin) and proteins (whey proteins, caseins, soy proteins, gelatin, and zein) have been employed for encapsulation of probiotic organisms as they are recognized as safe (GRAS) materials. To avoid the use of toxic ingredients, polymers for electrospinning mostly use water or ethanol as a solvent for food applications [24, 42].

The viability of the encapsulated probiotics is affected by the concentration, type, and characteristics of the encapsulating material. It is reported that more than fifty polymers have been converted to ultrafine fibers successfully by electrospinning with diameters ranging from <3 nm to 1 µm [36].

11.3.1.1   PROTEINS

Common protein-based encapsulating materials for probiotics are WPI [2], WPC [51, 55], zein [48], and gelatin [30], etc. Utilization of both WPC and WPI is essential to produce a resistant and coherent structure for encapsulation of probiotics [2]. Studies reported the feasibility of WPC along with pullulan to encapsulate *Bifidobacterium* by electrospinning [51]. The authors reported a greater enhancement in cell viability for WPC-based microcapsules as compared to pullulan-based structures. Librán et al. [48] used zein, along with other polymers like WPC, maltodextrin, etc., to encapsulate *B. longum* subsp. *infantis.* A lower protective ability as compared to other polymers was reported in zein-maltodextrin combination probably due to its excessive water intake.

Gómez-Mascaraque et al. [30] introduced gelatin-coated WPC microcapsules of *L. plantarum* by coaxial electrospraying, and concluded that gelatin-coated capsules were unable to protect the microorganisms from adverse conditions.

11.3.1.2 CARBOHYDRATES

Polysaccharides are promising vehicles for micro-and nanoencapsulation of probiotics as they are biodegradable, biocompatible, and could be modified to achieve the desired properties [29, 83]. Based on their origin, carbohydrate-based delivery systems can be classified as: plants (cellulose, starch, guar gum, and pectin), animals (chitosan), and algae (carrageenan and alginate), and microbes (cyclodextrins, xanthan, and dextran) [24]. Inulin [62, 80], fructooligosaccharides [27], maltodextrin [48], cyclodextrin [44], pullulan [51], alginate [47] and gum Arabic [94] are found to be satisfactory for electrospinning of probiotics.

Inulin can favor and stimulate the growth of intestinal *Bifidobacterium* and stimulate insulin secretion or reduce serum glucose [62]. Moayyedi et al. [54] used inulin as an encapsulating material and demonstrated that the addition of inulin protected *L. rhamnosus* against bile, NaCl, Penicillin G, and lysozyme. However, the application of inulin as a wall material was limited by its sticky and hygroscopic behaviors. These properties could be improved by enhancing the transition temperature (Tg) using high molecular weight whey proteins. Rosolen et al. [80] evaluated the efficiency of inulin for microencapsulation of *L. lactis* ssp. *lactis* R7 by spray drying. It was observed that microencapsulated *L. lactis* R7 showed better protection to adverse environment when inulin and cheese whey were combinedly for use as wall material.

Fructooligosaccharides (FOS) occur naturally in garlic, onion, asparagus, chicory, artichoke, banana, and among other plants. Rajam and Anandharamakrishnan [74] analyzed the effect of FOS as wall material for encapsulation of *L. plantarum* (MTCC 5422) by spray drying and concluded that its high hygroscopic nature and low molecular weight limited its use as encapsulating matrix. Addition of WPI or WPI to FOS facilitated aggregation and reduced the stickiness of particles. However, the low conductivity and viscosity of FOS make it tough to electrospun into nanofibers [27].

Lopez-Rubio et al. [50] used pullulan [18] for encapsulation of *Bifido-bacterium* by electrospinning, and reported that smooth and round capsules with ultrafine fibers were formed.

11.3.2 PARAMETERS INFLUENCING ELECTROSPINNING

Viscosity, elasticity, thermal conductivity, surface tension, etc., (feed solution parameters) and applied voltage, flow rate, and tip to collector distance (TCD) (process parameters) can influence electrospinning. Altering these parameters makes it possible to produce fibers with desired morphology and microstructure. Bead formation is mainly influenced by surface tension, viscosity of polymer solution and the net charge density carried by the liquid-jet. The conductivity of the solution and applied electrostatic field also influence net charge density carried by the moving jet. As the net charge density decreases, the beads become smaller and more spindle-like, thereby yielding fibers of smaller diameter [92].

The most critical parameter that influences the electrospinning process is the applied voltage. Generally, application of higher voltage ejects more fluid from the spinneret, yielding fibers of larger diameter. Škrlec et al. [87] studied the viability of *L. plantarum* after electrospinning at 10, 15, and 20 kV-applied voltage. *L. plantarum* showed the highest viability at 15 kV. Maintaining the viability during long-term storage is much more vital and important than achieving higher viability of probiotics immediately after electrospinning. The quantity of solution at the spinneret tip in electrospinning is governed by the flow rate. For a certain applied voltage, a corresponding flow rate maintains a stable Taylor cone. High flow rates result in fibers with large diameters, while low flow rates yield uniform electrospun fibers [7].

Flow rates of 0.3–5 mL/h [2, 50, 94, 87] were used in electrospinning for encapsulation of probiotics. As the diameter of the spinneret increases, plugging may occur at the spinneret tip. However, if the diameter is too small, it is difficult to form droplet from the feed solution [96]. The elongation of the polymer jet takes place between the spinneret tip and the collector. The distance from the tip to the collector (TCD) has a direct effect on the time taken by the polymer jet to reach the collector and the electrostatic field strength. Insufficient TCD may result in inadequate drying of fibers [29]. Commonly used TCD in the electrospinning of probiotics is 100–150 mm [2, 94].

11.3.3   CHARACTERISTICS OF ELECTROSPUN PROBIOTICS

It is important to evaluate and characterize the encapsulated probiotics in terms of their morphological, molecular, and crystallographic properties using scanning electron microscopy (SEM), Fourier transform infrared spectroscopy (FTIR) or optical microscopy, etc. In addition, the viability of microorganisms after electrospinning and during storage needs to be ascertained. Lopez-Rubio et al. [7] evaluated the SEM micrographs of *B. animalis* Bb12 encapsulated in pullulan and WPC capsules, and reported that round, smooth, and ultrafine capsules were obtained in case of pullulan with a diameter >1 μm. Fung et al. [10] reported that at 8% concentration of PVA, the SEM images displayed the formation of uniform, smooth, and bead free electrospun nanofibers.

FTIR can identify the functional groups of materials. It helps to understand how successfully the probiotic strains get encapsulated in the nanofiber mats. This technique aids in measuring the absorption of infrared radiation, which identify various molecular groups and structures. Škrlec et al. [87] used FTIR to identify the effect of drying on the bacterial proteins during electrospinning, and concluded that the presence of trehalose (lyoprotectant) enhanced probiotic survival in electrospun nanofibers up to 24 weeks.

The viability of microorganisms after the electrospinning process could be estimated by using the pour plate method. Alehosseini et al. [2] reported a survival rate of 99.4% after electrospinning and 84.2% for freeze-drying. This difference might be because of lower applied stress during electrospinning compared to freeze-drying. The authors also added that the bacteria had only 2 log reduction in the number of viable cells after electrospinning compared to freeze-dried cultures. Electrospraying did not affect the viability of probiotic cells, which showed greater efficiency than the other methods of preparation of dried cultures [50]. It was observed that the viability of *B. animalis* Bb12 after electrospinning showed a decrease of 1 log unit while the freeze-dried process had 3 log reduction.

Mohaisen et al. [55] reported 81.1% viability after electrospinning of *L. brevis*. The authors added that this viability percentage was significant, and the electrospinning process appeared to provide the least effect on inactivation of probiotic microorganisms.

11.3.4   PROBIOTIC ATTRIBUTES

The selection of strain is very important in the encapsulation of probiotics. It is done based on the *in-vitro* tolerance to physiological stresses, such as low pH, bile, and increased osmolarity. Probiotic bacterial cells are sensitive to these stresses, and encapsulation of probiotics cells is presumed to protect the bacterial cells from these adverse conditions in the gastrointestinal (GI) tract.

11.3.4.1   BILE SALT RESISTANCE

The probiotic strains possess the natural ability to hydrolyze bile salts. Though many bile salt hydrolases (BSH) have been identified and characterized, yet it is not yet clear if BSH activity is a favorable trait in probiotic microorganisms. BSH has been reported in *Bifidobacterium, Lactobacillus, Enterococcus,* etc., [9]. Lactobacilli are the largest contributors to BSH activity in the GI tract [45]. Zhang et al. [95] reported that probiotic microorganisms could be effectively safeguarded by encapsulation, thereby creating the necessity of bile salt to permeate through the wall materials enclosing the probiotic cells. For example, *L. acidophilus* encapsulated in alginate in the form of microspheres considerably enhanced its viability and survival against 1% bile salt [15].

11.3.4.2   ACID TOLERANCE

Probiotic bacteria must be tolerant to the acidic conditions in the stomach (pH 1.5–3) to survive through it. Acid tolerance of probiotics is evaluated by simple *in-vitro* tests, which can ascertain the quality of probiotic cultures. Zhang et al. [95] reported the effect of encapsulation of *B. bifidum* against adverse acidic environment in a simulated gastric fluid (SGF) and simulated intestinal fluid (SIF). Encapsulation of *B. bifidum* exerted a protective effect and enhanced its survivability under such conditions. Therefore, it is concluded that electrospinning of probiotic bacteria protects the viability and survivability in the GI environment, and maintain their stability under storage.

11.4   FORTIFICATION OF ENCAPSULATED PROBIOTICS

Several foods are fortified with probiotics to enhance their functional properties. However, the present trend is the use of encapsulated probiotics to offer protection of probiotics from varying levels of oxygen, temperature, freezing, and acidic conditions during processing and storage, and finally to maintain viability during GI transit [63].

11.4.1 DAIRY FOODS

Many functional dairy products fortified with probiotics have successfully been manufactured (Table 11.1). However, some situations could be inimical to the viability of probiotics, such as prohibitory compounds produced by other competitive starter cultures in fermented milk, thermal stresses occurring during pasteurization, freezing conditions prevailing in products such as frozen desserts and ice creams, inhibition by other additive components, such as salts, and flavors, and low water activity conditions in dehydrated products. Thus, the fortification of probiotics in encapsulated form is a better alternative to overcome these adverse conditions [14].

TABLE 11.1 Some Dairy Foods Fortified with Encapsulated Probiotics

Dairy Food	Encapsulation Method	Probiotic Strain(s) Fortified	Wall Material(s)	References
Cheddar cheese	Droplet extrusion and emulsion	B. longum	Sodium alginate and palmitoylated alginate	[3]
Iranian Doogh	Extrusion	L. acidophilus LA-5 B. lactis Bb-12	Alginate	[52]
Kasar cheese	Extrusion	L. acidophilus LA-5 B. bifidum BB-12	Alginate	[68]
Kefir	Extrusion	B. animalis	Sodium alginate	[31]
Yogurt	Extrusion	B. bifidum F-35	Whey/alginate	[58]
Yogurt-ice cream	Extrusion	L. acidophilus La-5	Sodium alginate	[1]

Studies reported that yogurt fortified with encapsulated probiotics exhibited increased survival of probiotic microorganisms during storage, and the therapeutic value of yogurt was enhanced [4]. A 7-week storage study of yogurt made with probiotic bacteria encapsulated with calcium-induced alginate-starch indicated an improvement in survivability by 1 and 2 log population of B. lactis and L. acidophilus, respectively, owing to the protection offered by microencapsulation. There was no significant effect on the physicochemical and organoleptic qualities of yogurt during storage. However, a slight variation in texture (smoothness) of yogurt was observed compared to that made with non-encapsulated probiotics [41].

Microentrapment in alginate beads proved to be a potential solution to protect probiotic lactic cultures against attacks of bacteriophages, while the cells immobilized in alginate gels exhibited low proteolytic activity in milk, thereby preventing off-flavors in dairy products [14]. Ice cream containing 1% resistant starch with free and encapsulated *B. lactis* (Bb-12) and *L. casei* (Lc-01) were prepared, and their survival was monitored for 180 days at –20°C. The results revealed that encapsulation could considerably enhance the survival rate of these probiotics in ice cream by 30% without any adverse effect on the organoleptic properties [33].

Partly skimmed Mozzarella cheese was fortified with *L. paracasei ssp. paracasei* LBC-1 in free and microencapsulated forms at levels of 10^7 and 10^8 CFU/g, respectively. Results indicated better survival of encapsulated strains throughout the pasta-filata process of cheese making [67]. Probiotics encapsulated in carrageenan or alginate beads had greater viability in various dairy-based frozen desserts and ice creams. Encapsulation enabled the use of cryoprotective compounds, such as glycerol in the beads, without having them added to the ice-cream mix [88].

Chios mastic gum (*Pistacia lentiscus var. chia*) was examined as an encapsulant for immobilization of *L. casei* ATCC393 for the development of functional fermented milk. These gums exhibited antimicrobial properties against many microorganisms without significantly affecting the probiotic cell counts during 8 weeks of storage [89]. The functional properties of yogurt drink were improved by the addition of *L. brevis,* which was encapsulated in sodium alginate by electrospinning. Results indicated enhanced viability of encapsulated probiotics as compared to free microorganisms, and the use of electrospinning led to a stable yet more functional food with enhanced health benefits [55].

11.4.2   OTHER FOODS

Probiotics are added to milk and milk products traditionally. Vegetable and fruit juices can also act as delivery media for probiotics [71]. Some of the non-dairy foods fortified with probiotic microorganisms are listed in Table 11.2.

11.4.3   CYTOTOXICITY AND BIOLOGICAL EFFECTS OF PROBIOTIC ENCAPSULATES

Though the beneficial effects of probiotics on human health are well known, many studies revealed that their stability and survival are very sensitive to processing and GI environment of the host. Cytotoxicity should be considered so that the selected polymer for encapsulation of probiotics and the technique should not be cytotoxic and non-antimicrobial to ensure that the host and the probiotic bacteria are not inhibited.

TABLE 11.2 Some Non-Dairy Foods Fortified with Encapsulated Probiotics

Food Product	Encapsulation Method	Probiotic Strain(s) Fortified	Wall Material(s)	References
Apple juice	Electrospinning	*L. brevis*	Sodium alginate	[55]
	Spray drying	*L. rhamnosus*	WPI or WPI+ modified resistant starch	[93]
Carrot juice	Extrusion	*L. casei*	Chitosan-Ca-alginate	[37]
Dry fermented sausages	Extrusion	*L. reuteri*	Alginate	[59]
Fruit juices	Freeze drying	*B. longum, B. breve*	Poly-γ-glutamic acid	[10]
Grape juice	Internal gelation	*L. acidophilus, B. bifidum*	Alginate	[56]
Mango juice	Gelation	*L. plantarum*	Calcium alginate-soy protein isolate	[72]
Orange juice	Electrospinning	*L. brevis*	Sodium alginate	[55]
Pomegranate juice	Extrusion	*L. plantarum*	Alginate beads with chitosan	[64]

In a study by Singh et al. [86], *L. rhamnosus* entrapped in cellulose particles was evaluated for its viability, cytotoxicity, and stability. The colon Caco-2 cell line was used to assess the cytotoxicity of the probiotics encapsulated with cellulose. The results indicated that the particles exhibited a very low toxic profile to the cell line, and the formulation made with cellulose presented the best bacterial survivability when exposed to simulated GI fluids. Similarly, novel probiotic formulations comprising three different probiotic strains namely *L. rhamnosus*, *L. plantarum*, and *B. animalis subsp. Lactis,* taken both individually as well as a mixture of them, in the form of chitosan-coated alginate microcapsules did not show any specific cytotoxic effect [20].

11.5 SUMMARY

There is a difficulty to electrospin many food-grade polymers, due to their solubility, insufficient molecular entanglement, poor viscoelastic characteristics, etc. However, electrospun encapsulates as a delivery vehicle for probiotics in industrial scale is still limited and needs to be researched. Similarly, many studies reported the stability and viability of electrospun probiotics during storage, but the loss of their functionality and their synergy during long-term storage need to be ascertained.

KEYWORDS

- **acid tolerance**
- **bile salt tolerance**
- **cytotoxicity**
- **electrospinning**
- **electrospraying**
- **probiotics**
- **wall materials**

REFERENCES

1. Ahmadi, A., Milani, E., Madadlou, A., & Mortazavi, S., (2014). Synbiotic yogurt-ice cream produced *via* incorporation of microencapsulated *Lactobacillus acidophilus* (la-5) and fructooligosaccharide. *Journal of Food Science and Technology, 51*(8), 1568–1574.
2. Alehosseini, A., Sarabi-Jamab, M., Ghorani, B., & Kadkhodaee, R., (2019). Electro-encapsulation of *Lactobacillus casei* in high-resistant capsules of whey protein containing transglutaminase enzyme. *LWT-Food Science and Technology, 102*, 150–158.
3. Amine, K. M., Champagne, C. P., & Raymond, Y., (2014). Survival of microencapsulated *Bifidobacterium longum* in cheddar cheese during production and storage. *Food Control, 37*, 193–199.
4. And, C. I., & Kailasapathy, K., (2005). Effect of co-encapsulation of probiotics with prebiotics on increasing the viability of encapsulated bacteria under *in vitro* acidic and bile salt conditions and in yogurt. *Journal of Food Science, 70*(1), 18–23.
5. Anekella, K., & Orsat, V., (2013). Optimization of microencapsulation of probiotics in raspberry juice by spray drying. *LWT-Food Science and Technology, 50*(1), 17–24.
6. Arslan-Tontul, S., & Erbas, M., (2017). Single and double-layered microencapsulation of probiotics by spray drying and spray chilling. *LWT-Food Science and Technology, 81*, 160–169.
7. Ballengee, J. B., & Pintauro, P. N., (2011). Morphological control of electrospun Nafion nanofiber mats. *Journal of the Electrochemical Society, 158*(5), 568–572.

8. Basholli-Salihu, M., Mueller, M., Salar-Behzadi, S., Unger, F. M., & Viernstein, H., (2014). Effect of lyoprotectants on B-glucosidase activity and viability of *Bifidobacterium infantis* after freeze-drying and storage in milk and low pH juices. *LWT-Food Science and Technology, 57*(1), 276–282.
9. Begley, M., Hill, C., & Gahan, C. G., (2006). Bile salt hydrolase activity in probiotics. *Applied and Environmental Microbiology, 72*(3), 1729–1738.
10. Bhat, A. R., Irorere, V. U., Bartlett, T., Hill, D., Kedia, G., Charalampopoulos, D., Nualkaekul, S., & Radecka, I., (2015). Improving survival of probiotic bacteria using bacterial poly-gamma-glutamic acid. *International Journal of Food Microbiology, 196,* 24–31.
11. Bhushani, A. J., & Anandharamakrishnan, C., (2014). Electrospinning and electrospraying techniques: Potential food-based applications. *Trends in Food Science and Technology, 38*(1), 21–33.
12. Brun-Graeppi, A. K. A. S., Richard, C., & Bessodes, M., (2011). Cell microcarriers and microcapsules of stimuli-responsive polymers. *Journal of Controlled Release, 149*(3), 209–224.
13. Burgain, J., Gaiani, C., Linder, M., & Scher, J., (2011). Encapsulation of probiotic living cells: From laboratory scale to industrial applications. *Journal of Food Engineering, 104*(4), 467–483.
14. Champagne, C. P., & Fustier, P., (2007). Microencapsulation for delivery of probiotics and other ingredients in functional dairy products: Chapter 19. In: Saarela, M., (ed.), *Functional Dairy Products* (Vol. 2, pp. 404–426). New York-USA: Woodhead Publishing.
15. Chandramouli, V., Kailasapathy, K., Peiris, P., & Jones, M., (2004). Improved method of microencapsulation and its evaluation to protect *Lactobacillus* spp. in simulated gastric conditions. *Journal of Microbiological Methods, 56*(1), 27–35.
16. Chávarri, M., Marañón, I., & Villarán, M. C., (2012). Encapsulation technology to protect probiotic bacteria: Chapter 23. In: Rigobelo, E., (ed.), *Probiotics* (pp. 501–540). IntechOpen.
17. Chen, M. J., & Chen, K. N., (2007). Applications of probiotic encapsulation in dairy products: Chapter 4. In: Jamileh, M. L., (ed.), *Encapsulation and Controlled Release Technologies in Food Systems* (pp. 83–107). New York, USA: Wiley-Blackwell.
18. Cheng, K. C., Demirci, A., & Catchmark, J. M., (2011). Pullulan: Biosynthesis, production, and applications. *Applied Microbiology and Biotechnology, 92*(1), 29–44.
19. Coghetto, C. C., Brinques, G. B., & Siqueira, N. M., (2016). Electrospraying microencapsulation of *Lactobacillus plantarum* enhances cell viability under refrigeration storage and simulated gastric and intestinal fluids. *Journal of Functional Foods, 24,* 316–326.
20. D'Orazio, G., & Di Gennaro, P., (2015). Microencapsulation of new probiotic formulations for gastrointestinal delivery: *In vitro* study to assess viability and biological properties. *Applied Microbiology and Biotechnology, 99*(22), 9779–9789.
21. De Vos, P., & Faas, M. M., (2010). Encapsulation for preservation of functionality and targeted delivery of bioactive food components. *International Dairy Journal, 20*(4), 292–302.
22. Dimitrellou, D., Kandylis, P., & Lević, S., (2019). Encapsulation of *Lactobacillus casei* ATCC 393 in alginate capsules for probiotic fermented milk production. *LWT-Food Science and Technology, 116,* 1–7.
23. Probiotics in Food, (2006). *Report of a Joint FAO/WHO Consultation on Evaluation of Health and Nutritional Properties of Probiotics in Food Including Powder Milk with*

Live Lactic Acid Bacteria (Vol. 85, pp. 1–56). Cordoba, Argentina-2001; FAO Food and Nutrition Paper.

24. Fathi, M., Martín, A., & McClements, D., (2014). Nanoencapsulation of food ingredients using carbohydrate-based delivery systems. *Trends in Food Science and Technology, 39*(1), 18–39.
25. Favaro-Trindade, C. S., Heinemann, R. J. B., & Pedroso, D. L., (2011). Developments in probiotic encapsulation. *CAB Reviews, 6*(4), 1–8.
26. Favaro-Trindade, C. S., Okuro, P. K., & Matos, Jr, F. E., (2015). Encapsulation via spray chilling/cooling/congealing: Chapter 5. In: Mishra, M., (ed.), *Handbook of Encapsulation and Controlled Release* (pp. 71–88). Boca Raton-USA: CRC Press.
27. Feng, K., Zhai, M. Y., Zhang, Y., & Linhardt, R. J., (2018). Improved viability and thermal stability of the probiotics encapsulated in a novel electrospun fiber mat. *Journal of Agricultural and Food Chemistry, 66*(41), 10890–10897.
28. Fung, W. Y., Yuen, K. H., & Liong, M. T., (2011). Agro-waste-based nanofibers as a probiotic encapsulant: Fabrication and characterization. *Journal of Agricultural and Food Chemistry, 59*(15), 8140–8147.
29. Ghorani, B., & Tucker, N., (2015). Fundamentals of electrospinning as a novel delivery vehicle for bioactive compounds in food nanotechnology. *Food Hydrocolloids, 51,* 227–240.
30. Gómez-Mascaraque, L. G., Ambrosio-Martín, J., Perez-Masiá, R., & Lopez-Rubio, A., (2017). Impact of acetic acid on the survival of *L. plantarum* upon microencapsulation by coaxial electrospraying. *Journal of Healthcare Engineering, 2017,* 1–6.
31. Gonzalez-Sanchez, F., & Azaola, A., (2010). Viability of microencapsulated *Bifidobacterium animalis* ssp. *lactis* BB12 in kefir during refrigerated storage. *International Journal of Dairy Technology, 63*(3), 431–436.
32. Heidebach, T., Först, P., & Kulozik, U., (2012). Microencapsulation of probiotic cells for food applications. *Critical Reviews in Food Science and Nutrition, 52*(4), 291–311.
33. Homayouni, A., Azizi, A., Ehsani, M. R., Yarmand, M. S., & Razavi, S. H., (2008). Effect of microencapsulation and resistant starch on the probiotic survival and sensory properties of synbiotic ice cream. *Food Chemistry, 111*(1), 50–55.
34. Hou, R. C. W., Lin, M. Y., Wang, M. M. C., & Tzen, J. T. C., (2003). Increase of viability of entrapped cells of *Lactobacillus delbrueckii* ssp. *bulgaricus* in artificial sesame oil emulsions. *Journal of Dairy Science, 86*(2), 424–428.
35. Hu, M. X., Li, J. N., Guo, Q., Zhu, Y. Q., & Niu, H. M., (2019). Probiotics biofilm-integrated electrospun nanofiber membranes: A new starter culture for fermented milk production. *Journal of Agricultural and Food Chemistry, 67*(11), 3198–3208.
36. Huang, Z. M., Zhang, Y. Z., Kotaki, M., & Ramakrishna, S., (2003). Review on polymer nanofibers by electrospinning and their applications in nanocomposites. *Composites Science and Technology, 63*(15), 2223–2253.
37. Ivanovska, T. P., & Petrushevska-Tozi, L., (2014). Optimization of synbiotic microparticles prepared by spray drying to development of new functional carrot juice. *Chemical Industry and Chemical Engineering Quarterly, 20*(4), 549–564.
38. Jaworek, A., & Sobczyk, A. T., (2008). Electrospraying route to nanotechnology: An overview. *Journal of Electrostatics, 66*(3/4), 197–219.
39. Jayalalitha, V., (2010). *Microencapsulation of Probiotics and Incorporation in Yoghurt* (p. 233). Doctoral dissertation; Tamil Nadu Veterinary and Animal Sciences University; Chennai.

40. John, R. P., Tyagi, R. D., Brar, S. K., Surampalli, R. Y., & Prévost, D., (2011). Bio-encapsulation of microbial cells for targeted agricultural delivery. *Critical Reviews in Biotechnology, 31*(3), 211–226.

41. Kailasapathy, K., (2006). Survival of free and encapsulated probiotic bacteria and their effect on the sensory properties of yoghurt. *LWT-Food Science and Technology, 39*(10), 1221–1227.

42. Kayaci, F., & Uyar, T., (2012). Encapsulation of vanillin/cyclodextrin inclusion complex in electrospun polyvinyl alcohol (PVA) nanowebs: Prolonged shelf-life and high temperature stability of vanillin. *Food Chemistry, 133*(3), 641–649.

43. Kechagia, M., & Basoulis, D., (2013). Health benefits of probiotics: A review. *ISRN Nutrition, 2013*, 1–7.

44. Keskin, N. O. S., Celebioglu, A., Sarioglu, O. F., Uyar, T., & Tekinay, T., (2018). Encapsulation of living bacteria in electrospun cyclodextrin ultrathin fibers for bioremediation of heavy metals and reactive dye from wastewater. *Colloids and Surfaces B: Biointerfaces, 161*, 169–176.

45. Klaenhammer, T. R., & Kullen, M. J., (1999). Selection and design of probiotics. *International Journal of Food Microbiology, 50*(1/2), 45–57.

46. Krasaekoopt, W., Bhandari, B., & Deeth, H., (2004). The influence of coating materials on some properties of alginate beads and survivability of microencapsulated probiotic bacteria. *International Dairy Journal, 14*(8), 737–743.

47. Laelorspoen, N., Wongsasulak, S., Yoovidhya, T., & Devahastin, S., (2014). Microencapsulation of *Lactobacillus acidophilus* in zein-alginate core-shell microcapsules *via* electrospraying. *Journal of Functional Foods, 7*, 342–349.

48. Librán, C. M., Castro, S., & Lagaron, J. M., (2017). Encapsulation by electrospray coating atomization of probiotic strains. *Innovative Food Science and Emerging Technologies, 39*, 216–222.

49. López-Rubio, A., & Sanchez, E., (2009). Encapsulation of living *Bifidobacteria* in ultrathin PVOH electrospun fibers. *Biomacromolecules, 10*(10), 2823–2829.

50. Lopez-Rubio, A., & Lagaron, J. M., (2012). Whey protein capsules obtained through electrospraying for the encapsulation of bio-actives. *Innovative Food Science and Emerging Technologies, 13*, 200–206.

51. López-Rubio, A., & Sanchez, E., (2012). Electrospinning is a useful technique for the encapsulation of living *Bifidobacteria* in food hydrocolloids. *Food Hydrocolloids, 28*(1), 159–167.

52. Meral, R., Alav, A., Karakas, C., Dertli, E., Yilmaz, M. T., & Ceylan, Z., (2019). Effect of electrospun nisin and curcumin loaded nanomats on the microbial quality, hardness, and sensory characteristics of rainbow trout fillet. *LWT-Food Science and Technology, 113*, 1–8.

53. Mills, S., Stanton, C., Fitzgerald, G. F., & Ross, R., (2011). Enhancing the stress responses of probiotics for a lifestyle from gut to product and back again. *Microbial Cell Factories, 10*(1), 1–15.

54. Moayyedi, M., & Eskandari, M. H., (2018). Effect of drying methods (electrospraying, freeze-drying and spray drying) on survival and viability of microencapsulated *Lactobacillus rhamnosus* ATCC 7469. *Journal of Functional Foods, 40*, 391–399.

55. Mohaisen, M. J. M., Yildirim, R. M., Yilmaz, M. T., & Durak, M. Z., (2019). Production of functional yogurt drink, apple, and orange juice using nano-encapsulated *L. Brevis* within sodium alginate-based biopolymers. *Science of Advanced Materials, 11*(12), 1788–1797.

56. Mokhtari, S., Jafari, S. M., & Khomeiri, M., (2019). Survival of encapsulated probiotics in pasteurized grape juice and evaluation of their properties during storage. *Food Science and Technology International, 25*(2), 120–129.

57. Mortazavian, A. M., & Ehsani, M. R., (2008). Viability of calcium-alginate-microencapsulated probiotic bacteria in Iranian yogurt drink during refrigerated storage and under simulated gastrointestinal conditions. *Australian Journal of Dairy Technology, 63*(1), 25–30.

58. Mousa, A., Liu, X. M., Chen, Y. Q., Zhang, H., & Chen, W., (2014). Evaluation of physicochemical, textural, microbiological, and sensory characteristics in set yogurt reinforced by microencapsulated *Bifidobacterium bifidum* F-35. *International Journal of Food Science and Technology, 49*(7), 1673–1679.

59. Muthukumarasamy, P., & Holley, R. A., (2006). Microbiological and sensory quality of dry fermented sausages containing alginate-microencapsulated *Lactobacillus reuteri*. *International Journal of Food Microbiology, 111*(2), 164–169.

60. Nagy, Z. K., Wagner, I., Suhajda, Á., & Tobak, T., (2014). Nanofibrous solid dosage form of living bacteria prepared by electrospinning. *Express Polymer Letters, 8*(5), 352–361.

61. Nieuwland, M., Geerdink, P., & Brier, P., (2013). Food grade electrospinning of proteins. *Innovative Food Science and Emerging Technologies, 20*, 269–275.

62. Niness, K. R., (1999). Inulin and oligofructose: What are they? *The Journal of Nutrition, 129*(7), 1402–1406.

63. Ningtyas, D. W., Bhandari, B., Bansal, N., & Prakash, S., (2019). The viability of probiotic *Lactobacillus rhamnosus* (non-encapsulated and encapsulated) in functional reduced-fat cream cheese and its textural properties during storage. *Food Control, 100*, 8–16.

64. Nualkaekul, S., Lenton, D., & Cook, M. T., (2012). Chitosan coated alginate beads for the survival of microencapsulated *Lactobacillus plantarum* in pomegranate juice. *Carbohydrate Polymers, 90*(3), 1281–1287.

65. Okuro, P. K., De Matos, J. F. E., & Favaro-Trindade, C. S., (2013). Technological challenges for spray chilling encapsulation of functional food ingredients. *Food Technology and Biotechnology, 51*(2), 171–182.

66. Oliveira, A. C., Moretti, T. S., & Boschini, C., (2007). Microencapsulation of *B. lactis* (BI 01) and *L. acidophilus* (LAC 4) by complex coacervation followed by spouted-bed drying. *Drying Technology, 25*(10), 1687–1693.

67. Ortakci, F., & Broadbent, J. R., (2012). Survival of microencapsulated probiotic *Lactobacillus paracasei* LBC-1e during manufacture of mozzarella cheese and simulated gastric digestion. *Journal of Dairy Science, 95*(11), 6274–6281.

68. Özer, B., Uzun, Y. S., & Kirmaci, H. A., (2008). Effect of microencapsulation on viability of *Lactobacillus acidophilus* LA-5 and *Bifidobacterium bifidum* BB-12 during kasar cheese ripening. *International Journal of Dairy Technology, 61*(3), 237–244.

69. Pedroso, D. L., Dogenski, M., & Thomazini, M., (2013). Microencapsulation of *Bifidobacterium animalis* subsp. *lactis* and *Lactobacillus acidophilus* in cocoa butter using spray-chilling technology. *Brazilian Journal of Microbiology, 44*(3), 777–783.

70. Pimentel-González, D. J., & Campos-Montiel, R. G., (2009). Encapsulation of *Lactobacillus rhamnosus* in double emulsions formulated with sweet whey as emulsifier and survival in simulated gastrointestinal conditions. *Food Research International, 42*(2), 292–297.

71. Prado, F. C., Parada, J. L., Pandey, A., & Soccol, C. R., (2008). Trends in non-dairy probiotic beverages. *Food Research International, 41*(2), 111–123.

72. Praepanitchai, O. A., Noomhorm, A., & Anal, A. K., (2019). Survival and behavior of encapsulated probiotics (*Lactobacillus plantarum*) in calcium-alginate-soy protein isolate-based hydrogel beads in different processing conditions (pH and temperature) and in pasteurized mango juice. *BioMed Research International, 2019*, 1–8.

73. Anonymous (2019). *Probiotics Market by Application, Ingredient (Bacteria, Yeast), Form (Dry, Liquid), End User, and Region-Global Forecast to 2023* (p. 169). Top Market Reports.   https://www.marketsandmarkets.com/Market-Reports/probiotic-market-advanced-technologies-and-global-market-69.html (accessed on 19 January 2021).

74. Rajam, R., & Anandharamakrishnan, C., (2015). Microencapsulation of *Lactobacillus plantarum* (MTCC 5422) with fructooligosaccharide as wall material by spray drying. *LWT-Food Science and Technology, 60*(2), 773–780.

75. Rao, A. V., Shiwnarain, N., & Maharaj, I., (1989). Survival of microencapsulated *Bifidobacterium pseudolongum* in simulated gastric and intestinal juices. *Canadian Institute of Food Science and Technology Journal, 22*(4), 345–349.

76. Rathore, S., & Desai, P. M., (2013). Microencapsulation of microbial cells. *Journal of Food Engineering, 116*(2), 369–381.

77. Reis, C. P., & Neufeld, R. J., (2006). Review and current status of emulsion/dispersion technology using an internal gelation process for the design of alginate particles. *Journal of Microencapsulation, 23*(3), 245–257.

78. Rezaei, A., Nasirpour, A., & Fathi, M., (2015). Application of cellulosic nanofibers in food science using electrospinning and its potential risk. *Comprehensive Reviews in Food Science and Food Safety, 14*(3), 269–284.

79. Rokka, S., & Rantamaki, P., (2010). Protecting probiotic bacteria by microencapsulation: Challenges for industrial applications. *European Food Research and Technology, 231(1)*, 1–12.

80. Rosolen, M. D., Bordini, F. W., & De Oliveira, P. D., (2019). Symbiotic microencapsulation of *Lactococcus lactis* subsp. *lactis* R7 using whey and inulin by spray drying. *LWT-Food Science and Technology, 115*, 1–7.

81. Salalha, W., Kuhn, J., Dror, Y., & Zussman, E., (2006). Encapsulation of bacteria and viruses in electrospun nanofibers. *Nanotechnology, 17*(18), 4675–4681.

82. Santivarangkna, C., Kulozik, U., & Foerst, P., (2007). Alternative drying processes for the industrial preservation of lactic acid starter cultures. *Biotechnology Progress, 23*(2), 302–315.

83. Schiffman, J. D., & Schauer, C. L., (2008). A review: Electrospinning of biopolymer nanofibers and their applications. *Polymer Reviews, 48*(2), 317–352.

84. Semyonov, D., Ramon, O., Kaplun, Z., & Levin-Brener, L., (2010). Microencapsulation of *Lactobacillus paracasei* by spray freeze-drying. *Food Research International, 43*(1), 193–202.

85. Shah, N. P., & Ravula, R. R., (2000). Microencapsulation of probiotic bacteria and their survival in frozen fermented dairy desserts. *Australian Journal of Dairy Technology, 55*(3), 139–144.

86. Singh, P., Medronho, B., & Dos, S. T., (2018). On the viability, cytotoxicity, and stability of probiotic bacteria entrapped in cellulose-based particles. *Food Hydrocolloids, 82*, 457–465.

87. Škrlec, K., Zupančič, Š., & Mihevc, S. P., (2019). Development of electrospun nanofibers that enable high loading and long-term viability of probiotics. *European Journal of Pharmaceutics and Biopharmaceutics, 136*, 108–119.

88. Sultana, K., & Godward, G., (2000). Encapsulation of probiotic bacteria with alginate-starch and evaluation of survival in simulated gastrointestinal conditions and in yoghurt. *International Journal of Food Microbiology, 62*(1–2), 47–55.

89. Terpou, A., Nigam, P. S., Bosnea, L., & Kanellaki, M., (2018). Evaluation of Chios mastic gum as antimicrobial agent and matrix-forming material targeting probiotic cell encapsulation for functional fermented milk production. *LWT-Food Science and Technology, 97*, 109–116.

90. Terpou, A., Papadaki, A., & Lappa, I. K., (2019). Probiotics in food systems: Significance and emerging strategies towards improved viability and delivery of enhanced beneficial value. *Nutrients, 11*(7), 1591–1598.

91. Thies, C., (2016). Encapsulation by complex coacervation: Chapter 3. In: Jamileh, M. L., (ed.), *Encapsulation and Controlled Release Technologies in Food Systems* (pp. 41–71). New York, USA: Wiley-Blackwell.

92. Walczak, Z. K., (2002). *Processes of Fiber Formation* (1st edn., p. 397). Oxford, United Kingdom: Elsevier.

93. Ying, D., Schwander, S., & Weerakkody, R., (2013). Microencapsulated *Lactobacillus rhamnosus* GG in whey protein and resistant starch matrices: Probiotic survival in fruit juice. *Journal of Functional Foods, 5*(1), 98–105.

94. Zaeim, D., & Sarabi-Jamab, M., (2018). Electrospray-assisted drying of live probiotics in acacia gum microparticles matrix. *Carbohydrate Polymers, 183*, 183–191.

95. Zhang, F., Li, X. Y., Park, H. J., & Zhao, M., (2013). Effect of microencapsulation methods on the survival of freeze-dried *Bifidobacterium bifidum*. *Journal of Microencapsulation, 30*(6), 511–518.

96. Zhao, S., Wu, X., Wang, L., & Huang, Y., (2004). Electrospinning of ethylecyanoethyl cellulose/tetrahydrofuran solutions. *Journal of Applied Polymer Science, 91*(1), 242–246.

97. Zuidam, N. J., & Shimoni, E., (2010). *Encapsulation Technologies for Active Food Ingredients and Food Processing* (p. 400). New York, USA: Springer-Verlag.

98. Zuidam, N. J., & Shimoni, E., (2010). Overview of microencapsulates for use in food products or processes and methods to take them: Chapter 2. In: Zuidam, N. J., & Nedovic, V., (eds.), *Encapsulation Technologies for Active Food Ingredients and Food Processing* (pp. 3–29). New York-USA: Springer-Verlag.

CHAPTER 12

DESIGN AND APPLICATIONS OF MECHANICAL EXPELLERS IN DAIRY, FOOD, AND AGRICULTURAL PROCESSING

RAJASEKHAR TELLABATI and REKHA RAVINDRA MENON

ABSTRACT

Expeller pressing is a mechanical method for extracting liquids like oil; water entrapped in the matrix of solid material, such as oil-bearing seeds, fish, food waste, and cattle waste. The extraction can be done by a solvent, supercritical fluid, and mechanical expelling. Mechanical expression with a suitable design is a good alternative in view of the growing demand of natural essential oils with food-grade quality. The principle of operation of hydraulic and screw expellers, design, and applications are discussed in this chapter. The design considerations optimization of the expeller for a particular application is narrated. Pretreatments, machine variables, yield performance, must satisfy the final product quality requirements.

12.1 INTRODUCTION

Extraction is a process in which the selective component is separated from the sample material. The extraction can be done by a solvent, supercritical fluid and mechanical expelling. The specifications, quantity, and quality of the obtained component vary with the method of extraction.

Mechanical expression of oil can be accomplished by subjecting the oilseeds under pressure that are placed between permeable barriers. Under such pressure, the compaction of oil seeds occurs, releasing the oil from matrix [10]. The different kinds of equipment for mechanical expression are hydraulic press, uniaxial or screw press. The hydraulic press is a batch method, while the screw press can be continuously operated with a slightly higher yield in case of oil expulsion.

Screw expeller is one of the mechanical oil extracting methods. The principle of extraction involves the crushing of seeds with the pressure generated between the screw and barrel surface. The crushed oil seeds are squeezed due to compaction, and the oil is expelled through barrel perforation while the cake is extruded through the die or annular space provided at the discharge end. Oil can be extracted from corn, sunflower, cottonseed, peanuts, palm kernel, rice bran, etc. The screw expeller has low oil recovery compared to solvent extraction method and can be improved by certain pretreatments.

The screw expelling method can be used on a small scale, for speciality products or as a pre-press stage in a large solvent extraction process. The performance of the screw press is affected by various operating parameters, such as temperature, pressure, speed of rotation of screw and other pretreatments like cooking, dehulling, flaking, etc. Screw expeller occupies less space, less capital required and is suitable for small-scale production [34].

This chapter explores the importance and applications of the mechanical expeller and its design aspects and its applications in dairy, food, and agricultural processing.

12.2 SOLID-LIQUID EXTRACTION

Solid-Liquid extraction can be done chemically with a suitable solvent and physically with the help of a mechanical expression using hydraulic or screw presses.

Solvent extraction is a process, where organic solvents (e.g., hydrocarbons, ketones, alcohols) are used to dissolve the targeted component embedded in the solid matrix (e.g., oil-rich seeds and nuts). This is a capital intensive method as refining of oil is required before utilization. The toxicity of solvents renders the spent unutilized as fodder. The explosion hazard due to the organic solvent is also inevitable. The factors that affect the extraction efficiency of the oil in solvent extraction depends on the type of solvent, contact time and temperature, pretreatments like cooking, the thickness of flake [50]. Solvent extraction is capable of extracting 98% of oil [36].

Supercritical fluid extraction is a process where solids or liquids extracted are held at pressure and temperature above the critical point of solvent [41]. Supercritical fluid has properties of both liquid as well as gas and can penetrate substances easily like gases and dissolve materials easily like liquids. The different steps are involved in supercritical fluid extraction (SFE) (Figure 12.1). The compressed CO_2 is passed over the grains at the rate of 40 g/min for 3 hours, and the system is held at 50°C and 30 MPa. The extracted oil or component is collected in a separate depressurized vessel, and CO_2 gas is sent to the diaphragm compressor for the next cycle [41].

Mechanical pressing is the method of compression to expel the liquid from the porous matrix. Pressing can be accomplished by hydraulic or pneumatic press and screw press. In batch method, hydraulic or pneumatic press can be utilized while for continuous method screw expeller can be used.

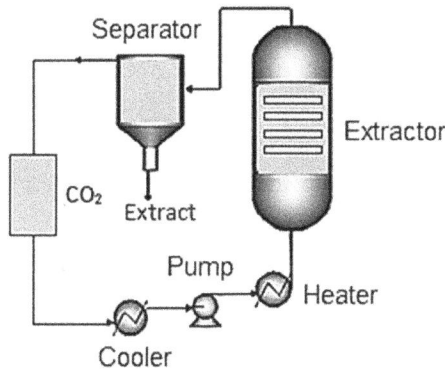

FIGURE 12.1   Steps involved in supercritical fluid extraction (SFE).

12.3   MECHANICAL EXTRACTION SYSTEMS

12.3.1   HYDRAULIC EXPELLER

Hydraulic press is used in the production of speciality oil, e.g., cocoa butter, olive oil, batch type whey removal from paneer, ghee recovery from ghee residue, etc. The pressure applied varies based on the product. Hydraulic presses are classified into open type and closed type [50]:

- In open type hydraulic press, the product is wrapped in cloths or other filtering fabric and pressure is applied directly on it. During pressing, a corrugated metal plate is placed in between each block.
- In closed type press, a steel cage with sufficient strength is designed to withstand the developed pressure. The pressure is applied by the

hydraulic piston to the desired level over the oil seeds or nuts that are placed in the steel cage [50]. Gradually as the pressure reaches the desired level, the oil from the seeds is expelled through the perforations of the steel cage.

A typical hydraulic press consists of a hydraulic unit, plunger, sieve plate or a cage, collection chamber, frame, support, etc. (Figure 12.2).

During the initial stage of hydraulic pressing, the voids filled with air are forced out due to compressive load, as the process continues at a critical point the oil starts flowing. In the second dynamic stage, the liquid displaces the air and air-fluid is extracted [38]. In the final stage, the pores are completely filled with expelled oil and the rate of expulsion is at the maximum (Figure 12.3).

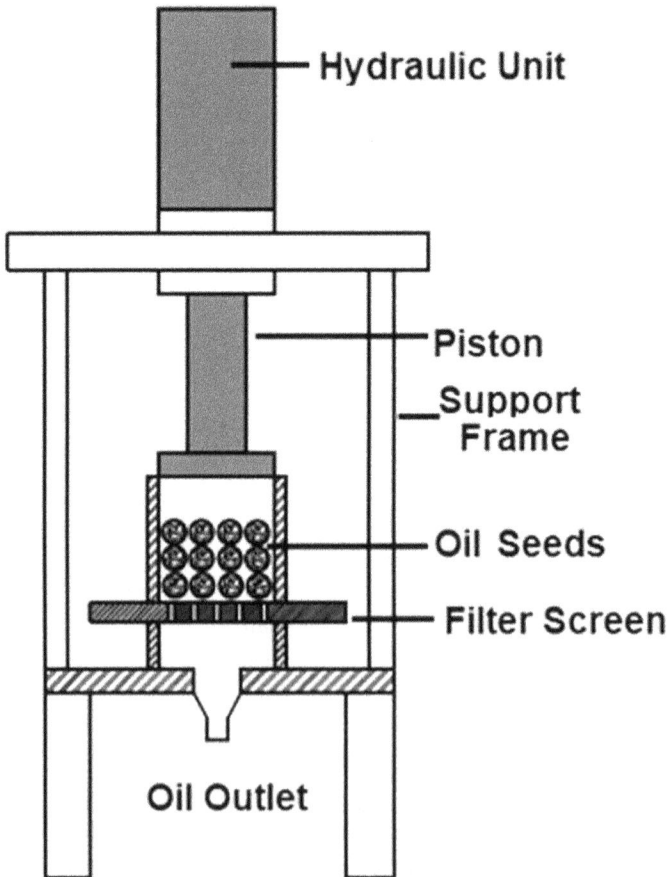

FIGURE 12.2   Schematic view of hydraulic expeller.

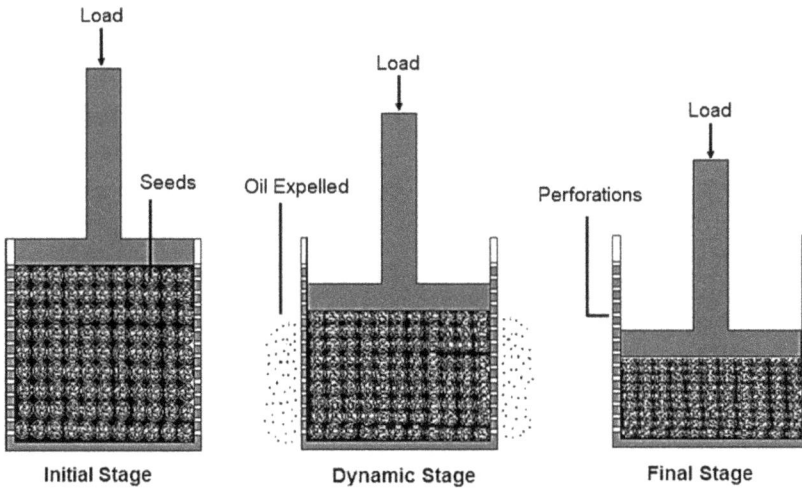

FIGURE 12.3   Different stages during hydraulic pressing.

12.3.2   SCREW PRESS EXPELLER

Screw expeller is used for extracting oil from raw materials, such as oilseeds, nuts, etc. The main parts of the screw expeller are drive, hopper, perforated barrel, main screw shaft, pressure cone/die, cutter, and supporting frame. The design of perforations and screw shaft is product specific. The compression ratio plays an important role to expel the liquid from the solid phase [4]. The pressure developed by screw crushes the oilseeds, nuts, and squeezing action expels the oil through perforation provided in the barrel and soil is collected in trough [40]. The residue mass is extruded as cake through the annular space or die. The recovery of oil varies with the design of screw expeller, pretreatments, and available oil content in the raw material. The extracted oil is used as a food product or for industrial purpose. The different oil products include peanut oil, coconut oil, rice bran oil, fish oil, corn oil, sunflower oil, jatropha oil, canola oil, etc.

12.4   DESIGN CONSIDERATIONS OF A MECHANICAL EXPELLER

12.4.1   DESIGN OF A SCREW EXPELLER

The design considerations include material of construction, design of screw, optimization of shaft length with respect to screw pitch, the rotational speed of the shaft, taper angle through the length of the shaft, and the dimensions of the barrel [22, 39]. The design and selection of screw shaft for expulsion and extrusion play a vital role in terms of performance, power consumption, wear, and tear of screw and characteristics of the conveyed material [1]. The screw shaft design depends on the raw material and the pressure to be developed during the process. The design and selection of screw, barrel, die geometries, and operating conditions are dependent on the characteristics of a particular product.

The simulation study on various screw designs indicated that screw with tapered shaft reduced cost of maintenance and improved life span [23]. Various designs include constant pitch screw with a straight shaft, constant pitch with tapered shaft, variable pitch with a straight shaft, and variable pitch with tapered shaft (Figure 12.4(a)). The screw with a tapered shaft and constant pitch had a life span of 5950 hours, which was double when compared to other designs [23].

CFD simulation of different dependent and independent variables during the development of pressure through the barrel due to screw pressing indicated a pressure rise of 1 MPa at the exit [13].

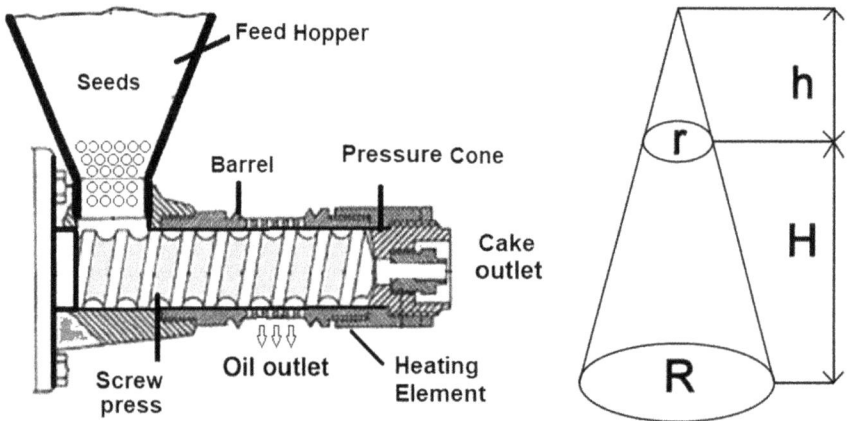

FIGURE 12.4   (a) Cross-sectional view of a screw press oil expeller; (b) design of a hopper.

12.4.1.1   HOPPER DESIGN

Hoppers with round or square outlets are designed with respect to the frustum volume of the pyramid. The hopper angle is considered for mass-flow rather than funnel flow. The volume of the hopper (Figure 12.4(b)) is obtained by deducting the volume of the smaller pyramid from the larger pyramid [35].

$$\text{Volume of Hopper, } V = \frac{1}{2}\pi\left(R^2 H - r^2 h\right) \tag{1}$$

where; V is the volume of the hopper (m³), R is an outer radius (m), H is the height of the cone (m), r is the inner radius (m); and h is an inner height (m).

12.4.1.2 DESIGN OF A SCREW SHAFT

The design of the screw shaft is critical as the torsion; compressive stresses are generated during the crushing of oilseeds. Sufficient squeezing force must be produced to reach the oil-point. The shaft can be designed based on maximum shear stress.

$$d_s = \left(\frac{16.T.f_s}{\tau_y.\pi}\right)^{\!1/3} \tag{2}$$

where; d_s is the diameter of screw shaft (m), T is torque (N.m), t_y is yield stress (N/m²) and f_s is the factor of safety.

12.4.1.3 COMPRESSION RATIO

The compression ratio in screw press expeller is defined as the ratio between volumes displaced in one rotation at feed end to the volume per rotation at discharge end. The compression curve varies along the length of the barrel in the screw press (Figure 12.5). It is divided in the feed zone, ram zone and plug zone [33].

The screw shaft can be preferably tapered screw conveyor with decreasing volumetric displacement from feed end to the discharge end. The ratio compression ratio affects the residual moisture or oil content in the pressed cake.

$$\text{Compression ratio} = \frac{Volume\ displaced\ per\ revolution\ at\ feed\ end}{Volume\ displaced\ per\ revolution\ at\ discharge\ end} \tag{3}$$

Design of successive decrease in screw depth in tapered screw shaft from the feed end to discharge end is given as below:

$$h_s = \frac{h_{fe} - h_{de}}{n-1} \tag{4}$$

where; h_{fe} is screw depth at feed end (m), h_{de} is screw depth at discharge end (m), n is the number of screw flights or turns and h_s is the successive decrease in screw depth (m).

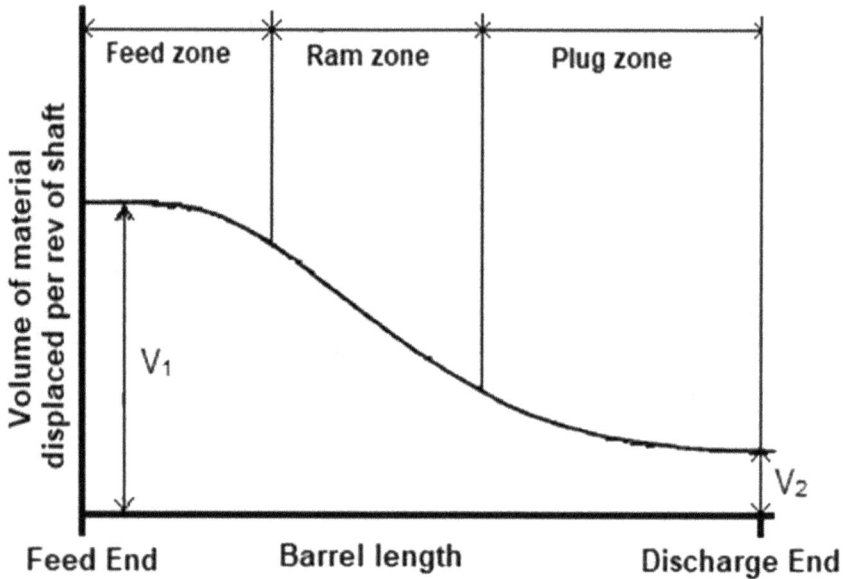

FIGURE 12.5 Compression curve along the length of the barrel in a screw press.

In case when moisture expulsion is criteria like whey removal from paneer, dewatering of solid waste, fish waste variable pitch screw shaft is necessary. Design of successive variation of screw pitch in the straight shaft with variable pitch is given by the equation:

$$p_{ss} = \frac{p_{fe} - p_{de}}{n-1} \tag{5}$$

where; p_{fe} is the pitch of the screw at feed end (m), p_{de} is pitch of the screw at discharge end (m), n is the number of screw flights or turns and p_{ss} is the successive decrease in pitch of the screw (m).

The maximum load that can be lifted by the screw shaft [46] during crushing of raw material is given by the equation:

$$W_s = \frac{T_{max}}{R} \cdot \left(\frac{\tan(\theta_{tp}) + \left(\frac{\mu}{\cos(\alpha)}\right)}{1 - \mu.\tan(\theta_{tp}).\cos(\alpha)} \right) \tag{6}$$

$$\alpha = \operatorname{atan}\left(\tan(\theta_{th}).\cos(\theta_{tp})\right) \tag{7}$$

where; W_s is load lifted by screw (N), T_{max} is maximum torque generated (N.m), R is radius of the screw (m), q_{tp} is taper angle (deg), q_{th} is thread angle (deg) and m is coefficient of friction.

12.4.1.4 BARREL DESIGN PRESSURE

The strength of the barrel should be sufficient to bear the pressure generated by the screw shaft during the oil expulsion. The maximum design pressure in the barrel can be estimated as follows:

$$P_c = \frac{2.t.\tau_a}{d_i} \tag{8}$$

where; P_c is maximum pressure inside the barrel (MPa), t is the thickness of the barrel (m), t is allowable stress (N/m²) and d_i is inside diameter of barrel (m).

12.4.1.5 CAPACITY OF THE EXPELLER

The theoretical design capacity of the screw expeller [6] can be calculated by the following expression:

$$Q_e = 60.\frac{\pi}{4}.\left(d_m^{\ 2} - d_s^{\ 2}\right).p_s.N_s.\varphi.\rho \tag{9}$$

where; Q_e is capacity of Screw expeller (kg/hr), d_m is mean diameter of the screw (m), d_s is mean diameter of the shaft (m), p_s is the pitch of the screw (m), N_s is speed of the screw (rpm), j is filling factor (0.8) and r is density of the product (Kg/m³).

12.4.1.6 POWER CALCULATION

Power required for screw expeller [35] can be estimated by torque and corresponding power relationship:

$$P_e = \frac{2.\pi.N_m.T}{60} \tag{10}$$

where; P_e is power required to run the screw expeller (Hp), N_m is the maximum speed of the screw (rpm) and T is maximum toque generated (N.m).

12.4.1.7 DIE GEOMETRY

In expellers, the spent cake is extruded through the annular space provided between the barrel and the pressure cone. The pressure cone assists in developing pressure inside the barrel. Considering the flow through the circular die is laminar, Newtonian the rate of flow through various shapes can be calculated using the Hagen Poiseuille equation:

$$Q = \frac{\pi R^4}{8L_d} \cdot \frac{\Delta P}{\mu_d} \tag{11}$$

where; DP is the pressure drop across die (MPa), μ_d is the viscosity of the dough at die (MPa.s), R is the radius of the die opening (m) and L_d is the length of the die (m).

12.4.2 STRUCTURAL ANALYSIS

The design and structural analysis has importance to optimize the cost of manufacturing and efficient expulsion of liquid from the matrix. Intensive pressures are generated to crush different kinds of seeds, etc., which lead to a rise in temperatures. The wearing of flights in screw press is predictable by recording wearing rate over time and interpretation of data with multiple regression analysis. Multiple regression double-log models were reported to fit the best to predict the wearing rate of screw [18].

Finite element analysis (FEA) with Solid Works and ANSYS aids in the determination of optimized clearance required between screw and barrel [34]. The cutting grooves inside the bars arranged in the form of barrel can improve efficiency for extracting oil from rapeseed, sesame seed, which are smaller in size. Static structural simulation reported that expansion of screw is more at higher temperatures at the feed throat, which causes narrowing [1]. It was recommended to cool the screw at the feed section for reducing the temperature at transition, metering section, and to increase the clearance at the throat region.

12.5 PARAMETERS FOR FOOD PROCESSING AND EXPELLER PERFORMANCE

12.5.1 PRESSURE

The pressure inside the expeller is governed by the force required to crush the given type of oil seeds. The higher the pressure applied, the higher is the yield of oil [10]. The pressure varies along the length of the expeller barrel (Figure 12.6). The pressure can be increased [44] with:

- Increase in taper of the screw shaft;
- Decreasing the pitch of the screw gradually as it progresses towards the discharge end;
- Decreasing the diameter of the barrel gradually towards the delivery end;
- Arranging the adjustable pressure cone or choke at the outlet or combination thereof.

The expelling efficiency can be increased with an increase in pressure at a constant temperature, initial moisture content and the time of heating during almond oil expression [8]. A typical behavior was observed that there was a decrease in expeller efficiency at a pressure range of 6.72 to 7.01 MPa due to compaction of almond cake blocking the expulsion of the oil. The oil yield in peanuts was increased with increase in pressure gradually up to 20 MPa and remained constant thereafter [2]. The boiling point of palm kernel and rapeseed are 3.91 MPa and 6.7–8.4 MPa, respectively [49].

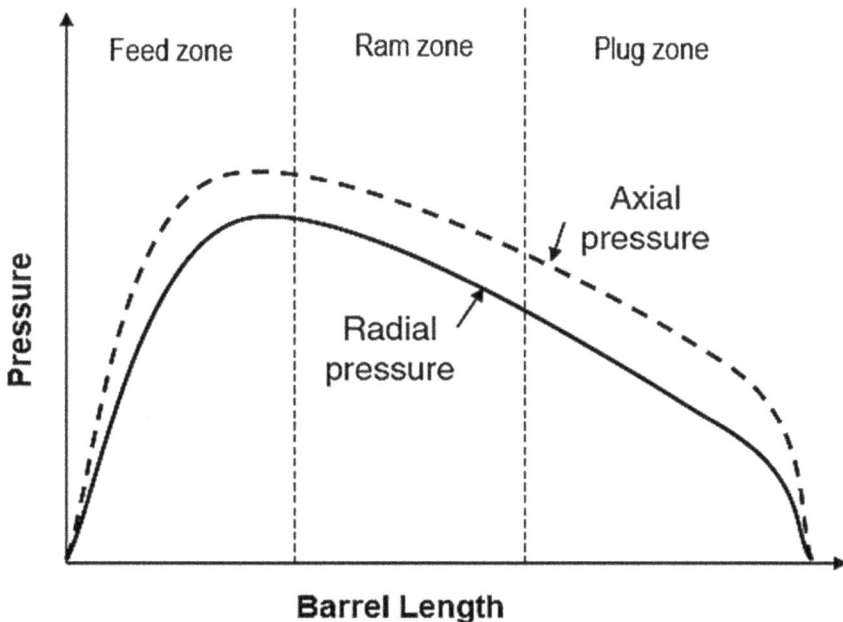

FIGURE 12.6 Pressure distribution along the length of barrel in screw expeller.

12.5.2 TEMPERATURE

The effect of temperature on the expelling efficacy has a corresponding increase until the cellular structure of the mass becomes soft enough to allow the liquid to flow out of the embedded cell structure. The rise in temperature lowers the oil viscosity that enhances the oil expulsion [10, 17]. The temperature had shown significant influence on the rate of pressing for almost all seeds around 100C where the protein gets coagulated, releasing the oil globules. A significant influence on the rate of pressing was not observed in the temperature range of 40 to 80°C [49]. During continuous pressing due to frictional forces, the mechanical energy is converted to heat energy, thereby increasing the temperature in the barrel. However, heating elements can also be provided to attain the required temperature around the barrel during compression.

12.5.3 SPEED OF SCREW ROTATION

The effect of speed of screw-on expeller performance is dependent on the design aspects of the shaft and screw. Screw speed has no direct impact on expeller performance [7]. The screw speed and screw geometry can be correlated with crushing velocity during continuous pressing [44]. A higher level of oil yield can be obtained at a lower rotational speed of the screw due to longer residence time in the expeller. However, minimum speed is to be maintained to generate sufficient pressure and torque [11]. The heat generation at higher speed is double the value of heat generated at lower speeds [37].

12.5.4 TYPE OF SEEDS

The effect of kind of feed varies with its composition and porosity. The seeds having a thin hull when compressed blocks the pore formation while the thicker hull is strong enough and generated pores during oil expulsion [11]. The pretreatment conditions impact the structure and initial moisture content of the seeds, which ultimately influence the expeller performance.

12.5.5 MOISTURE CONTENT OF SEEDS

The initial moisture content of the oil seeds greatly influences the oil expulsion. The increase in moisture content decreases the oil yield [47]. The quality of the oil is affected by the moisture content in the final product. Significant variation in pressing rate was observed with respect to the moisture content of the feed. The moisture content acts as a lubricant, which reduces the resistance while pressing enhances the pressing rate [42]. On the contrary, overall oil recovery is reduced due to lower residence time compared to feed having lower moisture. The initial moisture content and softening of protein can be achieved by cooking. The moisture content present in the sample can be calculated by weighing the product before and after drying.

$$\text{Moisture Content (\%)} = \left(\frac{m_i - m_f}{m_i} \right) \times 100 \tag{12}$$

where; m_i is the initial weight of product (g) and m_f is the final weight of the product (g).

12.5.6 EFFECT OF PRETREATMENT

Cooking is one of the common pretreatments to soften the protein in the seeds, thus improving the ease of oil expulsion. Cooking is generally characterized by time and temperature combination (Table 12.1). Undercooking or overcooking is not acceptable; hence optimum time-temperature combination can help to achieve higher efficiencies.

TABLE 12.1 Optimized Pretreatment Conditions for Rapeseed, Peanut, and Sunflower Oil Seeds

Parameter	Type of Oilseed		
	Rapeseed	Peanut	Sunflower
Cooking temperature (°C)	90	95–100	85
Oilseed flux (t/h)	10	10	8
Residence time (min)	30–35	60	40

Dehulling is necessary for seeds, such as almonds, peanuts, soya bean, linseed, etc., which comprises the significant portion. For peanut, the hull comprises 25% of the total mass. Dehulling helps in preventing the absorption of oil into the shell, which incurs losses. In order to attain the required protein concentration in the cake, soybean seeds are desheled. The yield of oil was 72% when dehulled seeds are subjected to screw pressing.

Flaking of seeds increases the surface area of seeds during solvent extraction, while in the case of screw pressing oil yield decreased from 80%

to 55% when fed at a temperature of 20°C. When the temperature of flaked seed is increased to 100°C there is no significant variation in the oil yield when compared to whole seeds. During hydraulic pressing, the flaked seeds yielded higher oil content [33]. The effect of different process parameter on expeller performance is given in Table 12.2.

TABLE 12.2 Influence of Process Parameters and Pretreatments on Screw Press Expeller Performance

Parameters	Pressure in Barrel	Temperature of Barrel	Cake Oil Content	Capacity
Choke opening (mm) ↓	↑	↑	↓	↓
Flaking	↑	↓	↑	↑
Heating (°C) ↑	↑	↑	↓	↑
Heating and Flaking ↑	↑	↑	↓	↑
Initial Moisture content ↑	↓	↓	↑	↓
Screw speed (rpm) ↓	↑	↓	↓	↓

NOTE: = ↑ Increase; ↓ = Decrease.

The variation in feed rate may lead to different level of compression, and ultimately it affects the oil yield. Optimum feed rate is required for the smooth operation of oil expeller as follows:

$$Feed\ Rate\left(\frac{Kg}{h}\right) = \frac{Quantity\ of\ Product\ fed\ (Kg)}{Time\ taken\ (h)} \tag{13}$$

12.6 MODELING OF SCREW EXPELLER

Mathematical modeling is a complex process in expelling of oil seeds. Empirical models based on cell structure are specific for a certain type of seeds and machine design. Terzagi one dimensional consolidation model is more appropriate for hydraulic compression and requires ample insight of properties of the material to be compressed. An extended Terzagi model known as Shirato model was employed in the expression of cocoa nibs [48]. Shirato model works on the ratio of consolidation with respect to properties of material, pressure, and time of pressure application. The ratio of consolidation can be utilized in comparing the rate of pressing in various seeds [49].

A linear programming models considering the input variable like screw speed, rate of feed, motor power, die geometry, pressure, and temperatures are suitable to optimize the yield of biodiesel from Jatropha seeds [29]. The oil extraction reported is 35.42% (on basis of seed weight) when compared to conventional rates of 28–32% when optimized machine variables are applied in the screw press.

The optimum values of operating variables were determined with computational fluid dynamics (CFD) in a screw expeller. Experimental validation reported extraction of oil by 98% of total oil content compared to 75% of oil content in the existing designs. A 12 Kg/h mass flow was achieved without backpressure [13].

The food extrusion models commonly applied are: response surface modeling (RSM), residence time distribution (RTD), FEA, comprehensive, and black-box modeling. Based on the characteristics of the product, the modeling or the analysis is selected [3]. Optimization mathematically is one of the best methods as it includes the selection of objective function in a defined domain. Design analysis of spares for their optimal performance enhances the efficiency and overall productivity [14].

12.7 DEVELOPMENTS IN MECHANICAL EXPELLERS

Oil can be mechanically expressed by hydraulic press, screw press expeller, and multi-layer press. The extraction efficiency depends on the process parameters, the size of the particle and other pre-processing parameters [33]. The screw expeller majorly consists of: (a) shaft with screw flight; (b) choke mechanism; (c) barrel with drainage perforations; (d) Drive/power transmission; and (e) cooling mechanism.

12.7.1 EXTRACTION OF VEGETABLE OIL

Sunflower seeds (*Helinanthus annues L.*) are rich in oil content, and they are a good source for cooking oil and can be used in the production of margarine, cosmetics, etc. The cake obtained after extracting oil is a good source of protein. The common method to obtain uncontaminated oil is by mechanical expelling [16]. Sunflower oil expelling machine was designed in accordance with the physicochemical properties of sunflower seeds. The unit consists of a screw auger with a decreasing pitch. The heating of seeds was accomplished by steam. It was reported that expelling efficiency of >70% was achieved.

The designed machine has a throughput of 502.64 kg/day and oil yield of 24.4 liters/hr (lph). The performance of screw expeller was affected by the composition of the sunflower seeds [31].

Oil extraction capacity was more in Lot I with 39.7% oil content and 11% moisture on a wet basis, while oil extraction efficiency was good in Lot II having 42.6% oil and 6% moisture on a wet basis. The effect of pre-heat treatment on sunflower seeds indicated that the efficiency of extraction was improved with the increase in temperature. A mathematical model that can predict the level of oil extraction along the length of the barrel in sunflower seeds during screw pressing was developed [30]. In order to establish a mathematic model, the design of the barrel was divided into four compartments to which a separate collection system was provided to determine the level of extraction along with the perforated barrel chamber. The oil yield was in the range of 20 to 20.8% with a nozzle size of 12 mm and 8 mm diameter at the outlet. The higher the nozzle diameter, the lower was the oil yield.

Palm kernel is obtained from palm tree (*Elaeis guineensis*). The composition of Palm kernel has oil (47–49%), crude protein (7–9%), cellulose (9%), ash (2%), non-nitrogenous matter (23–24%), and water (9%). Different mechanical methods currently in use are hydraulic pressing and screw pressing method for Palm kernel oil. In the batch method, the perforated boxes filled kernels wrapped in cloth are subjected to hydraulic pressing to obtain the oil. In a continuous process, screw press having a series of worms welded on shaft rotated in a barrel having perforations is being used. The rotation of the shaft generates a crushing and squeezing action that extracts oil out through the perforated barrel. The continuous screw extraction requires high initial cost but low running cost with little operating skill [20]. The operating capacities range from 3 to 1000 tons/day.

The compressive stress, rate of feeding and rpm of the screw are optimized based on the characteristics of the palm kernel [7]. The design of the expeller consists of the helical threaded screw with the stationary barrel. The volume is reduced gradually through the axis, which compresses the meal leading to oil expulsion. The expelled oil is drained through the perforation of the barrel and the cake is discarded from the annular orifice. The study revealed that the yield of oil is directly proportional to the compressive force applied to the barrel while the speed of the screw has marginal influence. At 30 MPa pressure with 150 Kg per hour feed rate and 110 rpm speed, the oil yield is 46.5% with 94.5% expression efficiency; while at 10 MPa pressure and 50 rpm with the feed rate of 150 Kg per hour, the minimum of 16.3% oil yield having 33.6% efficiency was recorded.

Jatropha (*Jatropha curcas*) is a potential source of biodiesel with seeds containing 30 to 40% of oil content. The oil extraction process involves a series of post-harvest unit operations wherein screw expression is one of these. The optimized pretreatment was cooking the seeds to 100°C for 10 min with the moisture content of 9.7% on dry matter basis [42]. The yield of oil was 73% with a slight insignificant reduction of rate of pressing. The homogenous feed develops uniform pressing conditions, which lead to improved performance [19].

The seed shells build a porous matrix during pressing that facilitate the expulsion of oil through the cake. The mass balance between the crude oil extracted and the pressed cake will determine the separation efficiency. The overall recovery of oil was about 70 to 80% with optimum specific energy requirement. The design employs a screw pressing principle for oil extraction. The rotating screw first breaks and interstitial voids would be diminished, as the process continues till a pressure point is reached, where the oil starts oozing out. Screw press differs in geometry with oil outlet and cake press restriction [11]. The designed screw press expeller for Jatropha consists of a barrel with 101 mm in diameter; and throughput of 120 kg/h powered with 7.5 KW electrical motor was used for oil extraction. The rpm regulation was done by variable frequency drive. Thermocouples 9K-type temperature probes were mounted at different points and connected to display modules [19].

A mechanical expression system for almond kernel was optimized for expression efficiency [8] using central composite design (CCD) in response surface methodology (RSM). The parameters under study in optimization were temperature, duration of heating, pressure, and moisture. The mechanical expression system used for the study consists of a cylinder-piston ring, a hydraulic pressing system with load cell at the bottom. The cylinder is perforated, wherein during compression, the oil is recovered in an oil-collecting pan. In order to maintain the required temperature, a 605 W electric heater was used.

The optimum of 76.4% expression efficiency was obtained at temperature, heating time, moisture, and pressure of 93°C, 17 min, 8.71%(wet basis) and 6.4 MPa, respectively during mechanical expression of oil from almond kernel. The size of the particle did not show significant effect on oil yield from almonds, while the pressure applied and pressing time had a significant effect [5]. Maximum yield about 48.4% was obtained at 116.6 KPa pressure with 12 min pressing time. Further increase in pressure up to 130 KPa yielded 50% of oil content beyond which there is no effect of pressure on oil yield.

Pistachio (*Pistacia vera*) oil was extracted with hydraulic and screw-type expellers to attain high quality of oil in terms of its physicochemical properties and sensory attributes [43]. A batch of 200 g of pistachios was used for hydraulic press and screw press. Ground pistachios were used in the hydraulic press in which seeds were fed directly in the screw press at different processing conditions. The extracted oil was subjected to centrifugal separation of residues. The fatty acid profile of final products exhibited a higher proportion of oleic acid (54.80 to 55.17 g/100 g) and linoleic acid (30.93 to 31.52 g/100 g). The sensory attributes of oils produced with hydraulic press did not show significant variation at different processing parameters. Pistachio oils produced with screw press exhibited superior sensory quality attributes.

The screw expeller was employed for extraction of canola oil with optimized screw speed, choke opening, pretreatments, and initial moisture content [47]. Increase in pressure increased the throughput and residual oil content in pressed cake while decreased tendency was seen with the reduction in opening of choke and screw shaft speed.

The optimum operating parameters and oil extraction were related in rice bran and parboiled rice bran obtained from paddy husk during the extraction of rice bran oil by screw expeller [45]. The oil yield is represented by the oil content in the cake of rice bran. The main components of the screw press were driving motor rated 1 hp and 220 V, screw-type was Maddock, the barrel consists of 16 bars in the form of cage in which the screw was placed, a feed hopper was mounted, reduction gearbox and the structure for supporting the whole setup.

The study of oil extraction at various screw speeds (8.5, 11.3, 14.1, 16.9 and 19.8 rpm) and compression clearances (1.0, 1.3, 1.5, 1.7 and 1.9 cm) reported that there was no oil extraction with 8.5 rpm of the screw with clearances of 1.0, 1.3 and 1.5 cm. The initial content of oil in rice bran and parboiled rice bran was 18.5 and 21.9%, respectively. The maximum obtained oil in cake after extraction is 4.17% at 1.7 cm clearance and 8.5 rpm for rice bran while for parboiled rice bran it was 8.20% at 1.3 cm clearance and 8.5 rpm.

The kneading temperature of 90°C was optimized for maximum extraction efficiency during the continuous extraction of soybean oil with screw expeller [36]. This kind of mechanical pressing technology was found best from small-scale farmers. The pressed cake can be used as a fertilizer and also as animal feed as it is free from other toxic substances with solvent extraction technique. The extraction of oil in screw press was 68% compared to 98% in solvent extraction technique.

Mechanical expression of oil from moringa seeds was optimized by RSM [21]. The optimum condition established with an oil yield 28.20% having moisture 11.3% at a pressure of 19.63 MPa for 27.17 min and temperature of 85.57°C. There is no significant deviation between experimental and predicted values using screw type oil expeller.

The design of coconut oil expeller consists a screw shaft, perforated barrel, hopper, outlets, mechanical drive, and supporting structure [9]. The maximum yield of coconut oil of 53.3% was obtained at an optimized feed rate of 3.5 Kg/h, moisture 7.2% (wet basis). The effectiveness of compression and quantity of oil yield depends on the moisture content of the feed.

Shea butter is utilized for various applications in pharmaceutical, cosmetic, and food industries. The mechanical extraction of Shea butter is by cold press method and it does not involve heating, which preserves the functional components during extraction [28].

12.7.2 EXTRACTION OF FISH OIL

The marine products can be pressed continuously with a screw press while hydraulic press is used for batch pressing. The product needs to be supplied constantly for screw press and can be used for dewatering and for extraction of fish oil. Hydraulic press is used when a small amount of material needs to be compressed and to achieve higher compaction level with low moisture content [26].

A mechanical fish oil-extracting machine consisting of a hopper, screw drum, cylindrical casing, cake, oil outlet and supporting frame for the setup was developed [15]. The concave screen with pressing drum having clearance of 5 mm between screw and barrel allows the squeezed oil to pass through and collected at the oil outlet. Crushing, grinding, pressing, and conveying are accomplished by casing and the screw drum. In this system, the extraction process can be done in a continuous manner.

Hydraulic pressed discards of Sardine (*Sardina pilchardus*) were optimized for liquor yield and to minimize the moisture content in the obtained cake by statistically designed experiment [24]. The discarded waste comprises of organic material with small-sized with low values and are mostly dumped into the sea. It was reported that the reduction in volume in obtained cake is about 40 to 45% with respect to the initial volume of the material. This ultimately reduces the refrigeration and space requirement during storage.

The yield of liquor is 13.45% w/w with respect to the quantity of sardine discards fed into the hydraulic press under 350 bar pressure. Heating to 55°C improved the oil yield facilitating the oil release by disruption of tissues on the contrary, this reduced the free fatty acids (FFA) content [25] during extracting oil from whole sardine. A minimum pressure of 60 bars reduced the oxidative stress and residence time, which enhanced the quality characteristics of the oil. Two-stage pressing at 60 bar improved the yield of the oil with good quality.

Expeller for oil extraction from catfish (*Clarias gariapinus*) consists of a feeding chamber, expelling system, discharge outlet, electric drive, and supporting frame structure [12]. The designed hopper is pyramidal in shape made with the galvanized iron sheets. The tapered screw design generated friction, which causes heating of the feed thus facilitates in the extraction of oil. The oil is collected through perforation underneath and cake is extruded out through the outlet. It was reported that 80% of chemical-free fish oil was recovered with the help of screw press expeller.

Pressure is one of the important factors that affect the oil yield and energy consumption for the extraction of oil from crude fish with a screw expeller [14]. Other factors include the speed of the screw, size of restriction, moisture content and pretreatment like cooking. Higher speeds of screw resulted in low yield and higher throughput. With reduced restriction size, the pressure generated was more resulting in higher oil yield. Extraction of crude fish oil by screw expeller was optimized with RSM.

12.7.3   DEWATERING OF DAIRY AND FOOD WASTES

A screw expeller was designed for dewatering the cattle dung obtained from biogas plant [32]. The perforated screen with 3 mm size gave minimum total solids in the liquid phase. The 40% of the total solids in cow dung slurry were in the dissolved form, which cannot be removed by screw expeller. The initial total solids of 9% in the slurry were concentrated to 26% solid residues. The effect of pressing velocity during dewatering showed that slower pressing velocity increased level of dewatering [27]. Higher velocities resulted higher-pressure drop hence lower dewatering. Dewatering did not increase when pressure was increased above 1.6 MPa.

12.8   SUMMARY

The mechanical expression can be done by hydraulic pressing and screw pressing. Hydraulic pressing is a batch method, while screw expelling is a continuous method. The throughput of the screw expeller is far higher than hydraulic pressing without compromising the expression efficiency. The design of screw expeller consists of a screw shaft, barrel with perforations or slits, drive, pressure cone, die, and structure to support the whole setup. The design of the screw shaft is critical in the unit as the desired pressure and compaction is developed by screw shaft in conjunction with choke at the outlet. The screw expeller can be designed with a single screw or twin screw. Design modeling with RSM and CFD can aid in optimizing the design and life of machine components. Extensive research is undergoing in design optimization for application screw expeller in different agricultural products like jatropha oil, sunflower oil, soya bean oil, peanut, etc., meat products like fish oil from different variants of fish, dewatering of fish waste, and dairy waste. The future challenges include the design optimization for specific applications, improving the expeller efficiency, and affordable design for small-scale industries.

KEYWORDS

- **dewatering**
- **extraction modeling**
- **hydraulic pressing**
- **mechanical extraction**
- **oil expeller**
- **screw expeller**
- **screw press**

REFERENCES

1. Abdel, G. W. E., Ebeid, S. J., & Fikry, I., (2015). Mechanical design aspects of single-screw extruders using finite element analysis. *International Journal of Engineering and Technical Research, 3*(7), 2454–4698.
2. Adeeko, K. A., & Ajibola, O. O., (1990). Processing factors affecting yield and quality of mechanically expressed groundnut oil. *Journal of Agricultural Engineering Research, 45*, 31–43.
3. Adekola, K. A., (2016). Engineering review food extrusion technology and its applications. *Journal of Food Science and Engineering, 6*, 149–168.

4. Adekola, K. A., (2014). Analytical engineering designs for twin-screw food extruder dies. *International Journal of Engineering Innovation and Research, 3*(5), 713–717.
5. Adesina, B. S., & Bankole, Y. O., (2013). Effects of particle size, applied pressure, and pressing time on the yield of oil expressed from almond seed. *Nigerian Food Journal, 31*(2), 98–105.
6. Adetola, O. A., Olajide, O. J., & Olalusi, A. P., (2014). Development of a screw press for palm oil extraction. *International Journal of Scientific and Engineering Research, 5*(7), 1416–1422.
7. Akinoso, R., Raji, A. O., & Igbeka, J. C., (2009). Effects of compressive stress, feeding rate and speed of rotation on palm kernel oil yield. *Journal of Food Engineering, 93*(4), 427–430.
8. Akubude, V. C., Maduako, J. N., & Egwuonwu, C. C., (2018). Effect of processing parameters on the expression efficiency of almond oil in a mechanical expression rig. *Agricultural Engineering International: CIGR Journal, 20*(1), 109–117.
9. Amretha, K. A. K., & Manshooba, P. M., (2017). Development and testing of a power-operated coconut oil expeller. *International Journal of Engineering Research and Application, 7*(7), 01–05.
10. Arisanu, A. O., (2013). Mechanical continuous oil expression from oil seeds: Oil yield and press capacity. In: *5th International Conference on "Computational Mechanics and Virtual Engineering" Held on 24–25th October 2013 at Brasov, Romania* (pp. 347–352).
11. Atto, E. A., Abiodun, O. M., & Francis, O. O., (2018). Jatropha oil extraction optimization through varied processing conditions using mechanical process. *International Journal of Innovative Research and Development, 7*(9), 227–241.
12. Ayuba, A. B., Amina, I. M., Okouzi, S. A., & Olungenga, B. O., (2018). Design and fabrication of an extruder for the extraction of fish oil from catfish. *International Journal of Latest Technology in Engineering, Management, and Applied Science, 7*(8), 2278–2540.
13. Bahadar, A., Khan, M. B., & Mehran, T., (2013). Design and development of an efficient screw press expeller for oil expression from *Jatropha curcas* Seeds: A computational flow dynamics study of expeller for performance analysis. *Industrial and Engineering Chemistry Research, 52*(5), 2123–2129.
14. Bako, T., Umogbai, V. I., & Garba, A. J., (2018). Optimization of crude fish oil extraction using mechanical screw expeller validated by response surface methodology. *Journal of Postharvest Technology, 6*(1), 69–82.
15. Bako, T., Umogbai, V. I., & Obetta, S. E., (2014). Development of a fish oil-extracting machine. *Asian Journal of Science and Technology, 5*(8), 431–436.
16. Bamgboye, A. I., & Adejumo, A. O. D., (2007). Development of a sunflower oil expeller. *Agricultural Engineering International: CIGR Journal, 9*(9), 1–7.
17. Bamgboye, A. I., & Adejumo, O. I., (2011). Effects of processing parameters of Roselle seed on its oil yield. *International Journal of Agricultural and Biological Engineering, 4*(1), 82–86.
18. Basil, E. O., (2015). Predicting flights wear in screw presses in palm oil mills. *International Journal of Engineering and Technology, 5*(2), 80–86.
19. Chapuis, A., Blin, J., Carre, P., & Lecomte, D., (2014). Separation efficiency and energy consumption of oil expression using a screw-press: The case of *Jatropha Curcas L.* seeds. *Industrial Crops and Products, 52*, 752–761.
20. Ezeoha, S. L., Akubuo, C. O., & Ani, A. O., (2012). Indigenous design and manufacture of palm kernel oil screw press in Nigeria: Problems and prospects. *International Journal of Applied Agricultural Research, 7*(2), 67–82.

21. Fakayode, O. A., & Ajav, E. A., (2016). Process optimization of mechanical oil expression from moringa (*Moringa oleifera*) seeds. *Industrial Crops and Products, 90,* 142–151.

22. Fang, Q., & Hanna, M., (2010). Extrusion systems: Design. In: *Biological Systems Engineering: Papers and Publications* (p. 311). University of Nebraska, Lincoln, USA.

23. Firdaus, M., Salleh, S. M., Nawi, I., & Ngali, Z., (2017). Preliminary design on screw press model of palm oil extraction machine. In: *IOP Conference Series: Materials Science and Engineering* (Vol. 165, No. 1, pp. 12–29). Bristol, UK: IOP Publishing.

24. Galvez, R. P., Chopoin, C., & Mastail, M., (2009). Optimization of liquor yield during the hydraulic pressing of sardine (*Sardina pilchardus*) discards. *Journal of Food Engineering, 92,* 66–71.

25. Garcia, M. P. J., & Medina, R. M., (2014). Optimization of oil extraction from sardine (*Sardina pilchardus*) by hydraulic pressing. *International Journal of Food Science and Technology, 49*(10), 2167–2175.

26. Garcia, M. P. J., & Galvez, R. P., (2017). Pressing in the food industry: Example of fish discards processing. *Reference Module in Food Science* (pp. 1–4). London, UK: Elsevier Inc.

27. Gregor, H., Rupp, W., Janoske, U., & Kuhn, M., (2013). Dewatering behavior of sewage screenings. *Waste Management, 33*(4), 907–914.

28. Iddrisu, A. M., Didia, B., & Abdulai, A., (2019). Shea butter extraction technologies: Current status and future perspective. *African Journal of Biochemistry Research, 13*(2), 9–22.

29. Ihsan, A., Ahmad, R., & Umer, M., (2013). Optimization of a screw press using linear programming techniques. *Advanced Materials Research, 816,* 1265–1269.

30. Ionescu, M., Voicu, G., Biris, S., Matache, M., & Stefan, M., (2015). *Mathematical Models for Expressing the Oil Extraction at Screw Presses* (Vol. 77, No. 3, pp. 249–260). UPB Scientific Bulletin, Series D. Bucharest, Romania.

31. Jacobsen, L. A., & Backer, L. F., (1986). Recovery of sunflower oil with a small screw expeller. *Energy in Agriculture, 5*(3), 199–209.

32. Kataria, M. B., George, P. M., Mehta, H., & Jivani, R. G., (2011). *Experiments for Designing a Screen of a Screw Press for Dewatering of Cattle Dung Slurry* (pp. 1–3). National Conference on Recent Trends in Engineering and Technology held on 13-14 May at B.V.M. Engineering College, V.V. Nagar, Gujarat, India.

33. Khan, L. M., & Hanna, M. A., (1983). Expression of oil from oilseeds: Review. *Journal of Agricultural Engineering Research, 28*(6), 495–503.

34. Khan, M. H. U., Mondal, D., & Hoque, S., (2016). *Design and Construction of Oil Expeller Press with Structural Analysis of Screw with ANSYS* (pp. 1–4). International Conference on Mechanical, Industrial, and Energy Engineering. Khulna, Bangladesh.

35. Khurmi, R. S., & Gupta, J. K., (2008). Pressure vessels and shafts. In: *Textbook of Machine Design* (14th edn., pp. 224–260). New Delhi, India: Eurasia Publishing House (PVT) Ltd.

36. Moses, D. R., (2014). Performance evaluation of continuous screw press for extraction soybean oil. *American Journal of Science and Technology, 1*(5), 238–242.

37. Omobuwajo, T. O., Ige, M. T., & Ajayi, O. A., (1997). Heat transfer between the pressing chamber and the oil and oilcake streams during screw expeller processing of palm kernel seeds. *Journal of Food Engineering, 31*(1), 1–7.

38. Owolarafe, O. K., & Osunleke, A. S., (2008). Mathematical modeling and simulation of the hydraulic expression of oil from oil palm fruit. *Biosystems Engineering, 101*(3), 331–340.

39. Oyinlola, A., Ojo, A., & Adekoya, L. O., (2004). Development of a laboratory model screw press for peanut oil expression. *Journal of Food Engineering, 64*(2), 221–227.

40. Patil, M. A. A., Gaikwad, M. S. D., & Kulkarni, M. R., (2017). A review on the oil expeller screw shaft. *International Research Journal of Engineering and Technology, 40*(93), 1654–1657.

41. Pradhan, R. C., Meda, V., Rout, P. K., Naik, S., & Dalai, A. K., (2010). Supercritical CO_2 extraction of fatty oil from flaxseed and comparison with screw press expression and solvent extraction processes. *Journal of Food Engineering, 98*(4), 393–397.

42. Pradhan, R. C., Mishra, S., Naik, S. N., Bhatnagar, N., & Vijay, V. K., (2011). Oil expression from Jatropha seeds using a screw press expeller. *Biosystems Engineering, 109*(2), 158–166.

43. Rabadan, A., Orti, M. A., Gomez, R., Alvarruiz, A., & Pardo, J. E., (2017). Optimization of pistachio oil extraction regarding processing parameters of screw and hydraulic presses. *LWT-Food Science and Technology, 83*, 79–85.

44. Savoire, R., Lanoiselle, J. L., & Vorobiev, E., (2013). Mechanical continuous oil expression from oilseeds: A review. *Food Bioprocess Technology, 6*, 1–16.

45. Sayasoonthorn, S., Kaewrueng, S., & Patharasathapornkul, P., (2012). Rice bran oil extraction by screw press method: Optimum operating settings, oil extraction level and press cake appearance. *Rice Science, 19*(1), 75–78.

46. Shigley, J. E., & Mischke, C. R., (2001). *Mechanical Engineering Design* (6th edn., pp. 193–206). New York, USA: McGraw-Hill Companies, Inc.

47. Vadke, V. S., & Sosulski, F. W., (1988). Mechanics of oil expression from canola. *Journal of the American Oil Chemist's Society, 65*(7), 1169–1176.

48. Venter, M. J., Kuipers, N. J. M., & De Haan, A. B., (2007). Modeling and experimental evaluation of high-pressure expression of cocoa nibs. *Journal of Food Engineering, 80*(4), 1157–1170.

49. Willems, P., Kuipers, N. J. M., & De Hann, A. B., (2008). Hydraulic pressing of oilseeds: Experimental determination and modeling of yield and pressing rates. *Journal of Food Engineering, 89*, 8–16.

50. Williams, M. A., (2005). Recovery of oils and fats from oilseeds and fatty materials. In: Shahidi, F., (ed.), *Bailey's Industrial Oil and Fat Products* (pp. 99–189). New Jersey-USA: John Wiley and Sons, Inc.

INDEX

C

D